Springer Theses

Recognizing Outstanding Ph.D. Research

Aims and Scope

The series "Springer Theses" brings together a selection of the very best Ph.D. theses from around the world and across the physical sciences. Nominated and endorsed by two recognized specialists, each published volume has been selected for its scientific excellence and the high impact of its contents for the pertinent field of research. For greater accessibility to non-specialists, the published versions include an extended introduction, as well as a foreword by the student's supervisor explaining the special relevance of the work for the field. As a whole, the series will provide a valuable resource both for newcomers to the research fields described, and for other scientists seeking detailed background information on special questions. Finally, it provides an accredited documentation of the valuable contributions made by today's younger generation of scientists.

Theses may be nominated for publication in this series by heads of department at internationally leading universities or institutes and should fulfill all of the following criteria

- They must be written in good English.
- The topic should fall within the confines of Chemistry, Physics, Earth Sciences, Engineering and related interdisciplinary fields such as Materials, Nanoscience, Chemical Engineering, Complex Systems and Biophysics.
- The work reported in the thesis must represent a significant scientific advance.
- If the thesis includes previously published material, permission to reproduce this must be gained from the respective copyright holder (a maximum 30% of the thesis should be a verbatim reproduction from the author's previous publications).
- They must have been examined and passed during the 12 months prior to nomination.
- Each thesis should include a foreword by the supervisor outlining the significance of its content.
- The theses should have a clearly defined structure including an introduction accessible to new PhD students and scientists not expert in the relevant field.

Indexed by zbMATH.

More information about this series at https://link.springer.com/bookseries/8790

Stefan Meinecke

Spatio-Temporal Modeling and Device Optimization of Passively Mode-Locked Semiconductor Lasers

Doctoral Thesis accepted by
Technische Universität Berlin, Berlin, Germany

Springer

Author
Dr. Stefan Meinecke
Institut für Theoretische Physik
Technische Universität Berlin
Berlin, Germany

Supervisor
Prof. Kathy Lüdge
Institut für Physik
Technische Universität Ilmenau
Ilmenau, Germany

ISSN 2190-5053 ISSN 2190-5061 (electronic)
Springer Theses
ISBN 978-3-030-96247-0 ISBN 978-3-030-96248-7 (eBook)
https://doi.org/10.1007/978-3-030-96248-7

This Springer imprint is published by the registered company Springer Nature Switzerland AG
The registered company address is: Gewerbestrasse 11, 6330 Cham, Switzerland

Supervisor's Foreword

The dissertation of Stefan Meinecke published in this book is dedicated to the numerical analysis of semiconductor devices that are intended for the generation of optical pulse trains via passive mode locking. With this mode-locking method, it is possible to generate short pulses by a saturable absorber and a gain section, which are passed through serially in the device. In contrast to active mode locking usually realized via periodic gain or loss modulation, pulse generation in passive mode locking occurs without additional active elements and is therefore very suitable for integrated on-chip applications. In addition, the required semiconductor structures can be manufactured very easily by epitaxial processes. This means that the understanding of these integrated structures and the associated optimization of the emission properties have great application potential, ranging from the field of optical data transmission to spectroscopy and metrology.

Central issues with regard to practical applications are the maximization of the stability and regularity of the emitted pulse trains as well as the minimization of the pulse width at the highest possible peak intensity. This is where the work of Stefan Meinecke comes in, as he develops modeling concepts that allow a deep understanding and thus control of the complex nonlinear light–matter interactions within the device.

In his thesis, he investigates the passively mode-locked semiconductor lasers by numerical methods with a strong focus on the development of efficient numerical tools. This is necessary, as the modeling of integrated optical devices requires a good spatiotemporal resolution of the propagating light and the induced charge carrier dynamics as well as an accurate description of the material properties. The goal is to allow for both larger parameter studies and quantitative predictions. As such, a large methodological breadth is applied, ranging from stochastic calculations, sophisticated time domain-based evaluation algorithms to analytical methods. Along with that, Stefan Meinecke develops visualization and evaluation techniques for the rich spatiotemporal laser dynamics, which facilitate the physical interpretation of the observed features.

The investigations in this thesis revolve around two specific semiconductor devices, namely a monolithically integrated three-section tapered quantum-dot laser and a V-shaped external cavity laser. In both cases, the simulations closely tie in with experimental results, which have been obtained in collaboration with the TU Darmstadt and the ETH Zurich. Based on the successful numerical reproduction of the experimental findings, the emission dynamics of both lasers can be understood in terms of the cavity geometry and the active medium dynamics. The latter, in particular, highlights the value of the developed simulation tools, since the fast charge carrier dynamics are generally not experimentally accessible during mode-locking operation. Lastly, the numerical models are used to perform laser design explorations and thus to derive recommendations for innovative and optimized devices.

Berlin, Germany
November 2021

Prof. Kathy Lüdge

Abstract

Passively mode-locked lasers produce regular optical pulse trains at high repetition rates and find many applications in science and technology. This thesis numerically investigates two specific devices: a monolithically integrated three-section tapered semiconductor quantum-dot laser and a V-shaped external cavity semiconductor laser. Both lasers are motivated and complemented by experiments. Special attention is given to the spatiotemporal intra-cavity electric field and active medium gain dynamics. The results are used to predict optimal laser configurations and operating conditions.

For both lasers, device-specific numerical models, which explicitly include the spatially inhomogeneous electric field and active medium evolution, are derived from first principles. In order to characterize the temporal pulse train stability, a review of the established long-term timing jitter estimation methods and their applications to simulated time series is presented. Additionally, a new computationally efficient pulse period fluctuations-based method is introduced. All methods are benchmarked with respect to their computational demands. The results are used to deduce usage recommendations for the various methods.

The three-section tapered quantum-dot laser produces competitive pulse performances in all measures. Both the pump current dependent sequence of observed emission states and the evolution of the performance figures can be reproduced by the numerical model. The results are explained in terms of the device geometry and the active medium dynamics. The analysis further unravels a pulse-shaping mechanism that is contrary to the published literature: Pulses broaden in the absorber and shorten in the gain sections. The numerical model is further used to map and understand the dynamics in the operation parameter space and to explore and identify an optimal device design with respect to the taper angle and the saturable absorber position.

The V-shaped external cavity laser stands out by exhibiting pulse-cluster mode-locking states at larger pump currents. The experimentally observed pump current-dependent sequence of emission states can be reproduced and understood in terms of the gain and net gain dynamics. The twofold interaction with the gain chip in the forward and backward directions is identified as the governing feature. The emission dynamics are consequently demonstrated to critically depend on the gain chip position. Favorable emission states distribute the pulse interactions with the gain chip equidistantly. Their stability is limited by positive net gain windows, which are characteristic for the given cavity configuration. Based on the net gain, analytic stability boundaries for the fundamental mode-locking state are derived and optimal cavity configurations are predicted.

Apart from the modeling and simulation techniques, this thesis presents evaluation and visualization methods, which enable the analysis of mode-locking and pulse-shaping mechanisms. Their application may facilitate the investigation and exploration of future generations of high-performance passively mode-locked lasers.

Publications Related to This Thesis

[1] D. Waldburger, C. G. E. Alfieri, S. M. Link, S. Meinecke, L. C. Jaurigue, K. Lüdge, and U. Keller. Multipulse instabilities of a femtosecond SESAM-modelocked VECSEL. *Opt. Express*, 26(17):21872–21886, August 2018. https://doi.org/10.1364/oe.26.021872.

[2] L. C. Jaurigue, B. Lingnau, S. Meinecke, and K. Lüdge. Passive mode-locking in a Vshaped cavity. *Proc. SPIE*, 1051505, 2018. https://doi.org/10.1117/12.2288036.

[3] S. Meinecke, L. Drzewietzki, C. Weber, B. Lingnau, S. Breuer, and K. Lüdge. Ultrashort pulse generation in a three section tapered passively mode-locked quantum-dot semiconductor laser. *Sci. Rep.*, 9:1783, 2019. https://doi.org/10.1038/s41598-018-38183-1.

[4] J. Hausen, S. Meinecke, B. Lingnau, and K. Lüdge. Pulse cluster dynamics in passively mode-locked semiconductor vertical-external-cavity surface-emitting lasers. *Phys. Rev. Appl.*, 11:044055, April 2019. https://doi.org/10.1103/physrevapplied.11.044055.

[5] J. Hausen, S. Meinecke, and K. Lüdge. Bifurcation scenario leading to multiple pulse emission in VECSELs. *Proc. SPIE*, 10901:109010F, 2019. https://doi.org/10.1117/12.2513751.

[6] M. Kolarczik, F. Böhm, U. Woggon, N. Owschimikow, A. S. Pimenov, M. Wolfrum, A. G. Vladimirov, S. Meinecke, B. Lingnau, L. C. Jaurigue, and K. Lüdge. *Coherent and incoherent dynamics in quantum dots and nanophotonic devices, Springer, ISBN 978-3-030-35655-2*, volume 194 of *Semiconductor Nanophotonics*, chapter 4, page 91. Springer Series in Solid-State Sciences, 2020. ISBN 978-3-030-35655-2. https://doi.org/10.1007/978-3-030-35656-9_4. Booktitle: Semiconductor Nanophotonics.

[7] J. Hausen, S. Meinecke, J. Javaloyes, S. V. Gurevich, and K. Lüdge. Phase-incoherent photonic molecules in V-shaped mode-locked vertical-external-cavity surface-emitting semiconductor lasers. *Phys. Rev. Appl.*, 14(044059), October 2020. https://doi.org/10.1103/physrevapplied.14.044059.

[8] S. Meinecke and K. Lüdge. Efficient timing jitter simulation for passively mode-locked semiconductor lasers. *App. Phys. Lett.*, 118(1):011104, 2021. https://doi.org/10.1063/5.0036928.

Acknowledgements

First and foremost, I would like to thank Kathy Lüdge for introducing me to the field of nonlinear dynamics and laser physics, providing a welcoming and stimulating work environment, supervising and supporting this thesis, and always being available for discussions. I would like to thank Benjamin Lingnau for helping me get started with many of the technical aspects of modeling and simulating semiconductor lasers. Furthermore, I would like to thank Jan Hausen and Felix Köster for proofreading this manuscript, countless scientific discussions, and enjoyable lunch and coffee breaks. I also thank Ulrike Wieland for the final proofreading of this thesis. Many thanks go to all the other current and former members of the AG Lüdge, among them especially André Röhm, Lina Jaurigue, David Schicke, Christoph Redlich, Larissa Bauer, and Roland Aust, for being supportive and providing a friendly work environment. Moreover, I would like to thank Andrea Schulze for promptly solving all arising bureaucratic issues. I would like to thank Stefan Breuer, Christoph Weber, and Dominik Auth for our fruitful experimental–theoretical collaboration on the mode-locked three-section quantum-dot laser, the many discussions on mode locking, and the ongoing exchange of ideas. I would like to thank Dominik Waldburger, Cesare Alfieri, and Ursula Keller for our experimental–theoretical collaboration on the V-shaped semiconductor laser and the in-depth discussion of the estimation and interpretation of the timing jitter in passively mode-locked lasers. I am grateful for the support provided by the SFB787 and the 'School of Nanophotonics,' which have offered many interesting seminars and workshops and allowed me to participate in multiple scientific conferences. Finally, I would like to thank Uwe Bandelow for taking the time to be the external reviewer on the examining board for this thesis.

Contents

Acronyms

AC	Autocorrelation
BW	Bandwidth
cw	Continuous wave (lasing)
DBR	Distributed Bragg reflector
DDE	Delay differential equation
DFT	Discrete Fourier transform
ES	Excited state
FFT	Fast Fourier transform
FML	Fundamental mode locking
FWHM	Full width at half maximum
GaAs	Gallium arsenide
GS	Ground state
HM	Half maximum
HML	Harmonic mode locking
InAs	Indium arsenide
InGaAs	Indium gallium arsenide
LEI	Leading-edge instability
lt	Long-term timing jitter
MP	Non-identical non-equidistant multi-pulse mode locking
OC	Output coupler
ODE	Ordinary differential equation
PC	Pulse cluster mode locking
PDE	Partial differential equation
pp	Pulse-to-pulse timing jitter
PS	Pulse switching
QD	Quantum dot
QSML	Q-switched mode locking
QW	Quantum well
rms	Root mean square
RO	Relaxation oscillation
SA	Saturable absorber

SESAM	Semiconductor saturable absorber mirror
TEI	Trailing-edge instability
TEP	Trailing-edge plateau
VECSEL	Vertical external cavity surface-emitting laser

Chapter 1
Introduction

1.1 Dynamical Systems and Computational Physics

The introduction of calculus by Newton and Leibniz in the mid-1600s has greatly transformed science by providing the means to quantitatively describe the time evolution of dynamical systems. Complemented by experimental observations, the newly advanced mathematical toolbox of the scientific method has lead to an ongoing explosion of knowledge and technologies. To highlight the historical significance, the newly developed tools of calculus, Newton's second law of motion, and the inverse-square law of gravitation represent the first unified quantitative theory for a wide range of physical phenomena [1]. This achievement is often expressed by the famous example that an apple, which falls from a tree, is governed by the same physical laws as the motion of celestial bodies.[1]

In particular, Newton's second law of motion is formulated as a differential equation, whose solution represents a time-dependent trajectory in the system's state space. This structural property reappears in the quantitative theories of natural phenomena, which have been developed in the subsequent centuries, such as Maxwell's equations, Gibb's fundamental relation of thermodynamics, and the Schrödinger equation. As pointed out by Feynman [2], it appears that all of natures fundamental laws are written in the language of differential equations.

The subset of the physical phenomena, which can be described by *linear* differential equations, is well understood since extensive methods to obtain analytic solutions have been developed in the past centuries. Critically, linear systems can be divided into subcomponents, solved separately, and be put back together, which is known as the superposition principle. This has lead, among others, to Fourier analysis, Laplace transforms and other integral transforms. In that sense, linear systems are the sum of their parts [3].

[1] Exemplary for an improved theory, Newton could demonstrate that Kepler's laws of planetary motion could be easily derived within the new framework.

S. Meinecke, *Spatio-Temporal Modeling and Device Optimization of Passively Mode-Locked Semiconductor Lasers*, Springer Theses,
https://doi.org/10.1007/978-3-030-96248-7_1

The analytic treatment of physical phenomena that require descriptions by *nonlinear* differential equations, on the other hand, is often very limited or even impossible. The extension of Newton's treatment of the two-body problem to a three-body problem, e.g., first appeared very difficult and later turned out to be impossible. Progress could only be achieved in the late 1800s by the work of Poincaré, who focused on qualitative rather than quantitative properties of the solutions. On that account, Poincaré developed a geometric approach to determine the asymptotic stability of equilibria and periodic orbits in the dynamical system's state space. This approach eventually evolved into the theory of nonlinear dynamics and stability analysis [4], which nowadays exhibits a scope far greater than physics [3]. However, the downside of this theory is that it only yields the asymptotic behavior of a given dynamical system and no transient dynamics, i.e., no complete trajectories.

Many of mankind's great problems are insufficiently understood, because either the determination of the asymptotic equilibria and their stability is prevented by the involved nonlinearities or the transient dynamics themselves are the object of interest. Exemplifying the former, controlled nuclear fusion for electricity generation has been on the roadmap for almost eight decades, yet still seems unattainable, as nonlinearities hinder the effective control and containment of the generated plasma [5–8]. Illustrating the latter, accurate weather forecasts require quantitatively exact trajectories, which propagate the atmospheric state that has been constructed from the data streams of weather stations [9–11]. The asymptotic system state is of no relevance in that case. Both problems are difficult to tackle, because they are both nonlinear and exhibit many degrees of freedom, i.e., they involve an infinite continuum of spatial variables. Typically, the resulting dynamics are thus complex both in space and time. From a theory and modeling perspective, the complexity in such cases is generated by the self-consistent coupling of the system's individual constituents. Produced by such mechanisms, complex and nonlinear spatio-temporal dynamics can be found in diverse fields ranging from general relativity to turbulent fluids, nonlinear waves, neural networks, atrial fibrillation, and lasers [3].

Significant progress regarding such problems has been made in the past decades due to the invention of high-speed computers [3, 12–14]. Numerical methods both provide the means to calculate individual trajectories by the direct integration of the differential equations and the search for and analysis of equilibria by path continuation techniques [15–17]. Given sufficient computing power, previously intractable problems can be numerically studied, in order to gain evidence and intuition to formulate new hypotheses. The advancements of semiconductor technologies, as described by Moore's law, thus have fueled computational physics by moving more and more problems into the realm of reasonable simulation costs. However, numerical methods often produce large amounts of data, which do not directly reveal the desired answers, but encode them in the simulated features of the considered system's state space. It is therefore often the case that the processing, visualization, and analysis of the simulation data is as challenging as the simulation itself. On that account, appropriate visualization techniques in particular are highly advantageous, in order to find and recognize patterns in the simulation data, which allow to make deductions and generalizations and to form new hypotheses.

This thesis studies the spatio-temporal dynamics of passively mode-locked semiconductor lasers by numerical methods. The complexity in those devices is generated by the self-consistent coupling of the intra-cavity electric field to the electrons of the optically active gain medium [18–20]. While both subsystems can be accurately described by linear partial differential equations, namely Maxwell's equations and the Schrödinger equation, their mutual interaction creates the characteristic semiconductor nonlinearity [20]. In order to not only evaluate the out-coupled optical pulse trains, but also study the underlying mechanisms, techniques to visualize and interpret the spatio-temporal pulse and gain dynamics are developed. The results are utilized to identify and predict optimal laser configurations and operating conditions.

1.2 Semiconductor Lasers

Only one year after lasing[2] was first demonstrated in 1960[3] with a solid-state ruby gain medium [25, 26], the concept of a semiconductor laser was published by Basov et al. [27]. They suggested to achieve the amplification of light by stimulated emission via the recombination of charge-carriers, which are injected across a semiconductor *p-n* junction. This idea was promptly picked up and one year later in 1962 working semiconductor lasers were independently demonstrated by multiple laboratories [28–31].

These early milestones lead to a rapid development, which resulted in lasers being nowadays ubiquitous in science and technology [19, 24, 32]. Semiconductor lasers in particular are among the building blocs of today's digital technology [33], e.g., as light sources of the global optical-fiber data-telecommunication network [34–38]. Even on other laser platforms, e.g., solid-state based active media, semiconductor lasers are often used as efficient optical pump sources [39]. Alongside lasers, semiconductor-based optical amplifiers [40–42], electro-optic modulators [43–45], and photodiodes [46] have been developed as further components of optoelectronic[4] technologies. The ever increasing demand for data transmission and computation capabilities requires and drives constant advancements of semiconductor laser technologies towards improved operation speeds and efficiencies, and reduced footprints [47]. Hence, the current 21st century has repeatedly been termed as the century of photonics [48–50] in analogy to the 20th century of electronics.

Fundamental Light-Matter Interactions in Semiconductors

The working principles of semiconductor lasers and other semiconductor-based optoelectronic devices rely on the interaction of light and matter. Being able to manipulate

[2] The word *laser* is an acronym for "light amplification by stimulated emission of radiation". The back-formed verb *to lase* has been adapted to describe the emission of coherent light from a laser.

[3] The first demonstration of a laser was preceded by a number of important discoveries and predictions [21–23]. See Ref. [24] for a detailed history.

[4] Optoelectronics describes the interface between electronic and photonic technologies.

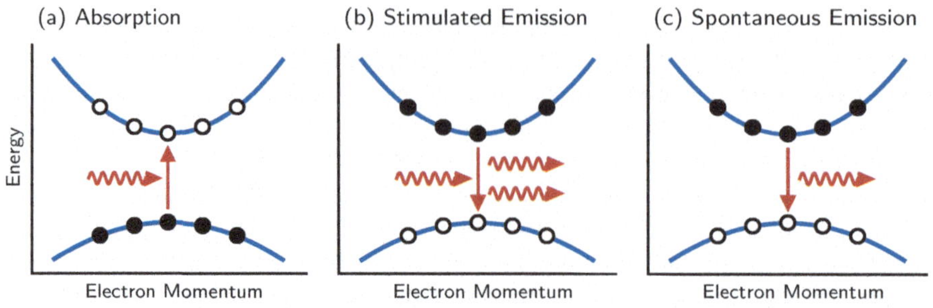

Fig. 1.1 Sketch of the fundamental light-matter interactions in direct band gap semiconductors. The panels from left to right illustrate the (stimulated) absorption, stimulated emission and spontaneous emission of a photon via an intra-band electronic transition. Full and empty circles indicate occupied and unoccupied electronic states in the valance band (bottom) and the conduction band (top). Based on [53]

one of those subsystems hence allows to manipulate the other one as well. The interaction is provided by optically active quantum-state transitions, which are provided by the electronic band structure in semiconductors materials [19, 20, 51, 52].

In thermodynamic equilibrium, i.e., without external driving, the electron-state occupation numbers are distributed according to the Fermi-Dirac distribution [19, 20]. Characteristic for semiconductors, the band structure exhibits an energy gap, which contains the chemical potential (Fermi energy). As a result, the respective lower band is almost fully occupied and is referred to as the *valance band*, while the upper band is almost empty and is referred to as the *conduction band*. The considered optical transitions then occur across the band gap between the conduction and the valance band.

The fundamental interaction processes between photons and the electronic semiconductor band structure are sketched in Fig. 1.1. The panels from left to right represent the (stimulated) absorption, stimulated emission and spontaneous emission of a photon via an intra-band electronic transition. Full and empty circles indicate occupied and unoccupied electronic states. In the case of (stimulated) absorption (a), a photon promotes an electron from the valance band to a vacant conduction-band state. Stimulated emission (b) describes the reverse process, where a photon triggers the transition of a conduction-band electron to an unoccupied valance-band state under the emission of a photon, which is identical to the initial photon. Lastly, the relaxation of a conduction-band electron to an empty valance-band state under the emission of a photon can also occur spontaneously as indicated in (c). In all cases, the energy $\hbar\omega$ of the participating photons must match the energy of the electronic transition $\epsilon_c - \epsilon_v$ for the process to occur.

The respective transition rates can be either directly calculated via a quantized theory [51, 54] or described via their Einstein coefficients $B^{c\leftrightarrow v}_{\text{stim,abs}}$ [52, 54]. In the latter case, the rates can be written as

$$\Gamma_{\text{stim}}^{\text{c}\to\text{v}} = B_{\text{stim}}^{\text{c}\to\text{v}} \rho_\text{c}(1-\rho_\text{v}) n_\text{ph}(\epsilon_\text{cv}) \tag{1.1}$$

for the stimulated emission process and

$$\Gamma_{\text{abs}}^{\text{v}\to\text{c}} = B_{\text{st}}^{\text{v}\to\text{c}} \rho_\text{v}(1-\rho_\text{c}) n_\text{ph}(\epsilon_\text{cv}) \tag{1.2}$$

for the stimulated absorption process. ρ_v and ρ_c denote the occupation numbers of the considered valance and conduction band states and $n_\text{ph}(\epsilon_\text{cv})$ the photon number corresponding to the transition energy ϵ_cv. Hence, both rates are proportional to the occupancy and vacancy of the initial and final electron state and the photon number with the matching energy. Either calculations with a quantized theory or thermodynamic equilibrium principles yield that the Einstein coefficients for stimulated absorption and emission are identical if the respective initial and final states are equally degenerate [51, 54].

To achieve a net amplification of light, the stimulated emission process must dominate the stimulated absorption process. This requires the electron distribution to be driven far out of equilibrium, such that the electron occupation numbers are larger in the conduction band than in the valance band, which is commonly referred to as a *population inversion* [18]. The generation of such an inversion can be achieved either by an optical or an electrical pump process. The latter in particular has been crucial for the success of semiconductor lasers since it enables the straightforward integration with electronics [19, 24, 52].

Semiconductor Heterostructures

The working principles of an electrically pumped semiconductor *p-n* junction are illustrated in Fig. 1.2, which sketches the electron band structure across a *p-n* junction without (a) and with (b) an applied forward bias. The upper edge of the valance band and the lower edge of the conduction band are indicated by solid blue lines and occupied states by gray shaded areas. The chemical potential μ is denoted by black dashed lines. The *p*- and *n*-doping is achieved by acceptor and donor impurities,

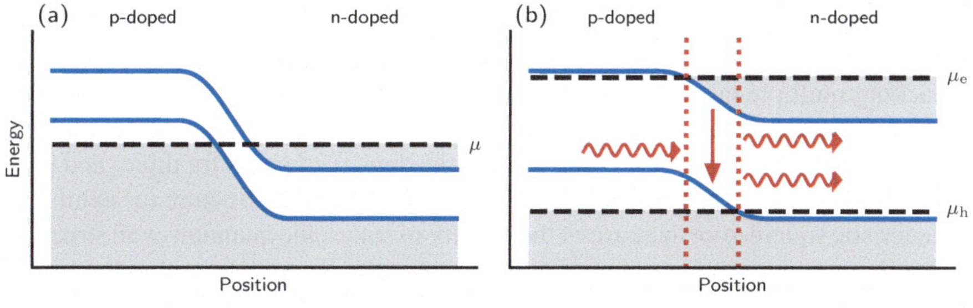

Fig. 1.2 Sketch of the electron band structure across a semiconductor *p-n* junction. μ indicates the chemical potential. **a** Without an applied bias and **b** with an applied forward bias, such that the energy levels are shifted and electrons and holes can recombine at the *p-n* junction. Based on [19]

which provide valance-band vacancies (holes) in the former and conduction-band occupancies in the latter case [19]. Compared to intrinsic semiconductors, doped semiconductors conduct electricity well.

In the case of no applied bias (a), the chemical potential μ is constant across the device and thus no net flow of charge-carriers within the device occurs. Moreover, charge-carriers recombine in the interface region of the *p*-*n* junction, such that a depletion region with no free charge-carriers is created. Hence, no optically active electron transitions occur between the bands.

Applying a forward bias alters the electron energies as sketched in (b) and drives the distribution out of thermal equilibrium [19]. However, fast carrier-carrier and carrier-phonon interactions establish quasi-Fermi distributions in both the valance and the conduction bands, which are described by their respective chemical potentials μ_h and μ_e as indicated in (b). A net flow of electrons and holes towards the interface region results and allows for their recombination across the band gap under the emission of photons. This process can be kept up by continuously injecting further electrons and holes into the *p*-*n* junction via electrodes, which are contacted to the *p*- and *n*-doped sides.

If such an electrically pumped semiconductor *p*-*n* junction is placed in between two mirrors, which form a resonator, a laser is constructed. However, the first lasers, which implemented such simple designs, could only be operated with short current pulses and at cryogenic temperatures, and exhibited large lasing threshold currents [24, 32, 33]. The first important step towards room temperature semiconductor lasers with continuous-wave (cw) emission was made by Kroemer in 1963 with the proposition of semiconductor heterostructures [55]. Due to the lack of mature fabrication technology, initial progress was slow and it took five years until this idea could be demonstrated [56, 57].[5] Replacing the simple *p*-*n* junction with multiple layers of different semiconductor compositions allowed for a better confinement of charge-carriers and photons and thus improved the lasing performance [19, 33].

The next significant step towards improved semiconductor lasers was achieved by the introduction of *quantum wells* [58, 59], which are thin active regions that provide charge-carrier quantum confinement in one spatial dimension [19]. Even though their thinness reduces the overlap between the lasing modes and the active medium, the overall optical gain could be increased via a larger density of states at lower energies, which results from the effective two-dimensional confinement structure [33]. The disadvantage of the reduced overlap can be furthermore mitigated by stacking multiple layers of quantum wells [60].

The difference to semiconductor bulk active regions is highlighted in Fig. 1.3, where the two leftmost panels (a) and (b) plot the density of states for three- and two-dimensional electron confinement structures. While the bulk structure (a) exhibits a characteristic square-root increase of the density of states, the quantum-well structure (b) shows a step-like increase that provides a non-zero density of states at the lowest energy. Since the occupation numbers follow quasi-Fermi distributions [19], the

[5] The development of practical room-temperature semiconductor lasers earned Kroemer and Alferov the Nobel Prize in 2000 [24].

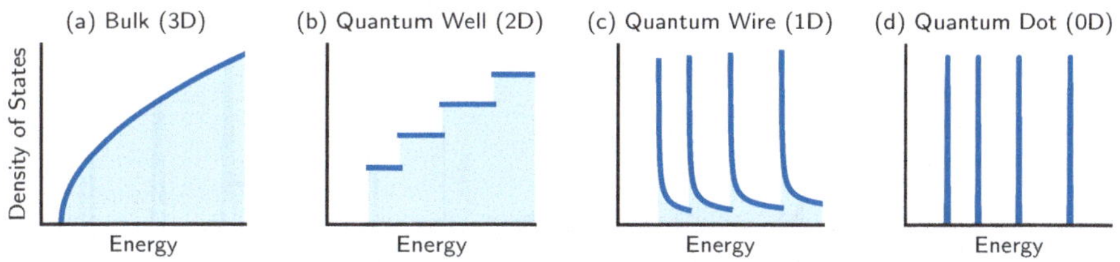

Fig. 1.3 Density of states for different dimensional confinements as a function of the energy. Lower dimensional structure exhibit a higher density of states at the lower energies

increased density of states at the lower energies produces a larger optical gain at small pump currents and thus greatly improves the efficiency of the laser.

The principle of charge-carrier quantum confinement can be continued to two and three spatial dimensions as shown in Fig. 1.3c, d. The corresponding one and zero-dimensional structures are commonly referred to as *quantum wires* [61] and *quantum dots* [62]. Both structures further concentrate the density of states to lower energies and thereby achieve strong optical gain at even smaller pump currents [63–66]. Semiconductor quantum-dot based lasers in particular have been shown to offer advantages such as high differential gain, ultra-fast gain recovery, broad gain spectra, and low-temperature sensitivity [36, 67–69].

1.3 Optical Pulse Trains and Frequency Combs

Optical frequency combs are optical spectra, which are characterized by equidistantly-spaced well-defined lines [39, 70]. If the individual lines are locked, i.e., have a fixed phase relationship, these combs find a multitude of applications in optical data communication [36, 71–73], metrology [70, 74–78], medicine [79–81], optical clocking [82, 83], and super-continuum generation [84–87].

Locked frequency combs are related by a Fourier transform to regular optical pulse trains in the time domain. Hence, their properties and generation mechanism can be equally analyzed and discussed in both the frequency and the time domain. Figure 1.4 shows sketches of a frequency comb (a) and the corresponding pulse train (b). By the Fourier transform, the comb spacing ν_c is related to the inverse period $1/T$ of the pulse train and the width of the comb (FWHM of the red envelope) is related to the inverse pulse width $1/\Delta_{pw}$. Therefore, the generation of ultra-short and regular optical pulses in the time domain produces a broad and stable comb in the frequency domain.

Various frequency-comb generation techniques have been developed in the past decades, which include electro-optic modulation of cw-lasers [88–90], gain and Q-switching of cw-lasers [70, 91–93], four-wave mixing [74, 94–96], and mode-locking of lasers [38, 39, 70, 97–104]. The last technique in particular has been proven to be deployable in a wide range of cases and is therefore further discussed in this work.

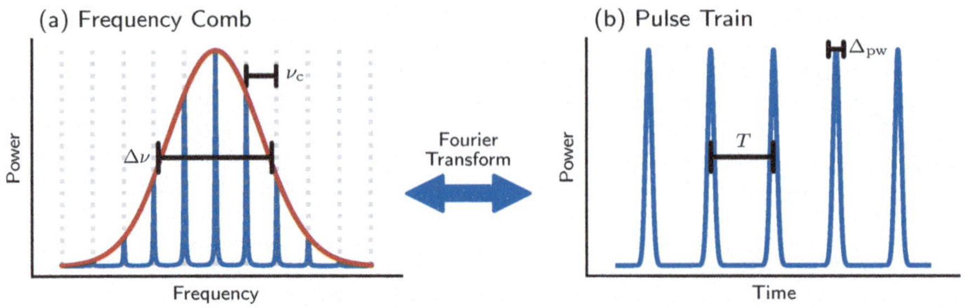

Fig. 1.4 Sketch of a frequency comb (**a**) and the corresponding regular pulse train (**b**). Both are related by a Fourier transform, which implies an inverse relationship between the characteristic measures. The comb spacing ν_c corresponds to the inverse pulse period $1/T$ and the comb width $\Delta\nu$ to the inverse pulse width $1/\Delta_{pw}$

Mode-Locking

The frequencies of the light emitted by a laser are both determined by the amplifying medium and the resonator. The latter imposes boundary conditions on the electric field, which only allow for the propagation of certain distinct and discrete modes. In the case of a Fabry-Perót cavity, e.g., the length of the resonator must be an integer-multiple of the respective modes' wavelengths. The bandwidth of the active medium, i.e., the gain bandwidth, then determines which resonator modes can be amplified. In stable cw-lasers, one mode, referred to as the maximum gain mode, wins the competition for the optical gain and suppresses the emission from all other modes [18]. In multi-mode lasers, on the other hand, mechanisms such as spatial-hole burning [18, 105] counteract the dominance of a maximum gain mode and thereby lead to the simultaneous emission from multiple modes. The individual modes are then coupled via the common gain medium but may nonetheless exhibit asynchronous dynamics. Hence, the emitted light, i.e., the superposition of all modes, generally does not exhibit any regular structure.

The emission characteristics are, however, drastically changed if a fixed relationship among all the relative phases ϕ_n of the participating modes can be established. In that case, the modes are locked and experience synchronized phase dynamics. Determined by the inverse mode-spacing $1/\nu_C$, all modes then periodically reestablish the same interference pattern. Accordingly, the periodicity of the interference pattern manifests in the emitted light intensity. If the phase relationship of the modes is furthermore described by the same constant relative-phase difference for all neighboring modes,[6] all modes periodically exhibit the same instantaneous phase and interfere constructively.

To illustrate the generation of short optical pulses, the implications of three different relative-phase relationships are demonstrated in Fig. 1.5. The top row shows the gain (red envelope) and the relative phases (black dots) of the cavity modes. The

[6] This condition translates to relative-phase relationship that is linear in the optical frequency and propagation constant.

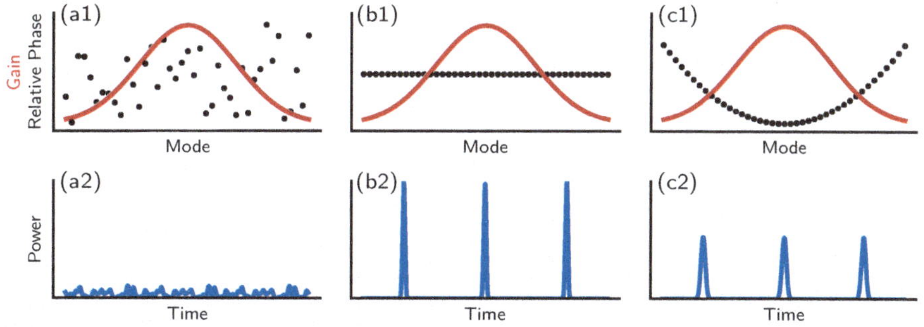

Fig. 1.5 Phase synchronization of the longitudinal cavity modes. The top row presents the gain (red envelope) and the relative phases (black dots) of the cavity modes. The bottom row plots the time dependence of the respective power, i.e., the squared modulus of their superposition. Three cases for the relative-phase relationships are considered: **a** random, **b** linear (constant), and **c** nonlinear (quadratic)

modes are represented by plane waves with the amplitudes A_n, which are proportional to the gain, the optical mode frequencies ω_n, which linearly increase with n, and the relative phases ϕ_n. The respective time dependent powers are then computed according to

$$P(t) = |E(t)|^2 = \left| \sum_n A_n e^{i(\omega_n t + \phi_n)} \right|^2 \tag{1.3}$$

and shown in the bottom row of Fig. 1.5. The left column considers randomly-distributed relative phases ϕ_n (a1), which produce a low-power time series (a2) without distinct features. Note, however, that although random, the phase differences are constant in time and thus the power time series is periodic with the inverse mode spacing $\Delta\nu$. The middle column presents a linear relative-phase relationship with equal relative phases ϕ_n. The resulting time dependent power (b2) exhibits the desirable short and high peak power pulses. Lastly, the third column presents a quadratic relative-phase relationship (c1). In that case, the relative phases ϕ_n never perfectly line up and thus produce pulses (c2), which are well defined but not as narrow and strong as in (b). Such pulses are commonly referred to as chirped [39, 103] and can be compressed retroactively by compensating the deviations from a linear phase relationship by additional optics.

Note that all presented examples technically present mode-locked frequency combs, since the relative phases ϕ_n of the individual modes remain constant in time. The term *mode-locking*, however, typically rather refers to the cases (b) and (c), where the relative-phase relationship promotes the generation of short optical pulses. Hence, mode-locking is often directly associated with the generation of ultra-short optical pulses.

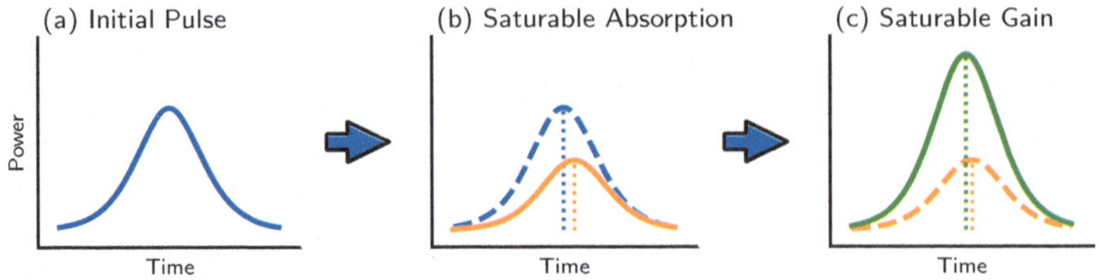

Fig. 1.6 Sketch of a proposed pulse-shaping mechanism in passively mode-locked lasers. **a** Initial pulse profile (blue line), **b** pulse profile after the interaction with the saturable absorber (orange line), and **c** pulse profile after the interaction with the saturable gain (green line). Dotted horizontal lines indicate the positions of the pulse maxima. The saturation of the gain and absorber causes an asymmetric amplification/absorption and thereby reshapes the pulse profile. Upon being subject to non-resonant and out-coupling losses, the process periodically repeats

Mode-Locking Techniques

Up to this point, the discussion of mode-locking has been entirely descriptive and did neither suggest nor explain any mechanism to achieve the synchronization of the individual resonator modes. In order to accomplish the locking of the modes, various techniques can be applied and can be distinguished into the categories of active and passive mode-locking.

Active mode-locking is achieved by externally modulating either the electric field or the amplifying medium [18, 39, 106] with a frequency that matches the mode spacing ν_c. This procedure produces sidebands, which excite and lock the neighboring cavity modes and thereby creates mode-locked pulse trains. Active mode-locking thus critically requires additional external components that facilitate the modulation of the field.

Passive mode-locking, on the other hand, employs no external components but places additional optical components within the laser cavity [100, 102, 103, 107, 108], which allow for a self-modulation of the electric field.[7] Mostly, saturable absorbers are used for this purpose since they exhibit the following properties[8]: Low intensity light does not saturate the absorber and can thus be continuously absorbed. High intensity pulses, on the other hand, are able to quickly saturate the absorption and pass without further losses afterwards. Consequently, such configurations favor short pulses, where the light intensity is strongly localized compared to homogeneous cw-light. Since the modulation of the electric field is not limited by external electronics, passively mode-locked laser are able to generate much shorter pulses with widths down to only a few femtoseconds [39, 100, 103].

[7] The self-modulation of the field that eventually leads to the synchronization of the laser modes can also be seen in the broader context of self-organization and pattern formation in dynamic systems [3, 109].

[8] Passive mode-locking can also be achieved via other optical nonlinearities. In the case of Kerr-lens mode-locking for example, the Kerr effect in the gain medium refocuses the electric field based on the incident intensity and thus facilitates mode-locking [39, 110].

The pulse-shaping mechanism in passively mode-locked lasers as proposed and discussed by in Refs. [38, 100, 101, 103, 107, 108, 111–115] can be best understood in the time domain and is sketched in Fig. 1.6. The left panel (a) presents an initial pulse shape. Upon arriving in the saturable absorber (b), the optical pulse excites transitons until the absorber approximately saturates and the remaining pulse can pass. This process reshapes and repositions the pulse as the absorption is strongest at the leading edge and weakest at the trailing edge of the pulse. In the subsequent interaction with the saturable gain (c), a similar but inverted process occurs. The amplification is strongest at the leading edge and weakest at the trailing edge as the gain saturates along the pulse propagation. Incorporating the non-resonant propagation and out-coupling losses, this procedure repeats and the final pulse shape (green line in (c)) represents the initial pulse shape of the succeeding cycle. If the absorber has a lower saturation energy than the gain, this process repeats until an equilibrium is reached and a stable train of mode-locked pulses can be observed. In the referenced literature, the interaction with the absorber is typically associated with a pulse shortening and the interaction with the gain with a pulse broadening. In the equilibrium of stable mode-locking, the interplay of both stabilizes the pulse shape. Especially in extended lasers cavities, the saturation nonlinearities, which are introduced by the gain and absorber, must furthermore counteract the effect of dispersion, to guarantee stable mode-locking [39, 100, 103].

In summary, mode-locking has been described in terms of the lasing modes' synchronization properties. While the lasing modes are best described in frequency space, the mechanism that invokes the locking of the modes is best described in the time domain. In that context, one must be careful with the usage of the term *modes*, since the concept of resonator modes, as introduced in the beginning of this section, is not well-defined in the presence of strong optical nonlinearities. The effect of the latter is a modified group-velocity dispersion, which generally results in a deviation of the observed pulse period from the cold-cavity round-trip time [39, 103, 115, 116]. Hence, the observed frequency comb spacing does not match the spacing of the (empty) resonator modes. The comb structure of the optical spectrum, which is produced by regular pulse trains, could indeed be solely interpreted as a property of the Fourier transform. It is nonetheless reasonable to speak of *mode-locking*, since the observed mode spacing is close to the empty-resonator mode spacing, which is a distinguishing feature when compared to, e.g., Q-switched lasers.

1.4 Thesis Outline

This thesis numerically investigates the emission dynamics of passively mode-locked semiconductor lasers with a special focus on the spatial laser cavity configuration and the pulse-train stability.

Chapter 2 introduces the fundamentals of semi-classical laser theory by deriving the spatially-inhomogeneous Maxwell-Bloch equations in Sects. 2.2 and 2.3. These equations are further adapted to suit the simulation of mode-locked lasers. On that account, two efficient numerical integration schemes are introduced in Sects. 2.4 and 2.5.

Chapter 3 introduces the long-term timing jitter as a measure to quantify the temporal pulse train stability. Multiple estimation techniques and their application to numerically simulated time series are demonstrated and discussed in Sects. 3.2–3.5. The various methods are benchmarked with respect to their computational cost in Sect. 3.6, to obtain condition dependent usage recommendations.

Chapter 4 investigates the mode-locking dynamics of a monolithically integrated three-section tapered semiconductor quantum-dot laser. The introduction of the device in Sect. 4.1 is followed by the detailed derivation of the numerical model in Sect. 4.2. The study of the emission dynamics starts off with an in-depth look into the basic emission states in Sect. 4.3 and then presents the comparison between experimental and simulated results. Building onto that, the emission dynamics, pulse performance, and timing jitter are characterized in the operation parameter space in Sects. 4.5 and 4.6, respectively. Lastly, further device design optimization with respect to the taper angle and the saturable absorber position are explored in Sects. 4.7 and 4.8, respectively, and optimal configurations are suggested.

Chapter 5 studies the emission dynamics of a V-shaped external cavity mode-locked semiconductor laser. Following the introduction of the device and the accompanying experiment in Sect. 5.1, a minimalist delay-differential equation model is derived in Sect. 5.2. The emission characteristics for symmetric V-shaped cavity configuration are investigated in Sect. 5.3 and discussed and interpreted in terms of the gain and net-gain dynamics. The impact of geometrically asymmetric laser cavity configurations is explored in Sect. 5.4. The obtained results further allow for a discussion of the experimental results in Sect. 5.4.2.

Lastly, Chap. 6 presents a summary of this thesis and gives an outlook on possible directions for future work.

References

1. Feynman RP, Leighton RB, Sands M (2011) The Feynman lectures on physics, vol. I: the new millennium edition: mainly mechanics, radiation, and heat, vol 1. Basic Books
2. Feynman RP (2017) The character of physical law. MIT Press
3. Strogatz SH (2014) Nonlinear dynamics and chaos. CRC Press
4. Kuznetsov YA, Sirovich L (1998) Elements of applied bifurcation theory, 2nd edn. In: Applied mathematical sciences. Springer-Verlag, New York, USA. 0-387-98382-1
5. Freidberg JP (2008) Plasma physics and fusion energy. Cambridge University Press
6. Lorenzini R, Martines E, Piovesan P, Terranova D, Zanca P, Zuin M, Alfier A, Bonfiglio D, Bonomo F, Canton A (2009) Self-organized helical equilibria as a new paradigm for ohmically heated fusion plasmas. Nat Phys 5(8):570–574
7. Markidis S, Lapenta G (2010) Multi-scale simulations of plasma with iPIC3D. Math Comput Simulation 80(7):1509–1519

8. Pedersen TS, Otte M, Lazerson S, Helander P, Bozhenkov S, Biedermann C, Klinger T, Wolf RC, Bosch HS (2016) Confirmation of the topology of the wendelstein 7–x magnetic field to better than 1: 100,000. Nat Commun 7(1):1–10
9. Lorenz EN (1963) Deterministic nonperiodic flow. J Atmos Sci 20:130
10. Roulstone I, Norbury J (2013) Invisible in the Storm: the role of mathematics in understanding weather. Princeton University Press
11. Ahrens CD, Henson R (2021) Meteorology today: an introduction to weather, climate, and the environment. Cengage Learning
12. Pang T (1999) An introduction to computational physics. Am J Phys 67(1):94–95. https://doi.org/10.1119/1.19198
13. Thijssen J (2007) Computational physics. Cambridge University Press
14. Press WH, Flannery BP, Teukolsky SA, Vettering WT (2007) Numerical recipes, 3rd edn. Cambridge University Press, Cambridge
15. Doedel EJ, Oldeman BE (2009) Auto-07P: Continuation and bifurcation software for ordinary differential equations. Concordia University, Montreal, Canada
16. Sieber J, Engelborghs K, Luzyanina T, Samaey G, Roose D (2014) DDE-BIFTOOL manual—bifurcation analysis of delay differential equations. arXiv: 1406.7144, June 2014
17. Uecker H, Wetzel D, Rademacher J (2014) pde2path–a matlab package for continuation and bifurcation in 2d elliptic systems. Numer Math Theor Meth Appl 7(1):58–106. https://doi.org/10.1017/s1004897900000295
18. Haken H (1986) Laser light dynamics, 1st edn., vol II. North Holland. ISBN 978-0444860217
19. Chow WW, Koch SW (1999) Semiconductor-laser fundamentals, 1st edn. Springer-Verlag, Berlin Heidelberg, Berlin. 978-3-662-03880-2. https://doi.org/10.1007/978-3-662-03880-2
20. Haug H, Koch SW (2004) Quantum theory of the optical and electronic properties of semiconductors, 2nd edn. World Scientific, Singapore
21. Perot A, Fabry C (1899) On the application of interference phenomena to the solution of various problems of spectroscopy and metrology. Astrophys J 9:87. https://doi.org/10.1086/140557
22. Einstein A (1917) Zur Quantentheorie der Strahlung. Phys Z 18:121–128
23. Schawlow AL, Townes CH (1958) Infrared and optical masers. Phys Rev 112(6):1940
24. Hecht J (2010) Short history of laser development. Opt Eng 49(9):1. https://doi.org/10.1117/1.3483597
25. Maiman TH (1960) Stimulated optical radiation in ruby. Nature 187:493–494
26. Maiman TH, Hoskins RH, D'Haenens IJ, Asawa CK, Evtuhov V (1961) Stimulated optical emission in fluorescent solids. II. spectroscopy and stimulated emission in ruby. Phys Rev 123:1151. https://doi.org/10.1103/physrev.123.1151
27. Basov NG (1961) Possibility of using indirect transistions to obtain negative temperatures in semiconductors. J Exp Theor Phys 39:1033
28. Hall RN, Fenner GE, Kingsley JD, Soltys TJ, Carlson RO (1962) Coherent light emission from GaAs junctions. Phys Rev Lett 9(9):366–368
29. Holonyak JN, Bevacqua SF (1962) Coherent (visible) light emission from $Ga(As_{1-x}P_x)$ junctions. Appl Phys Lett 1(4):82–83. https://doi.org/10.1063/1.1753706
30. Nathan MI, Dumke WP, Burns G, Dill JFH, Lasher G (1962) Stimulated emission of radiation from GaAs p-n junctions. Appl Phys Lett 1(3):62–64. https://doi.org/10.1063/1.1777371
31. Quist TM, Rediker RH, Keyes RJ, Krag WE, Lax B, McWhorter AL, Zeigler HJ (1962) Semiconductor maser of GaAs. Appl Phys Lett 1(4):91–92. https://doi.org/10.1063/1.1753710
32. Bimberg D (2012) Vom hässlichen Entlein zum Schwan – vor fünfzig Jahren wurde der Halbleiterlaser erfunden. Phys J
33. Coleman JJ (2012) The development of the semiconductor laser diode after the first demonstration in 1962. Semicond Sci Technol 27(9):090207
34. Knox WH (2000) Ultrafast technology in telecommunications. IEEE J Sel Top Quantum Electron 6(6):1273–1278. https://doi.org/10.1109/2944.902178
35. Keeler GA, Nelson BE, Agarwal D, Debaes C, Helman NC, Bhatnagar A, Miller DAB (2003) The benefits of ultrashort optical pulses in optically interconnected systems. IEEE J Sel Top Quantum Electron 9(2):477–485. https://doi.org/10.1109/jstqe.2003.813317

36. Bimberg D, Fiol G, Kuntz M, Meuer C, Lämmlin M, Ledentsov NN, Kovsh AR (2006) High speed nanophotonic devices based on quantum dots. Phys Stat Sol A 203(14):3523–3532. https://doi.org/10.1002/pssa.200622488
37. Kuntz M, Fiol G, Lämmlin M, Meuer C, Bimberg D (2007) High-speed mode-locked quantum-dot lasers and optical amplifiers. Proc IEEE 95(9):1767–1778. ISSN 0018-9219. https://doi.org/10.1109/jproc.2007.900949
38. Rafailov EU, Cataluna MA, Avrutin EA (2011) Ultrafast lasers based on quantum dot structures. WILEY-VCH, Weinheim. 978-3-527-40928-0
39. Ducasse A, Rullière C, Couillaud B (2005) Methods for the generation of ultrashort laser pulses: mode-locking. In: Rullière C (ed) Femtosecond laser pulses—principles and experiments. Springer. ISBN 0-387-01769-0
40. Adams MJ, Collins JV, Henning ID (1985) Analysis of semiconductor laser optical amplifiers. IEE Proc Optoelectron 132(1):58–63. ISSN 0267-3932. https://doi.org/10.1049/ip-j.1985.0012
41. O'Mahony MJ (1988) Semiconductor laser optical amplifiers for use in future fiber systems. J Light Technol 6(4):531–544. ISSN 0733-8724. https://doi.org/10.1109/50.4035
42. Schöll E (1988) Dynamic theory of picosecond optical pulse shaping by gain-switched semiconductor laser amplifiers. IEEE J Quantum Electron 24(2):435–442
43. Kaminow IP, Turner EH (1966) Electrooptic light modulators. Appl Opt 5(10):1612–1628. https://doi.org/10.1364/ao.5.001612
44. Islam MN, Hillman RL, Miller DAB, Chemla DS, Gossard AC, English JH (1987) Electroabsorption in GaAs/AlGaAs coupled quantum well waveguides. Appl Phys Lett 50(16). https://doi.org/10.1063/1.97930
45. Zucker JE, Hendrickson TL, Burrus CA (1988) Electro-optic phase modulation in GaAs/AlGaAs quantum well waveguides. Appl Phys Lett 52(12). https://doi.org/10.1063/1.99236
46. Rosencher E, Vinter B (2002) Optoelectronics. Cambridge University Press
47. Bimberg D, Arsenijević D, Larisch G, Li H, Lott JA, Moser P, Schmeckebier H, Wolf P (2014) Green nanophotonics for future datacom and ethernet networks. Proc SPIE 9134:913402. https://doi.org/10.1117/12.2053027
48. Lifante G (2003) Integrated photonics: fundamentals. Wiley Online Library
49. Pauzauskie PJ, Yang P (2006) Nanowire photonics. Mater Today 9(10):36–45
50. Thylén L, Wosinski L (2014) Integrated photonics in the 21st century. Photon Res 2(2):75–81
51. Fox M (2007) Quantum optics: an introduction, vol 15. In: Oxford master series in physics. Oxford University Press
52. Coldren LA, Corzine SW, Mashanovitch M (2012) Diode lasers and photonic integrated circuits. Wiley series in microwave and optical enginieering, 2nd edn. Wiley
53. Lingnau B (2015) Nonlinear and nonequilibrium dynamics of quantum-dot optoelectronic devices. Springer International Publishing, Switzerland, Springer theses. 978-3-319-25803-4. https://doi.org/10.1007/978-3-319-25805-8
54. Loudon R (2000) The quantum theory of light. Oxford University Press, Sept 2000. ISBN 0198501765
55. Kroemer H (1963) A proposed class of hetero-junction injection lasers. Proc IEEE 51(12):1782–1783. https://doi.org/10.1109/proc.1963.2706
56. Alferov ZI, Andreev VM, Korolkov VI, Portnoi EL, Tretyakov DN (1968) Coherent radiation of epitaxial heterojunction structures in the AlAs-GaAs system. Fiz. Tekh. Poluprovodn. 2(10):1545–1547
57. Rupprecht H, Woodall JM, Pettit GD (1967) Efficient visible electroluminescence at 300K from $Ga_{1-x}Al_xAs$ p-njunctions grownby liquid-phase epitaxy. Appl Phys Lett 11(3):81–83
58. Dingle R, Wiegmann W, Henry CH (1974) Quantum states of confined carriers in very thin $Al_xGa_{1-x}As$-GaAs-$Al_xGa_{1-x}As$ heterostructures. Phys Rev Lett 33:27–830. https://doi.org/10.1103/physrevlett.33.827
59. Dingle R, Henry CH (1976) Quantum effects in heterostructure lasers. United States Patent No. 3982207, Sept 1976

60. Dupuis R, Dapkus PD, Chin R, Holonyak N Jr, Kirchoefer SW (1979) Continuous 300K laser operation of single-quantum-well $Al_xGa_{1-x}As$-GaAs heterostructure diodes grown by metalorganic chemical vapor deposition. Appl Phys Lett 34(4):265–267
61. Lelarge F, Dagens B, Renaudier J, Brenot R, Accard A, van Dijk F, Make D, Gouezigou OL, Provost JG, Poingt F, Landreau J, Drisse O, Derouin E, Rousseau B, Pommereau F, Duan GH (2006) Recent advances on InAs/InP quantum dash based semiconductor lasers and optical amplifiers operating at 1.55 μm. IEEE J Sel Top Quantum Electron 13(1):111. ISSN 1077-260X. https://doi.org/10.1109/jstqe.2006.887154
62. Bimberg D, Grundmann M, Ledentsov NN (1999) Quantum dot heterostructures. Wiley, New York. 978-0-471-97388-1
63. Arakawa Y, Sakaki H (1982) Multidimensional quantum well laser and temperature dependence of its threshold current. Appl Phys Lett 40:939
64. Asada M, Miyamoto Y, Suematsu Y (1986) Gain and the threshold of three-dimensional quantum-box lasers. IEEE J Quantum Electron 22(9):1915–1921. ISSN 0018-9197
65. Bimberg D (2008) Semiconductor Nanostructures, 1st edn. Springer-Verlag, Berlin Heidelberg, Berlin. 978-3-540-77898-1
66. Chow WW, Jahnke F (2013) On the physics of semiconductor quantum dots for applications in lasers and quantum optics. Prog Quantum Electron 37:109–184. https://doi.org/10.1016/j.pquantelec.2013.04.001
67. Chow WW, Koch SW (2005) Theory of semiconductor quantum-dot laser dynamics. IEEE J Quantum Electron. 41:495–505. https://doi.org/10.1109/jqe.2005.843948
68. Lüdge K (2012) Modeling quantum dot based laser devices. In: Lüdge K (ed) Nonlinear laser dynamics—from quantum dots to cryptography, chap. 1. Wiley-VCH Weinheim, Weinheim, pp 3–34. ISBN 9783527411009
69. Norman JC, Mirin RP, Bowers JE (2021) Quantum dot lasers-history and future prospects. J Vac Sci Technol A 39(2):020802. https://doi.org/10.1116/6.0000768
70. Diddams SA (2010) The evolving optical frequency comb. JOSA B 27(11):B51–B62
71. Rafailov EU, Cataluna MA, Sibbett W (2007) Mode-locked quantum-dot lasers. Nat Photon 1(7):395–401. https://doi.org/10.1038/nphoton.2007.120
72. Schmeckebier H, Fiol G, Meuer C, Arsenijević D, Bimberg D (2010) Complete pulse characterization of quantum dot mode-locked lasers suitable for optical communication up to 160 Gbit/s. Opt Express 18(4):3415. https://doi.org/10.1364/oe.18.003415
73. Zeb K, Lu Z, Liu J, Mao Y, Liu G, Poole PJ, Rahim M, Pakulski G, Barrios P, Jiang W, Zhang X (2021) InAs/InP quantum dash buried heterostructure mode-locked laser for high capacity fiber-wireless integrated 5G new radio fronthaul systems. Opt Express 29(11):16164–16174. https://doi.org/10.1364/oe.424504
74. Udem T, Holzwarth R, Hänsch TW (2002) Optical frequency metrology. Nature 416:233–237. https://doi.org/10.1038/416233a
75. Keller U, Tropper AC (2006) Passively modelocked surface-emitting semiconductor lasers. Phys Rep 429:67–120
76. Klenner A, Schilt S, Südmeyer T, Keller U (2014) Gigahertz frequency comb from a diode-pumped solid-state laser. Opt Express 22(25):31008. https://doi.org/10.1364/oe.22.031008
77. Coddington I, Newbury N, Swann W (2016) Dual-comb spectroscopy. Optica 3(4):414–426. https://doi.org/10.1364/optica.3.000414
78. Link SM, Maas DJHC, Waldburger D, Keller U (2017) Dual-comb spectroscopy of water vapor with a free-running semiconductor disk laser 356:1164. https://doi.org/10.1126/science.aam7424
79. Loesel FH, Niemz MH, Bille JF, Juhasz T (1996) Laser-induced optical breakdown on hard and soft tissues and its dependence on the pulse duration: experiment and model. IEEE J Quantum Electron 32(10):1717–1722
80. Juhasz T, Loesel FH, Kurtz RM, Horvath C, Bille JF, Mourou G (1999) Corneal refractive surgery with femtosecond lasers. IEEE J Sel Top Quantum Electron 5(4):902–910
81. Nagy Z, Takacs A, Filkorn T, Sarayba M (2009) Initial clinical evaluation of an intraocular femtosecond laser in cataract surgery. J Refract Surg 25(12):1053–1060

82. Delfyett PJ, Hartman DH, Ahmad SZ (1991) Optical clock distribution using a mode-locked semiconductor laser diode system. IEEE J. Lightwave Technol. 9(10):1646–1649
83. Mule AV, Glytsis EN, Gaylord TK, Meindl JD (2002) Electrical and optical clock distribution networks for gigascale microprocessors. IEEE Trans Very Large Scale Integr VLSI Syst 10(5):582–594
84. Mayer AS, Klenner A, Johnson AR, Luke K, Lamont MRE, Okawachi Y, Lipson M, Gaeta AL, Keller U (2015) Frequency comb offset detection using supercontinuum generation in silicon nitride waveguides. Opt. Express 23(12):15440–15451
85. Klenner A, Mayer AS, Johnson AR, Luke K, Lamont MRE, Okawachi Y, Lipson M, Gaeta AL, Keller U (2016) Gigahertz frequency comb offset stabilization based on supercontinuum generation in silicon nitride waveguides. Opt. Express 24(10):11043–11053
86. Waldburger D, Link SM, Mangold M, Alfieri CGE, Gini E, Golling M, Tilma BW, Keller U (2016) High-power 100 fs semiconductor disk lasers. Optica 3(8):844–852. https://doi.org/10.1364/optica.3.000844
87. Waldburger D, Mayer AS, Alfieri CGE, Nürnberg J, Johnson AR, Ji X, Klenner A, Okawachi Y, Lipson M, Gaeta AL, Keller U (2019) Tightly locked optical frequency comb from a semiconductor disk laser. Opt Express 27(3):1786. https://doi.org/10.1364/oe.27.001786
88. Nakajima Y, Inaba H, Hosaka K, Minoshima K, Onae A, Yasuda M, Kohno T, Kawato S, Kobayashi T, Katsuyama T (2010) A multi-branch, fiber-based frequency comb with millihertz-level relative linewidths using an intra-cavity electro-optic modulator. Opt Express 18(2):1667–1676
89. Rueda A, Sedlmeir F, Kumari M, Leuchs G, Schwefel HGL (2019) Resonant electro-optic frequency comb. Nature 568(7752):378–381
90. Zhang M, Buscaino B, Wang C, Shams-Ansari A, Reimer C, Zhu R, Kahn JM, Lončar M (2019) Broadband electro-optic frequency comb generation in a lithium niobate microring resonator. Nature 568(7752):373–377
91. Sharma U, Kim CS, Kang JU (2004) Highly stable tunable dual-wavelength q-switched fiber laser for dial applications. IEEE Photon Technol Lett 16(5):1277–1279
92. Anandarajah PM, Maher R, Xu YQ, Latkowski S, O'carroll J, Murdoch SG, Phelan R, O'Gorman J, Barry LP (2011) Generation of coherent multicarrier signals by gain switching of discrete mode lasers. IEEE Photon J 3(1):112–122
93. Zhou R, Latkowski S, OCarroll J, Phelan R, Barry LP, Anandarajah PM, (2011) 40 nm wavelength tunable gain-switched optical comb source. Opt Express 19(26):B415–B420
94. Ye J, Schnatz H, Hollberg L (2003) Optical frequency combs: from frequency metrology to optical phase control. IEEE J Sel Top Quantum Electron 9(4):1041–1058. ISSN 1077-260X. https://doi.org/10.1109/jstqe.2003.819109
95. Schliesser A, Picque N, Hänsch TW (2012) Mid-infrared frequency combs. Nat Photon 6(7):440–449. ISSN 1749-4885. https://doi.org/10.1038/nphoton.2012.142
96. Lingnau B, Lüdge K (2018) Rabi-oscillation-enhanced frequency conversion in quantum-dot semiconductor optical amplifiers. Opt Quantum Electron 50(2):111. ISSN 1572-817X. https://doi.org/10.1007/s11082-018-1380-9
97. Lamb WE Jr (1964) Theory of an optical maser. Phys Rev 134(6A):A1429
98. Hargrove LE, Fork RL, Pollack MA (1964) Locking of He-Ne laser modes induced by synchronous intracavity modulation. Appl Phys Lett 5(1):4–5
99. Ippen EP, Shank CV, Dienes GJ (1972) Passive mode locking of the cw dye laser. Appl Phys Lett 21(8):348–350
100. Ippen EP (1994) Principles of passive mode locking. Appl Phys B 58:159–170. https://doi.org/10.1007/bf01081309
101. Keller U, Weingarten KJ, Kärtner FX, Kopf D, Braun B, Jung ID, Fluck R, Hönninger C, Matuschek N, Aus-der AuJ (1996) Semiconductor saturable absorber mirrors (SESAM's) for femtosecond to nanosecond pulse generation in solid-state lasers. IEEE J Sel Top Quantum Electron 2:435
102. Avrutin EA, Marsh JH, Portnoi EL (2000) Monolithic and multi-gigahertz mode-locked semiconductor lasers: constructions, experiments, models and applications. IEE Proc Optoelectron 147(4):251. https://doi.org/10.1049/ip-opt:20000282

103. Haus HA (2000) Mode-locking of lasers. IEEE J Sel Top Quantum Electron 6(6):1173–1185. https://doi.org/10.1109/2944.902165
104. Hillbrand J, Auth D, Piccardo M, Opačak N, Gornik E, Strasser G, Capasso F, Breuer S, Schwarz B (2020) In-phase and anti-phase synchronization in a laser frequency comb. Phys Rev Lett 124:023901. https://doi.org/10.1103/physrevlett.124.023901
105. Bardella P, Columbo L, Gioannini M (2017) Self-generation of optical frequency comb in single section quantum dot Fabry-Perot lasers: a theoretical study. Opt Express 25(21):26234–26252
106. Haus HA (1975a) A theory of forced mode locking. IEEE J Quantum Electron 11(7):323–330. ISSN 0018-9197
107. Haus HA (1975) Theory of mode locking with a slow saturable absorber. IEEE J Quantum Electron 11(9):736. https://doi.org/10.1109/jqe.1975.1068922
108. Haus HA (1975) Theory of mode locking with a fast saturable absorber. J Appl Phys 46(7):3049–3058. https://doi.org/10.1063/1.321997
109. Haken H (1987) Synergetics, 2nd edn. Springer, Berlin
110. Brabec T, Spielmann C, Curley PF, Krausz F (1992) Kerr lens mode locking. Opt Lett 17(18):1292–1294
111. New G (1974) Pulse evolution in mode-locked quasi-continuous lasers. IEEE J Quantum Electron 10(2):115. ISSN 0018-9197. https://doi.org/10.1109/jqe.1974.1145781
112. Derickson DJ, Helkey RJ, Mar A, Karin JR, Wasserbauer JG, Bowers JE (1992) Short pulse generation using multisegment mode-locked semconductor lasers. IEEE J Quantum Electron 28(10):2186–2202
113. Kärtner FX, Keller U (1995) Stabilization of solitonlike pulses with a slow saturable absorber. Opt Lett 20(1):16–18
114. Thompson MG, Rae AR, Xia M, Penty RV, White IH (2009) InGaAs quantum-dot mode-locked laser diodes. IEEE J Quantum Electron 15(3):661–672
115. Jaurigue LC (2017) Passively mode-locked semiconductor lasers: dynamics and stochastic properties in the presence of optical feedback. Springer thesis edition. Springer, Cham. ISBN 978-3-319-58874-2. https://doi.org/10.1007/978-3-319-58874-2
116. Vladimirov AG, Rachinskii D, Wolfrum M (2011) Modeling of passively mode-locked semiconductor lasers. In: Lüdge K (ed) Nonlinear laser dynamics—from quantum dots to cryptography, reviews in nonlinear dynamics and complexity, chap. 8. Wiley-VCH, Weinheim, pp 183–213. ISBN 978-3-527-41100-9

Chapter 2
Semi-classical Laser Theory

Following the discussion of elementary aspects of light-matter interaction and lasers in the introduction in Sect. 1.2, this chapter builds the common theory foundation for the mode-locked laser models, which are presented in detail in the respective chapters. Along a brief introduction to laser modeling hierarchies in Sect. 2.1, the traveling-wave description of the electric field along the optical axis is derived in Sect. 2.2 and the spatially inhomogeneous Maxwell-Bloch equations are derived in Sect. 2.3. In the next step, two schemes for the self-consistent integration of the derived equations are presented in Sects. 2.4 and 2.5.

2.1 Introduction

The classical treatment of light and its interaction with matter, i.e., Maxwell's equations and classical mechanics, works well for a variety of macroscopic cases. Up to this day, the design and optimization of a multitude of electric technologies, such as antennas, waveguides and electric circuits, are performed with great success within this classical framework. On the other hand, a number of phenomena, such as black-body radiation and the photo-effect, can not be described by such means. Those very two examples specifically have sparked the development of quantum mechanics and eventually lead to a full quantum theory of light and its interaction with matter.

In between those two cases, a semi-classical framework is often utilized for the description of lasers and optical amplifiers in which the electric field is predominantly created by stimulated emission. Semi-classical in this context refers to a quantum-mechanical treatment of the active medium and a classical treatment of the electric field.

S. Meinecke, *Spatio-Temporal Modeling and Device Optimization of Passively Mode-Locked Semiconductor Lasers*, Springer Theses,
https://doi.org/10.1007/978-3-030-96248-7_2

The classical field approximation is mostly sufficient for macroscopic devices with large photon numbers [1]. In a fully quantized system, the stimulated emission produces a coherent state [2, 3], also known as the Glauber state, that is subject to a Poissonian photon number distribution. This distribution abides to the law of large numbers and thus the relative fluctuations become negligible at large photon numbers. For that reason, the expectation value of the coherent state sufficiently describes the electric field, which can therefore be replaced by a classical field. On a microscopic level, the light-matter interaction is mediated by microscopic electric dipole moments that are induced by the electric field. These microscopic dipole moments result from excitations of the solid-state semiconductor crystal, which can only be appropriately described by quantum mechanics [1, 4]. Their sum results in a macroscopic polarization of the medium, which then enters the dielectric displacement field.

The complexity of the semi-classical light-matter interaction is generated by the self-consistent coupling of the equations of motion for the electric field to those for the active medium. The procedure is sketched in Fig. 2.1: The classical electric field $\mathcal{E}(\boldsymbol{r}, t)$ induces microscopic dipoles $\boldsymbol{d}_\alpha$ in the active medium. Statistical summation of those dipoles leads to a macroscopic medium polarization $\mathcal{P}(\boldsymbol{r}, t)$, which in turn drives the classical electric field $\mathcal{E}'(\boldsymbol{r}, t)$ via Maxwell's equations. Self-consistency is achieved by imposing the condition $\mathcal{E}(\boldsymbol{r}, t) = \mathcal{E}'(\boldsymbol{r}, t)$. The resulting equations are known as the Maxwell-Bloch equations [1, 4–6].

Since the semi-classical approximation does not include quantum fluctuations of the electric field by definition, spontaneous emission processes can not be properly accounted for. If fluctuations are of interest, they can be introduced phenomenologically via stochastic Langevin forces [7–9]. Hence, the semi-classical partial and ordinary differential equations are transformed into stochastic partial and ordinary equations. This approach is chosen in this thesis for the simulation of the timing jitter of mode-locked lasers, where spontaneous emission noise causes fluctuations of the pulse timing positions [10].

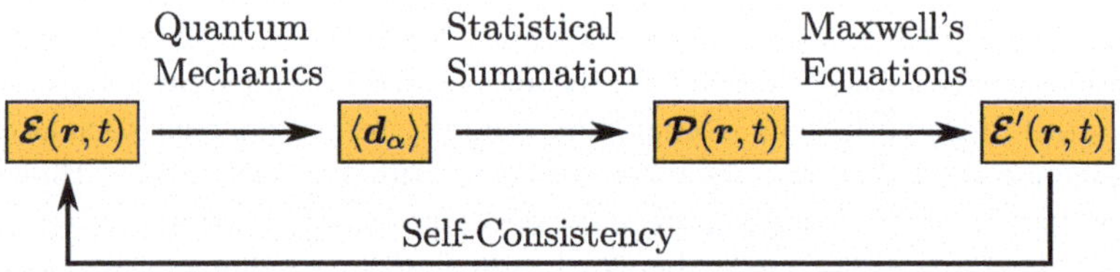

Fig. 2.1 Self-consistent semi-classical coherent light-matter interaction. A classical electric field $\mathcal{E}(\boldsymbol{r}, t)$ induces microscopic dipoles $\boldsymbol{d}_\alpha$ in a quantum mechanically described active medium. Statistical summation leads to a macroscopic medium polarization $\mathcal{P}(\boldsymbol{r}, t)$, which in turn drives the classical electric field via Maxwell's equations. Based on [1]

2.2 Traveling-Wave Equation for the Electric Field

The derivation of the equations of motion for the electric field dynamics sets out with Maxwell's macroscopic equations in matter. The governing equations for the dielectric displacement field $\mathcal{D}$, the magnetic field $\mathcal{B}$, the electric field $\mathcal{E}$, and the magnetizing field $\mathcal{H}$ are given by

$$\nabla \cdot \mathcal{D} = \rho_{\text{free}} \tag{2.1}$$

$$\nabla \cdot \mathcal{B} = 0 \tag{2.2}$$

$$\nabla \times \mathcal{E} = -\frac{\partial}{\partial t}\mathcal{B} \tag{2.3}$$

$$\nabla \times \mathcal{H} = \boldsymbol{j}_{\text{free}} + \frac{\partial}{\partial t}\mathcal{D}, \tag{2.4}$$

where ρ_{free} and $\boldsymbol{j}_{\text{free}}$ denote free charges and free charge currents. In these equations, the electric field $\mathcal{E}$ is related to the dielectric displacement $\mathcal{D}$ via

$$\mathcal{D} = \epsilon_{\text{b}}\mathcal{E} + \mathcal{P}, \tag{2.5}$$

where ϵ_{b} is the background permittivity of the host medium, which is related to the linear response of the medium and $\mathcal{P}$ is the induced polarization (dipoles per unit volume). The magnetic field $\mathcal{B}$ is related to the magnetizing field $\mathcal{H}$ via

$$\mathcal{B} = \mu\mathcal{H}, \tag{2.6}$$

where μ is the permeability of the host medium. Since μ in semiconductor materials is typically approximated by $\mu \approx \mu_0$ [1], the background permittivity ϵ_{b} is given by $\epsilon_{\text{b}} = n_{\text{b}}^2\epsilon_0$ where n_{b} is the background refractive index and ϵ_0 is the vacuum permittivity.

To further proceed, the curl of Eq. (2.3) is combined with the partial time derivative of Eq. (2.4) to yield the relation

$$\nabla \cdot (\nabla \cdot \mathcal{E}) - \nabla^2\mathcal{E} = -\mu_0\frac{\partial}{\partial t}\boldsymbol{j}_{\text{free}} - \mu_0\epsilon_{\text{b}}\frac{\partial^2}{\partial t^2}\mathcal{E} - \mu_0\frac{\partial^2}{\partial t^2}\mathcal{P}. \tag{2.7}$$

With neither free charges ρ_{free} nor free currents $\boldsymbol{j}_{\text{free}}$, the approximation $\nabla \cdot \mathcal{E} \approx 0$, and the relation $\mu_0\epsilon_{\text{b}} = n_{\text{b}}^2/c_0^2$, this expression simplifies to the inhomogeneous wave equation

$$\nabla^2\mathcal{E} - \frac{n_{\text{b}}^2}{c^2}\frac{\partial^2}{\partial t^2}\mathcal{E} = \mu_0\frac{\partial^2}{\partial t^2}\mathcal{P}, \tag{2.8}$$

where the induced polarization $\mathcal{P}$ drives the evolution of the electric field $\mathcal{E}$. The polarization $\mathcal{P}$ is exclusively composed of the optical transitions of the active medium, whereas the background refractive index n_b accounts for the linear polarization of the host medium.

In the next step, a coordinate system is chosen in which the optical axis of the laser cavity defines the z-axis along which the electric field is further decomposed into forward (+) and backward (−) traveling plane waves and the envelope functions $\Psi_{\mathcal{E}}^{\pm}$. The according ansatz thus reads

$$\mathcal{E}(\boldsymbol{r},t) = \boldsymbol{e}\frac{1}{2}\left[\Psi_{\mathcal{E}}^{+}(\boldsymbol{r},t)e^{ikz} + \Psi_{\mathcal{E}}^{-}(\boldsymbol{r},t)e^{-ikz}\right]e^{-i\omega t} + \text{c.c.}, \tag{2.9}$$

where $\boldsymbol{e}$ denotes the unit vector in the direction of the field polarization, k the propagation constant and c.c. abbreviates *complex conjugated.* The same ansatz

$$\mathcal{P}(\boldsymbol{r},t) = \boldsymbol{e}\frac{1}{2}\left[\Psi_{\mathcal{P}}^{+}(\boldsymbol{r},t)e^{ikz} + \Psi_{\mathcal{P}}^{-}(\boldsymbol{r},t)e^{-ikz}\right]e^{-i\omega t} + \text{c.c.} \tag{2.10}$$

is made for the polarization. Inserting both the expressions for the electric field Eq. (2.9) and the polarization Eq. (2.10) into the wave equation (2.8) and separating the forward (+) components yields the equation

$$\begin{aligned}
&\left[(\nabla^2\Psi_{\mathcal{E}}^{+})e^{ikz} + 2ik(\partial_z\Psi_{\mathcal{E}}^{+})e^{ikz} - k^2\Psi_{\mathcal{E}}^{+}e^{ikz}\right]e^{-i\omega t} + \text{c.c.}\\
&-\frac{n_b^2}{c^2}\left[\partial_t^2\Psi_{\mathcal{E}}^{+}e^{ikz} + 2i\omega\partial_t\Psi_{\mathcal{E}}^{+}e^{ikz} - \omega^2\Psi_{\mathcal{E}}^{+}e^{ikz}\right]e^{-i\omega t} + \text{c.c.}\\
&= \mu_0\left[\partial_t^2\Psi_{\mathcal{P}}^{+}e^{ikz} + 2i\omega\partial_t\Psi_{\mathcal{P}}^{+}e^{ikz} - \omega^2\Psi_{\mathcal{P}}^{+}e^{ikz}\right]e^{-i\omega t} + \text{c.c.}.
\end{aligned} \tag{2.11}$$

Using the dispersion relation $k^2 = \omega^2 n_b^2/c^2 = \omega^2/v_g^2$, this equation simplifies to

$$\begin{aligned}
&\left[(\nabla^2\Psi_{\mathcal{E}}^{+})e^{ikz} + 2ik(\partial_z\Psi_{\mathcal{E}}^{+})e^{ikz} - \frac{1}{n_g^2}\left(\partial_t^2\Psi_{\mathcal{E}}^{+}e^{ikz} + 2i\omega\partial_t\Psi_{\mathcal{E}}^{+}e^{ikz}\right)\right]\\
&e^{-i\omega t} + \text{c.c.} = \mu_0\left[\partial_t^2\Psi_{\mathcal{P}}^{+}\mathrm{e}^{\mathrm{i}kz} + 2\mathrm{i}\omega\partial_t\Psi_{\mathcal{P}}^{+}\mathrm{e}^{\mathrm{i}kz} - \omega^2\Psi_{\mathcal{P}}^{+}\mathrm{e}^{\mathrm{i}kz}\right]\mathrm{e}^{-\mathrm{i}\omega t} + \text{c.c.}.
\end{aligned} \tag{2.12}$$

To proceed, this equation is multiplied by $e^{-ikz+i\omega t}$ and the slowly varying envelope approximation (SVEA) [6, 11] is applied. The latter exploits that the amplitude envelope functions $\Psi_{\mathcal{E},\mathcal{P}}$ change both on a much slower time scale and a much shorter length scale than the optical oscillations of the electric field, i.e.,

$$|\partial_t\Psi_{\mathcal{E},\mathcal{P}}^{+}(\boldsymbol{r},t)| \ll \omega|\Psi_{\mathcal{E},\mathcal{P}}^{+}(\boldsymbol{r},t)| \text{ and } |\partial_z\Psi_{\mathcal{E},\mathcal{P}}^{+}(\boldsymbol{r},t)| \ll k|\Psi_{\mathcal{E},\mathcal{P}}^{+}(\boldsymbol{r},t)|. \tag{2.13}$$

Hereinafter, terms that oscillate with $e^{-2ikz+2i\omega t}$, which result from the complex conjugated part of the equation and higher derivatives, can therefore be dropped. Hence, only the first-order derivatives of $\Psi_{\mathcal{E}}^{+}$ remain and the traveling-wave equation for the slowly varying envelope functions $\Psi_{\mathcal{E},\mathcal{P}}^{+}$ of the forward-traveling components

$$2ik\partial_z\Psi^{+}_{\mathcal{E}}(\boldsymbol{r},t) + \frac{1}{v_{\mathrm{g}}^2}2i\omega\partial_t\Psi^{+}_{\mathcal{E}}(\boldsymbol{r},t) = \mu_0\omega^2\Psi^{+}_{\mathcal{P}}(\boldsymbol{r},t) \tag{2.14}$$

is obtained. Performing the same calculations for the backward (−) components $\Psi^{-}_{\mathcal{E},\mathcal{P}}$ simply flips the sign in front of the partial derivative ∂_z. Thus the final form of the traveling-wave equation is given by

$$\left[\pm\partial_z + \frac{1}{v_{\mathrm{g}}}\partial_t\right]\Psi^{\pm}_{\mathcal{E}}(\boldsymbol{r},t) = \frac{i\omega}{2\epsilon_{\mathrm{b}}v_{\mathrm{g}}}\Psi^{\pm}_{\mathcal{P}}(\boldsymbol{r},t). \tag{2.15}$$

Given the validity of the assumptions, the corresponding solutions $\Psi^{\pm}_{\mathcal{E},\mathcal{P}}$ fully specify the electric field $\boldsymbol{\mathcal{E}}$ and the polarization $\boldsymbol{\mathcal{P}}$ according to Eqs. (2.9) and (2.10).

It is often convenient to further separate the transverse spatial dependence of the envelope functions from their longitudinal and time dependence. Thus, the electric field envelope functions are written as

$$\Psi^{\pm}_{\mathcal{E}}(\boldsymbol{r},t) = m^{\pm}_{\mathcal{E}}(\boldsymbol{r})E^{\pm}(z,t), \tag{2.16}$$

where $m_{\mathcal{E}}(\boldsymbol{r})$ is the time invariant mode profile and $E^{\pm}(z,t)$ can be interpreted as the instantaneous field strength along the optical axis z. With the help of the electric field mode profile $m_{\mathcal{E}}(\boldsymbol{r})$, the envelope functions corresponding to the polarization are expressed as

$$\Psi^{\pm}_{\mathcal{P}}(\boldsymbol{r},t) = m^{\pm}_{\mathcal{E}}(\boldsymbol{r})m^{\pm}_{\mathcal{P}}(\boldsymbol{r})P^{\pm}(z,t), \tag{2.17}$$

where $m_{\mathcal{P}}$ denotes the time-invariant spatial polarization profile and $P^{\pm}(z,t)$ the instantaneous polarization strength along the optical axis z. The polarization profile $m_{\mathcal{P}}$ is defined by the active region geometry and acts like an indicator function that evaluates to one in optically active regions and to zero otherwise.

To utilize the newly introduced spatial profiles, Eq. (2.15) is multiplied by $\Psi^{\pm^*}_{\mathcal{E}}$. The resulting equation is complemented by its complex conjugate and integrated over its transverse directions $r_{\perp}$:

$$\int \mathrm{d}r_{\perp}\,\Psi^{\pm^*}_{\mathcal{E}}\left[\pm\partial_z + \frac{1}{v_{\mathrm{g}}}\partial_t\right]\Psi^{\pm}_{\mathcal{E}} + \text{c.c.} = \int \mathrm{d}r_{\perp}\,\frac{\omega}{2\epsilon_{\mathrm{b}}v_{\mathrm{g}}}\left[i\Psi^{\pm^*}_{\mathcal{E}}\Psi^{\pm}_{\mathcal{P}} + \text{c.c.}\right]. \tag{2.18}$$

Inserting the mode profile definitions Eqs. (2.16) and (2.17) into this equation and assuming that changes of the electric field profile are negligible compared to changes of the instantaneous electric field strength, i.e., $(\partial_z m_{\mathcal{E}})E^{\pm} \ll m_{\mathcal{E}}(\partial_z E^{\pm})$, yields

$$\begin{aligned}&\left[\pm\partial_z + \frac{1}{v_{\mathrm{g}}}\partial_t\right]|E^{\pm}|^2\int \mathrm{d}r_{\perp}\,m^{\pm}_{\mathcal{E}}(\boldsymbol{r})^2\\ &= \frac{\omega}{2\epsilon_{\mathrm{b}}v_{\mathrm{g}}}\left[iE^{\pm^*}P^{\pm} + \text{c.c.}\right]\int \mathrm{d}r_{\perp}\,\mathrm{m}^{\pm}_{\mathcal{E}}(\boldsymbol{r})^2\,\mathrm{m}^{\pm}_{\mathcal{P}}(\boldsymbol{r}),\end{aligned} \tag{2.19}$$

where $|E^{\pm}(z,t)|^2$ is proportional to the intensity (photon density) of the forward/backward-traveling field. With the introduction of the transverse confinement factor [9, 12, 13]

$$\Gamma_{\perp}^{\pm}(z) = \frac{\int \mathrm{d}r_{\perp}\, \mathrm{m}_{\mathcal{E}}^{\pm}(\boldsymbol{r})^2\, \mathrm{m}_{\mathcal{P}}^{\pm}(\boldsymbol{r})}{\int \mathrm{d}r_{\perp}\, \mathrm{m}_{\mathcal{E}}^{\pm}(\boldsymbol{r})^2} \tag{2.20}$$

this equation can be written as

$$\left[\pm\partial_z + \frac{1}{v_{\mathrm{g}}}\partial_t\right]|E^{\pm}|^2 = \frac{\omega\Gamma_{\perp}^{\pm}(z)}{2\epsilon_{\mathrm{b}}v_{\mathrm{g}}}\left[iE^{\pm^*}P^{\pm} + \mathrm{c.c.}\right], \tag{2.21}$$

which represents a traveling-wave equation for the intensity $|E^{\pm}(z,t)|^2$. By reversing the construction of this equation, the traveling-wave equation for the instantaneous electric field envelope function $E^{\pm}$ is found:

$$\left[\pm\partial_z + \frac{1}{v_{\mathrm{g}}}\partial_t\right]E^{\pm}(z,t) = \frac{i\omega\Gamma_{\perp}^{\pm}(z)}{2\epsilon_{\mathrm{b}}v_{\mathrm{g}}}P^{\pm}(z,t). \tag{2.22}$$

Note that the complete transverse spatial dependence of the electric field mode has been moved to the transverse confinement factor $\Gamma_{\perp}^{\pm}(z)$. The detour in its definition via the field intensity instead of the field amplitude is required to match the conventional definition via the field intensity. The confinement factor can be understood as the spatial overlap of the electric field mode with the active medium geometry. This notion is also often introduced phenomenologically to the equations of motion for the electric field [7, 8, 14–16].

The instantaneous electric field strength $E^{\pm}(z,t)$ has been constructed such that the integral of $|E^{+}(z,t)|^2 + |E^{-}(z,t)|^2$ along the longitudinal direction z corresponds to the integral of $|\mathcal{E}|^2$ over the mode volume. Thus, $E^{\pm}$ can be used without further conversions to calculate the energy density of the electric field within the laser cavity. On that account, the instantaneous electric field strength $E^{\pm}$ will be simply referred to as the electric field. With a similar argument, the instantaneous polarization strength $P^{\pm}$ will be simply referred to as the polarization.

Discussion of the Traveling-Wave Equation Approximations

The derivation of the traveling-wave equation (2.22) requires two critical approximations, which shall be briefly discussed with respect to their applicability to a mode-locked semiconductor laser. For the sake of this discussion, a mode-locked laser that emits 500 fs pulses at a mean photon energy of $\hbar\omega \approx 1.0\,\mathrm{eV}$ is considered.

Firstly, the slowly varying envelope approximation assumes that optical oscillations are much faster than the envelope functions. The chosen photon energy corresponds to an optical frequency of $\nu \approx 240\,\mathrm{THz}$ and yields the optical cycle length $1/\nu \approx 4.2\,\mathrm{fs}$ as well as the wavelength $\lambda \approx 340\,\mathrm{nm}$ within a semiconductor medium with the refractive index $n \approx 3.7$. A 500 fs pulse, on the other hand, roughly extends over $\approx 40\ \mu$m within the laser cavity. Hence, such a pulse contains approximately 120 optical cycles, which puts the slowly varying envelope approximation on firm ground.

Secondly, the separation of the electric field components $\Psi_{\mathcal{E}}^{\pm}$ into a time invariant mode profile $m_{\mathcal{E}}^{\pm}$ and the instantaneous electric field strength $E^{\pm}$ is strictly valid only for stable single-mode emission [1, 9, 13]. A complete multi-mode description can be achieved by a linear superposition of cavity modes with corresponding mode profiles. In the traveling-wave equation framework, however, the dynamical mode structure of the electric field is contained within the envelope functions $E^{\pm}(z, t)$. Thus, a reasonable decomposition into a mode profile and an electric field strength has to restrict the optically excited modes to narrow region around the mean optical frequency. Regarding the considered laser, 500 fs pulses correspond to spectral width of $\delta\nu \approx 2\,\text{THz}$, which is more than two orders of magnitude less than the mean optical frequency $\nu \approx 240\,\text{THz}$. Therefore, acceptable results can be expected from this approximation as well.

2.3 Maxwell-Bloch Equations

The previous section showed that the evolution of the electric field is driven by the polarization (dipole density) of the optically active transitions of the charge-carrier system. The polarization itself, however, is induced by the incident electric field. Their mutual interaction thus determines the global dynamics of the laser. Hence, this section derives the dynamical equations for the active medium charge-carrier system, which arise due to the coupling to the classical electric field. The relevant constituents of the semiconductor medium for that matter are the electrons in the conduction and the valance band, which are treated quantum mechanically in the second quantization framework. This is done in the free-carrier approximation, which assumes that carrier-carrier and carrier-phonon interactions primarily cause the carrier distribution to relax towards quasi-equilibrium distributions [1]. Hence, carriers behave like an ideal Fermi-gas that only interacts with the electric field. The corresponding free carrier Hamiltonian in the electron-hole picture [1, 2] reads

$$\begin{aligned} H &= H_{\text{kin}} + H_{\text{c-f}} \\ &= \sum_{\alpha} \left(\epsilon_{c,\alpha} c_{\alpha}^{\dagger} c_{\alpha} + \epsilon_{v,\alpha} v_{-\alpha}^{\dagger} v_{-\alpha} \right) + \sum_{\alpha\alpha'} \left[\hbar\Omega_{\alpha\alpha'}(\mathcal{E}) c_{\alpha}^{\dagger} v_{-\alpha'}^{\dagger} + \text{h.c.} \right], \end{aligned} \tag{2.23}$$

where c and v denote the fermionic electron and hole annihilation operators that satisfy the usual commutation relations. The first sum belongs to the kinetic Hamiltonian H_{kin} and contains the single-particle energies ϵ_c and ϵ_v. The index α describes a suitable set of quantum numbers, e.g., momentum and spin, and runs over all electron and hole states. The index $-\alpha$ implies an inverted momentum and spin compared to α. The second sum belongs to the carrier-field interaction Hamiltonian $H_{\mathrm{c-f}}$ and describes the electron-hole coupling via the interaction matrix element $\hbar\Omega_{\alpha\alpha'}(\boldsymbol{\mathcal{E}})$ [2].

The modeling of the mode-locked lasers requires the full spatio-temporal description of the charge-carrier system, since strongly localized pulses propagate within the laser cavity. The local interaction with the optical pulses produces strong saturation effects in the carrier distribution functions. Hence, the interaction has to be described by the spatially inhomogeneous Maxwell-Bloch equations [17–20].

Localized Charge-Carrier Wave Functions

The derivation of the spatially inhomogeneous Maxwell-Bloch equations is first presented for active media with charge-carrier wave functions, which are spatially localized. In this case, the optical transitions are inherently localized and thereby trivially induce the spatial dependence. Among nanostructured semiconductor devices, complete spatial localization occurs in quantum dots, which exhibit an atom-like behavior. For such systems, the interaction matrix element reads [2, 13]

$$\hbar\Omega_{\alpha\alpha'}(\boldsymbol{\mathcal{E}}) = \langle\alpha|_c - e_0\boldsymbol{r}\cdot\boldsymbol{\mathcal{E}}(\boldsymbol{r},t)| - \alpha'\rangle_v = -e_0\int \mathrm{d}^3r\, \Psi^{c^*}_{\alpha}(\boldsymbol{r})\boldsymbol{r}\cdot\boldsymbol{\mathcal{E}}(\boldsymbol{r},t)\Psi^{v}_{-\alpha'}(\boldsymbol{r}), \tag{2.24}$$

where $\Psi^c_\alpha(\boldsymbol{r})$ and $\Psi^v_{-\alpha'}(\boldsymbol{r})$ are the electron and hole wave functions in position space. If the electric field is approximately constant along the spatial extend of the wave functions, i.e., the optical wavelength is much larger than the spatial overlap of $\Psi^c_\alpha(\boldsymbol{r})$ and $\Psi^v_{-\alpha'}(\boldsymbol{r})$, the electric field can be pulled out of the integral and the matrix element reads

$$\hbar\Omega_{\alpha\alpha'}(\boldsymbol{\mathcal{E}}) \approx \langle\alpha|_c - e_0\boldsymbol{r}| - \alpha'\rangle_v\cdot\boldsymbol{\mathcal{E}}(\boldsymbol{r}_{\alpha\alpha'},t) = \boldsymbol{\mu}_{\alpha\alpha'}\cdot\boldsymbol{\mathcal{E}}(\boldsymbol{r}_{\alpha\alpha'},t) \tag{2.25}$$

where $\mu_{\alpha\alpha'}$ is called the dipole interaction transition moment. The electric field has to be evaluated at the center of mass $\boldsymbol{r}_{\alpha\alpha'}$ of the states α and $-\alpha'$. Using this definition and assuming that only electron and hole states with identical momentum and spin produce a non-zero dipole moment, the field-carrier interaction Hamiltonian can be written as

$$H_{\mathrm{c-f}} = \sum_\alpha\left[\boldsymbol{\mu}_\alpha c^\dagger_\alpha v^\dagger_{-\alpha} + \mathrm{h.c.}\right]\cdot\boldsymbol{\mathcal{E}}(\boldsymbol{r}_\alpha,t) = \int \mathrm{d}^3r\hat{\boldsymbol{\mathcal{P}}}(\boldsymbol{r})\cdot\boldsymbol{\mathcal{E}}(\boldsymbol{r},t) \tag{2.26}$$

with the shorthand $\alpha = \alpha\alpha'|_{\alpha=\alpha'}$. The operator of the macroscopic active medium polarization [1] then reads

$$\hat{\boldsymbol{\mathcal{P}}}(\boldsymbol{r}) = \sum_\alpha\left[\boldsymbol{\mu}_\alpha c^\dagger_\alpha v^\dagger_{-\alpha} + \boldsymbol{\mu}^*_\alpha c_\alpha v_{-\alpha}\right]\delta(\boldsymbol{r}-\boldsymbol{r}_\alpha), \tag{2.27}$$

where the delta function ensures the proper spatial dependence. The interaction Hamiltonian thus represents the electric field and active medium polarization interaction. The link between the classical electric field description and the quantum mechanical description of the charge-carrier system is generated by the expectation value $\langle . \rangle$ of the polarization operator

$$\langle \hat{\mathcal{P}}(\boldsymbol{r}) \rangle = \mathcal{P}(\boldsymbol{r}, t), \tag{2.28}$$

where $\mathcal{P}(\boldsymbol{r}, t)$ represents the polarization that drives the classical wave equation (2.8).

To practically evaluate the polarization, the electron and hole creation and annihilation operators are used to define the following system variables

$$\rho_\alpha^{\mathrm{e}} = \langle c_\alpha^\dagger c_\alpha \rangle \tag{2.29}$$

$$\rho_\alpha^{\mathrm{h}} = \langle v_{-\alpha}^\dagger v_{-\alpha} \rangle \tag{2.30}$$

$$\tilde{p}_\alpha = \langle v_{-\alpha} c_\alpha \rangle \tag{2.31}$$

via the quantum mechanical expectation value. ρ_α^{e} and ρ_α^{h} denote the occupation numbers of electrons and holes in the respective states α and $-\alpha$ and $\tilde{p}_\alpha$ denotes the intra-band microscopic polarization amplitude between those two states.

The time evolution of those system variables is computed by first applying the Heisenberg equation of motion to the respective enclosed operators to produce the following ordinary differential equations

$$\frac{\mathrm{d}}{\mathrm{d}t} c_\alpha^\dagger c_\alpha = \frac{i}{\hbar} \left(\boldsymbol{\mu}_\alpha c_\alpha^\dagger v_{-\alpha}^\dagger + \boldsymbol{\mu}_\alpha^* c_\alpha v_{-\alpha} \right) \boldsymbol{\mathcal{E}}(\boldsymbol{r}_\alpha, t) \tag{2.32}$$

$$\frac{\mathrm{d}}{\mathrm{d}t} v_{-\alpha}^\dagger v_{-\alpha} = \frac{i}{\hbar} \left(\boldsymbol{\mu}_\alpha c_\alpha^\dagger v_{-\alpha}^\dagger + \boldsymbol{\mu}_\alpha^* c_\alpha v_{-\alpha} \right) \boldsymbol{\mathcal{E}}(\boldsymbol{r}_\alpha, t) \tag{2.33}$$

$$\frac{\mathrm{d}}{\mathrm{d}t} v_{-\alpha} c_\alpha = -\frac{i}{\hbar} \left(\epsilon_{c,\alpha} + \epsilon_{v,\alpha} \right) v_{-\alpha} c_\alpha - \frac{i}{\hbar} \boldsymbol{\mu}_\alpha \left(c_\alpha^\dagger c_\alpha + v_{-\alpha}^\dagger v_{-\alpha} - 1 \right) \boldsymbol{\mathcal{E}}(\boldsymbol{r}_\alpha, t). \tag{2.34}$$

Further defining the transition frequencies $\omega_\alpha = (\epsilon_{c,\alpha} + \epsilon_{v,\alpha})/\hbar$, applying expectation values and writing these equations in terms of the previously defined system variables yields the equations

$$\frac{\mathrm{d}}{\mathrm{d}t} \rho_\alpha^{\mathrm{e,h}}(\boldsymbol{r}_\alpha, t) = \frac{i}{\hbar} \left(\boldsymbol{\mu}_\alpha \tilde{p}_\alpha^* - \boldsymbol{\mu}_\alpha^* \tilde{p}_\alpha \right) \boldsymbol{\mathcal{E}}(\boldsymbol{r}_\alpha, t) \tag{2.35}$$

$$\frac{\mathrm{d}}{\mathrm{d}t} \tilde{p}_\alpha(\boldsymbol{r}_\alpha, t) = -i\omega_\alpha \tilde{p}_\alpha - \frac{i}{\hbar} \boldsymbol{\mu}_\alpha \left(\rho_\alpha^{\mathrm{e}} + \rho_\alpha^{\mathrm{h}} - 1 \right) \boldsymbol{\mathcal{E}}(\boldsymbol{r}_\alpha, t), \tag{2.36}$$

where the electron (e) and hole (h) dynamics are summarized in one equation, since they are both subject to the same interaction with the polarization and electric field. The spatial coordinate $\boldsymbol{r}$ in these equations is strictly speaking a function of the index α and has to be evaluated accordingly. The macroscopic polarization can then be computed as

$$\mathcal{P}(\boldsymbol{r},t) = \left\langle 2\sum_{\alpha} \boldsymbol{\mu}_{\alpha}^{*} \tilde{p}_{\alpha}(\boldsymbol{r}_{\alpha},t)\delta(\boldsymbol{r}-\boldsymbol{r}_{\alpha}) \right\rangle_{\boldsymbol{r}}, \tag{2.37}$$

where $\langle . \rangle_{\boldsymbol{r}}$ denotes spatial average over a small region around $\boldsymbol{r}$, i.e., a convolution with suitable localized function. The included delta function then only selects microscopic polarization amplitudes $\tilde{p}_{\alpha}$ that contribute to the spatial coordinate $\boldsymbol{r}$. The obtained Eqs. (2.35), (2.36) and (2.37) can be utilized to self-consistently solve the full wave equation (2.8).

The goal of this subsection, however, is to provide the source term for the traveling-wave equation (2.22) of the scalar instantaneous electric field strength $E^{\pm}(z,t)$. Hence, similar to the electric field, the microscopic polarization amplitudes $\tilde{p}_{\alpha}(\boldsymbol{r}_{\alpha},t)$ are decomposed into forward and backward-traveling fields according to

$$\tilde{p}_{\alpha}(\boldsymbol{r}_{\alpha},t) = \left(p_{\alpha}^{+}(\boldsymbol{r}_{\alpha},t)e^{ikz}m_{\mathcal{E}}^{+}(\boldsymbol{r}_{\alpha}) + p_{\alpha}^{-}(\boldsymbol{r}_{\alpha},t)e^{-ikz}m_{\mathcal{E}}^{-}(\boldsymbol{r}_{\alpha})\right)e^{-i\omega t}, \tag{2.38}$$

which defines the new variables $p_{\alpha}^{+}(\boldsymbol{r}_{\alpha},t)$ and $p_{\alpha}^{-}(\boldsymbol{r}_{\alpha},t)$ that represent slowly varying envelope functions. Inserting the decompositions of the microscopic polarization amplitude and the electric field into Eq. (2.36) leads to the equation

$$\begin{aligned}
&\left(\frac{\mathrm{d}}{\mathrm{d}t}p_{\alpha}^{+}(\boldsymbol{r}_{\alpha},t)\right)e^{ikz} + \left(\frac{\mathrm{d}}{\mathrm{d}t}p_{\alpha}^{-}(\boldsymbol{r}_{\alpha},t)\right)e^{-ikz} \\
&= -i(\omega_{\alpha}-\omega)\left[p_{\alpha}^{+}e^{ikz} + p_{\alpha}^{-}e^{-ikz}\right] \\
&\quad - \frac{i\mu_{\alpha}}{2\hbar}\left(\rho_{\alpha}^{\mathrm{e}} - \rho_{\alpha}^{\mathrm{h}} - 1\right)\left[\frac{1}{2}\left(E^{+}e^{ikz} + E^{-}e^{-ikz}\right)e^{-i\omega t} + \mathrm{c.c.}\right],
\end{aligned} \tag{2.39}$$

where the mode profiles $m_{\mathcal{E}}$ have canceled and μ_{α} denotes the projection of the transition dipole moment onto the polarization direction of the electric field. To further proceed, the rotating-wave approximation [1, 5] is applied, which assumes that fast rotating terms $\propto e^{2i\omega t}$ and $\propto e^{\mp 2ikz}$ average out on the timescales of $p_{\alpha}^{\pm}$ and can therefore be neglected. This leads to a decoupling of the forward (+) and backward (−) direction and produces the two separate (±) equations

$$\frac{\mathrm{d}}{\mathrm{d}t}p_{\alpha}^{\pm}(\boldsymbol{r}_{\alpha},t) = -i\Delta\omega_{\alpha}p_{\alpha}^{\pm}(\boldsymbol{r}_{\alpha},t) - \frac{i\mu_{\alpha}}{2\hbar}\left(\rho_{\alpha}^{\mathrm{e}}(\boldsymbol{r}_{\alpha},t) + \rho_{\alpha}^{\mathrm{h}}(\boldsymbol{r}_{\alpha},t) - 1\right)E^{\pm}(z_{\alpha},t), \tag{2.40}$$

where $\Delta\omega_{\alpha} = \omega_{\alpha} - \omega$ denotes the detuning of the optical transition $\Delta\omega_{\alpha}$ from the chosen mode frequency ω. Similarly, the decompositions of the microscopic polarization amplitude and the electric field are inserted into the equation for the occupation numbers Eq. (2.35) and the rotating-wave approximation is applied to yield the equation

$$\frac{d}{dt}\rho_\alpha^{e,h}(\boldsymbol{r}_\alpha, t)$$
$$= \frac{\mu_\alpha}{\hbar}\text{Im}\left(E^+(z_\alpha, t)p_\alpha^{+^*}(\boldsymbol{r}_\alpha, t)|m_\mathcal{E}^+(\boldsymbol{r}_\alpha)|^2 + E^-(z_\alpha, t)p_\alpha^{-^*}(\boldsymbol{r}_\alpha, t)|m_\mathcal{E}^-(\boldsymbol{r}_\alpha)|^2\right), \tag{2.41}$$

which, unlike the polarization equation, still contains the mode profiles $m_\mathcal{E}^+$ and $m_\mathcal{E}^-$. In the next step, the equations for the polarization amplitudes (2.40) and occupation numbers (2.41) are averaged along the transverse directions $\boldsymbol{r}_\perp$ and along a small section of the longitudinal direction z. It is furthermore assumed that both the electric field and the microscopic polarization amplitude are approximately constant along the transverse directions within the active region, such that the mode profiles can be chosen $|m_\mathcal{E}^\pm(\boldsymbol{r})|^2 = 1$ for positions $\boldsymbol{r}$ within the active region. The resulting equations then read

$$\frac{d}{dt}\rho_\beta^{e,h}(z, t) = \frac{\mu_\beta}{\hbar}\text{Im}\left(E^+(z, t)p_\beta^{+^*}(z, t) + E^-(z, t)p_\beta^{-^*}(z, t)\right) \tag{2.42}$$

$$\frac{d}{dt}p_\beta^\pm(z, t) = -i\Delta\omega_\beta p_\beta^\pm(z, t) - \frac{i\mu_\beta}{2\hbar}\left(\rho_\beta^e(z, t) + \rho_\beta^h(z, t) - 1\right)E^\pm(z, t), \tag{2.43}$$

where the index β denotes a set of suitable quantum numbers that describe the localized states at the longitudinal coordinate z. β thus still carries the transverse spatial dependencies of the localized optically active transitions. Note that the spatial dependencies $\boldsymbol{r}$ and z only appear parametrically in these equations. The macroscopic polarization that drives the traveling-wave equation (2.22) can then be written as

$$P^\pm(z, t) = \frac{2}{A_\perp^{\text{act}}\text{dz}}\sum_\beta \mu_\beta^* p_\beta^\pm(z, t), \tag{2.44}$$

where $A_\perp^{\text{act}}\text{dz}$ represents the normalization volume, which contains the transverse cross-section of the active region $A_\perp^{\text{act}}$ and an infinitesimal section dz along the longitudinal direction z. For practical numerical implementations, dz transitions into the finite discretization length Δz and thus produces a finite normalization volume. The index β then runs over all optically active localized states within that volume.

On a final note, the presented treatment of the active medium has omitted carrier-carrier and carrier-phonon interactions. Their addition to the free-carrier Hamiltonian Eq. (2.23) leads in two kinds of modifications in the resulting equations: Firstly, first-order many-body effects produce changes in the transition energies and polarization amplitudes, which are known as the band-gap renormalization and Coulomb enhancement [1, 4, 21, 22]. Secondly, the interactions lead to a scattering of the involved charge-carriers, i.e., a redistribution among the considered states. Those processes moreover lead to loss of coherence, i.e., a dephasing, of the microscopic polarization amplitudes [23–26] with the combined characteristic time T_2. The latter effects can be included by phenomenologically adding the term R_β^{scat} to the occupation number equation and the term $-T_{2,\beta}^{-1}p^\pm$ to polarization amplitude equations.

With an additional term R_β^{sp} that represents losses due to spontaneous emission, the final equations read

$$\frac{\mathrm{d}}{\mathrm{d}t}\rho_\beta^{\mathrm{e,h}}(z,t) = \frac{\mu_\beta}{\hbar}\mathrm{Im}\left(E^+ p_\beta^{+^*} + E^- p_\beta^{-^*}\right) + R_\beta^{\mathrm{scat}} + R_\beta^{\mathrm{sp}} \tag{2.45}$$

$$\frac{\mathrm{d}}{\mathrm{d}t}p_\beta^\pm(z,t) = -i\Delta\omega_\beta p_\beta^\pm - \frac{i\mu_\beta}{2\hbar}\left(\rho_\beta^{\mathrm{e}} + \rho_\beta^{\mathrm{h}} - 1\right)E^\pm - T_{2,\beta}^{-1}p_\beta^\pm. \tag{2.46}$$

Coupled to the traveling-wave equation (2.22), those equations comprise a complete spatially inhomogeneous semi-classical description of a semiconductor laser with localized charge-carrier wave functions. The specifics of the considered laser must then be encoded via the device geometry and internal charge-carrier dynamics. In numerical implementations, each discretization point z is subject to a local set of matter equations.

Delocalized Charge-Carrier Wave Functions

If the charge-carrier wave functions are delocalized within the active medium, the derivation of the Maxwell-Bloch equations becomes more involved, but nevertheless produces equations, which are qualitatively identical to the equations obtained for localized wave functions [17–20, 27–29].

Note that the delocalization may apply to all spatial dimensions (bulk active media), to two spatial dimensions (quantum-well active media), or to only one spatial dimension (quantum-wire active media). The localized dimensions in the two latter cases still naturally induce a spatial dependence. Hence, quantum-well based surface emitting devices may be described by Eqs. (2.45) and (2.46), since the electric field propagation direction matches the localization dimension of the active medium.

The following subsection presents the derivation of the spatially inhomogeneous Maxwell-Bloch equations for fully delocalized charge-carrier wave functions. The procedure follows [28], but neglects many body effects and adapts the equations to the one-dimensional and scalar traveling-wave equation (2.22). The starting point is given by the off-diagonal density matrix elements

$$\rho_{\boldsymbol{k},\boldsymbol{k}'}^{\mathrm{e}} = \langle c_{\boldsymbol{k}}^\dagger c_{\boldsymbol{k}'}\rangle \tag{2.47}$$

$$\rho_{\boldsymbol{k},\boldsymbol{k}'}^{\mathrm{h}} = \langle v_{\boldsymbol{k}}^\dagger v_{\boldsymbol{k}'}\rangle \tag{2.48}$$

$$\tilde{p}_{\boldsymbol{k},\boldsymbol{k}'} = \langle v_{-\boldsymbol{k}} c_{\boldsymbol{k}'}\rangle, \tag{2.49}$$

where $\boldsymbol{k}$ represents the charge-carrier wave vector. If the active medium provides localization within at least one spatial dimension, $\boldsymbol{k}$ is restricted to the respective delocalized subspace and must be complemented by additional suitable quantum numbers that describe the states in the localized dimension. Spatially dependent distribution and microscopic polarization amplitude functions, i.e., Wigner representations, are obtained via the Wigner transform [28, 29]

$$\rho^{\mathrm{e,h}}(\boldsymbol{k},\boldsymbol{r}) = \sum_{\boldsymbol{q}} e^{-i\boldsymbol{q}\boldsymbol{r}} \rho^{\mathrm{e,h}}_{\boldsymbol{k}-\frac{1}{2}\boldsymbol{q},\boldsymbol{k}+\frac{1}{2}\boldsymbol{q}} \tag{2.50}$$

$$\tilde{p}(\boldsymbol{k},\boldsymbol{r}) = \sum_{\boldsymbol{q}} e^{-i\boldsymbol{q}\boldsymbol{r}} \tilde{p}_{\boldsymbol{k}-\frac{1}{2}\boldsymbol{q},\boldsymbol{k}+\frac{1}{2}\boldsymbol{q}}, \tag{2.51}$$

which corresponds to a Fourier transform with respect relative center of momentum. In order to formulate the semi-classical light-matter interaction, the electric field is expanded into spatial Fourier modes

$$\mathcal{E}(\boldsymbol{r},t) = \sum_{\boldsymbol{q}} e^{-i\boldsymbol{q}\boldsymbol{r}} \mathcal{E}_{\boldsymbol{q}}(t), \tag{2.52}$$

where $\mathcal{E}_{\boldsymbol{q}}(t)$ denotes the time dependent complex Fourier coefficients. The light-matter interaction Hamiltonian can then be written as

$$H_{\mathrm{c-f}} = \sum_{\boldsymbol{k},\boldsymbol{q}} \left[\boldsymbol{\mu}_{\boldsymbol{k}} \cdot \mathcal{E}_{\boldsymbol{q}}(t)\, c^{\dagger}_{\boldsymbol{k}+\frac{1}{2}\boldsymbol{q}} v^{\dagger}_{-\boldsymbol{k}+\frac{1}{2}\boldsymbol{q}} + \mathrm{h.c.}\right], \tag{2.53}$$

where $\boldsymbol{\mu}_{\boldsymbol{k}}$ denotes the interband transition dipole moment.[1] Using the Wigner representation of the microscopic polarization amplitude Eq. (2.51) allows to calculate the expectation value of the interaction Hamiltonian and to identify the spatially dependent macroscopic polarization

$$\mathcal{P}(\boldsymbol{r},t) = \frac{1}{V} \sum_{\boldsymbol{k}} \left[\boldsymbol{\mu}_{\boldsymbol{k}} \tilde{p}(\boldsymbol{k},\boldsymbol{r},t) + \mathrm{c.c.}\right], \tag{2.54}$$

which enters the wave equation (2.8). V denotes the normalization volume of the crystal. The temporal evolution of the Wigner representation of the distribution functions and the microscopic polarization amplitudes is obtained via the Heisenberg equation of motion and yields to lowest order [28] the already known optical Bloch equations

$$\frac{\mathrm{d}}{\mathrm{d}t}\rho^{\mathrm{e,h}}(\boldsymbol{k},\boldsymbol{r},t) = \frac{i}{\hbar}\left[\boldsymbol{\mu}_{\boldsymbol{k}}\tilde{p}^{*}(\boldsymbol{k},\boldsymbol{r},t) - \boldsymbol{\mu}^{*}_{\boldsymbol{k}}\tilde{p}(\boldsymbol{k},\boldsymbol{r},t)\right]\mathcal{E}(\boldsymbol{r},t) \tag{2.55}$$

$$\frac{\mathrm{d}}{\mathrm{d}t}\tilde{p}(\boldsymbol{k},\boldsymbol{r},t) = -i\omega_{\boldsymbol{k}}\tilde{p}(\boldsymbol{k},\boldsymbol{r},t) - \frac{i}{\hbar}\boldsymbol{\mu}_{\boldsymbol{k}}\left(\rho^{\mathrm{e}}(\boldsymbol{k},\boldsymbol{r},t) + \rho^{\mathrm{h}}(\boldsymbol{k},\boldsymbol{r},t) - 1\right)\mathcal{E}(\boldsymbol{r},t), \tag{2.56}$$

where $\omega_{\boldsymbol{k}}$ denotes the transition frequency. Higher-order expansion terms produce carrier transport terms, e.g., carrier diffusion in the first-order expansion [28]. In order to couple the dynamics to the traveling-wave equation, the microscopic polarization amplitude is written in terms of slowly varying forward and backward-traveling envelope functions

[1] In order to compute the transition dipole moments in the dipole approximation, the delocalization charge-carrier wave functions must be expanded into Bloch modes [4].

$$\tilde{p}(\boldsymbol{k}, \boldsymbol{r}, t) = \left(p^{+}(\boldsymbol{k}, \boldsymbol{r}, t)e^{ikz}m_{\mathcal{E}}^{+}(\boldsymbol{r}) + p^{-}(\boldsymbol{k}, \boldsymbol{r}, t)e^{-ikz}m_{\mathcal{E}}^{-}(\boldsymbol{r})\right) e^{-i\omega t}, \tag{2.57}$$

where $m_{\mathcal{E}}^{\pm}$ denotes the mode profiles of the forward and backward-traveling electric field. Inserting this equation into the dynamical equations (2.55) and (2.56), applying the rotating-wave approximation [1, 11], and integrating out the transverse direction yields the equations

$$\frac{\mathrm{d}}{\mathrm{d}t}\rho^{\mathrm{e,h}}(\boldsymbol{k}, z, t) = \frac{\mu_{\boldsymbol{k}}}{\hbar}\mathrm{Im}\left(E^{+}(z, t)p^{+^*}(\boldsymbol{k}, z, t) + E^{-}(z, t)p^{-^*}(\boldsymbol{k}, z, t)\right) \tag{2.58}$$

$$\frac{\mathrm{d}}{\mathrm{d}t}p^{\pm}(\boldsymbol{k}, z, t) = -i\Delta\omega_{\boldsymbol{k}}p^{\pm}(\boldsymbol{k}, z, t) - \frac{i\mu_{\boldsymbol{k}}}{2\hbar}$$
$$\left(\rho^{\mathrm{e}}(\boldsymbol{k}, z, t) + \rho^{\mathrm{h}}(\boldsymbol{k}, z, t) - 1\right) E^{\pm}(z, t). \tag{2.59}$$

Those equations are furthermore to be complemented by respective charger-carrier scattering and relaxation and polarization dephasing terms. Note that similar to the fully localized charge-carrier wave function case, the spatial dependence only appears parametrically in those equations due to the aforementioned lowest-order expansion of the dynamical equations. The macroscopic polarization that drives the traveling-wave equation (2.22) can then be written as:

$$P^{\pm}(z, t) = \frac{2}{V}\sum_{\boldsymbol{k}}\mu_{\boldsymbol{k}}^{*}p^{\pm}(\boldsymbol{k}, z, t). \tag{2.60}$$

In conclusion, the spatially inhomogeneous Maxwell-Bloch equations for localized and those for delocalization charge-carrier wave functions are completely isomorphic at the presented modeling depth. The practical differences between the two are embodied by the respective band structure of the active media, which are encoded in the corresponding quantum numbers β and $\boldsymbol{k}$.

2.4 Integration Scheme for the Traveling-Wave Equation

The traveling-wave equation (2.22) is a first-order partial differential equation (PDE) in the variables of the longitudinal position z and the time t. This represents a substantial reduction in complexity compared to the full wave equation Eq. (2.8), which is a second-order PDE in all three spatial dimensions and time. However, the self-consistent coupling to the charge-carrier system of the active medium generally produces a polarization, i.e., a source term that nonlinearly depends on both the electric field and the carrier occupation numbers. This nonlinearity typically defies the analytic treatment of a laser described by a traveling-wave equation model.

On that account, the analysis often resorts to the direct integration of the governing equations with a given initial system state. For the past three decades, finite-difference methods have been successfully applied to simulate the dynamics of mode-locked semiconductor lasers [16, 30–40].

However, the finite-difference methods often turn out to be too computationally expensive to perform statistical analyses of stochastic traveling-wave models and to perform large parameter studies in the case of mode-locked lasers. To account for short pulses and thus broad gain spectra, a sufficiently small time discretization is required, which implies a corresponding small spatial discretization to fulfill the Courant-Friedrichs-Lewy stability criterion. The resulting degrees of freedom from the spatial discretization then produce large computational demands [41].

Nevertheless, the computational costs can be drastically reduced by transforming the partial differential traveling-wave equation into sets of delay-coupled algebraic equations with a much coarser spatial discretization [41]. The procedure to derive the respective equations is presented and discussed in the following: With the method of characteristics, the traveling-wave equation

$$\left[\partial_z \pm \frac{1}{v_{\mathrm{g}}}\partial_t\right] E^{\pm}(z,t) = \pm S^{\pm}(z,t) \tag{2.61}$$

can be transformed into the ordinary differential equation

$$\frac{\mathrm{d}}{\mathrm{d}\zeta} E^{\pm}(z(\zeta), t(\zeta)) = \pm S^{\pm}(z(\zeta), t(\zeta)) \tag{2.62}$$

along its characteristic curve $(z(\zeta), t(\zeta))$, which is parameterized by the variable ζ. The field source term $S^{\pm}$ summarizes resonant effects mediated by the polarization $P^{\pm}$ and non-resonant waveguide losses. Note that the $\pm$, which accounts for the directionality and was previously placed in front of the spatial derivative ∂_z, has been moved to the time derivative ∂_t and the field source term. Formally evaluating the total derivative with respect to ζ yields

$$\frac{\mathrm{d}}{\mathrm{d}\zeta} E^{\pm}(z(\zeta), t(\zeta)) = \partial_z E^{\pm}\frac{\mathrm{d}}{\mathrm{d}\zeta} z(\zeta) + \partial_t E^{\pm}\frac{\mathrm{d}}{\mathrm{d}\zeta} t(\zeta) = \pm S^{\pm}(z(\zeta), t(\zeta)), \tag{2.63}$$

which recovers the partial differential equation in the variables z and t. By comparison to the traveling-wave equation (2.61), the total derivatives of $z(\zeta)$ and $t(\zeta)$ with respect to ζ, and consequently the characteristic curve $(z(\zeta), t(\zeta))$ itself, are found to be

$$\frac{\mathrm{d}}{\mathrm{d}\zeta} z(\zeta) = 1 \quad \rightarrow \quad z(\zeta) = z' + \zeta = \zeta \tag{2.64}$$

$$\frac{\mathrm{d}}{\mathrm{d}\zeta} t(\zeta) = \pm\frac{1}{v_{\mathrm{g}}} \quad \rightarrow \quad t(\zeta) = t' \pm \frac{\zeta}{v_{\mathrm{g}}}. \tag{2.65}$$

z' and t' are the constants of integration, which can be fixed by an initial state in the labframe. Choosing $z' = 0$ results in a rescaled time $t' = t \mp \frac{z}{v_{\mathrm{g}}}$, which corresponds to a co-moving frame in which the time t' is constant for an observer that moves along the electric field with the group velocity v_{g}.

In the next step, the ordinary differential equation (2.62) is integrated along ζ according to the fundamental theorem of calculus. With the explicit knowledge of the characteristic curve $(z(\zeta), t(\zeta))$, the result can be evaluated afterwards in the labframe z and t. Note that the integration along ζ corresponds to an integration over z and t at the same time, but with integration limits only in terms of z. The integration is first performed in the forward (+) direction along a small section Δz:

$$\int_{z-\Delta z}^{z} \mathrm{d}\zeta \frac{\mathrm{d}}{\mathrm{d}\zeta} E^{+}(z(\zeta), t(\zeta)) = \int_{z-\Delta z}^{z} \mathrm{d}\zeta\, S^{+}(z(\zeta), t(\zeta)) \tag{2.66}$$

$$\Leftrightarrow \int_{z-\Delta z}^{z} \mathrm{d}\zeta \frac{\mathrm{d}}{\mathrm{d}\zeta} E^{+}\left(\zeta, t' + \frac{\zeta}{v_{\mathrm{g}}}\right) = \int_{z-\Delta z}^{z} \mathrm{d}\zeta\, S^{+}\left(\zeta, t' + \frac{\zeta}{v_{\mathrm{g}}}\right) \tag{2.67}$$

$$\Leftrightarrow \left[E^{+}\left(z+\zeta, t' + \frac{z}{v_{\mathrm{g}}} + \frac{\zeta}{v_{\mathrm{g}}}\right)\right]_{-\Delta z}^{0} = \int_{-\Delta z}^{0} \mathrm{d}\zeta\, S^{+}\left(z+\zeta, t' + \frac{z}{v_{\mathrm{g}}} + \frac{\zeta}{v_{\mathrm{g}}}\right). \tag{2.68}$$

The last equality has shifted the position z from the integral limits to the integrands. With the introduction of the propagation time $\tau = \Delta z / v_{\mathrm{g}}$, the forward-moving electric field $E^{\pm}(z, t)$ at the position z can be expressed as

$$\Leftrightarrow E^{+}(z, t) = E^{+}(z - \Delta z, t - \tau) + \int_{-\Delta z}^{0} \mathrm{d}\zeta\, S^{+}\left(z+\zeta, t + \frac{\zeta}{v_{\mathrm{g}}}\right) \tag{2.69}$$

$$\approx E^{+}(z - \Delta z, t - \tau) + \frac{\Delta z}{2}\left[S^{+}(z, t) + S^{+}(z - \Delta z, t - \tau)\right], \tag{2.70}$$

where the integral over the source term has been approximated with the trapezoidal rule. This approximation works well as long as the source term does not vary significantly over the integration range. Convergence can be easily achieved by reducing the integration interval Δz, since the error of the trapezoidal approximation scales with $\Delta z^3 \partial_\zeta^2 S^{\pm}$.

The integration of the backward (−) direction along an interval Δz is set up similarly to the forward (+) direction, but with the integration domain from $z + \Delta z$ backward to z, since the goal is to find an expression for $E^{-}(z, t)$:

$$\int_{z+\Delta z}^{z} \mathrm{d}\zeta \frac{\mathrm{d}}{\mathrm{d}\zeta} E^{-}(z(\zeta), t(\zeta)) = -\int_{z+\Delta z}^{z} \mathrm{d}\zeta\, S^{-}(z(\zeta), t(\zeta)). \tag{2.71}$$

Note the flipped sign of the source term due to the backward direction. Repeating the same steps as for the forward direction produces the same result, but with a flipped sign in front of the integration interval Δz:

$$E^-(z,t) = E^-(z+\Delta z, t-\tau) + \int_0^{\Delta z} \mathrm{d}\zeta\, S^-\left(z+\zeta, t-\frac{\zeta}{v_\mathrm{g}}\right) \tag{2.72}$$

$$\approx E^-(z+\Delta z, t-\tau) + \frac{\Delta z}{2}\left[S^-(z,t) + S^-(z+\Delta z, t-\tau)\right]. \tag{2.73}$$

Finally, both the forward (+) and the backward (−) direction can be expressed in the formula

$$E^\pm(z,t) \approx E^\pm(z \mp \Delta z, t-\tau) + \frac{\Delta z}{2}\left[S^\pm(z,t) + S^\pm(z \mp \Delta z, t-\tau)\right], \tag{2.74}$$

which represents the foundation for a general spatial discretization scheme of the laser cavity.

To implement the integration of the traveling-wave equation for a given laser cavity, the longitudinal direction z is discretized into N sections with the length Δz_k and the index $k = 1, 2, \ldots, N$. The electric field and the source term are discretized likewise and the resulting variables $E_k^\pm$ and $S_k^\pm$ represent the mean field and mean source term within the sections Δz_k, which are considered to be localized in the middle of the respective sections z_k. Figure 2.2 presents a sketch of the spatial discretization scheme. The resulting system of delay-algebraic propagation equation for the electric field then reads

$$E_k^\pm(t) \approx E_{k\mp1}^\pm(t-\tau_k^\pm) + \frac{\delta_k^\pm}{2}\left[S_k^\pm(t) + S_{k\mp1}^\pm(t-\tau_k^\pm)\right], \tag{2.75}$$

where the shorthands

$$\delta_k^\pm = \frac{\Delta z_k + \Delta z_{k\mp1}}{2} \tag{2.76}$$

$$\tau_k^\pm = \frac{\Delta z_k + \Delta z_{k\mp1}}{2v_\mathrm{g}} \tag{2.77}$$

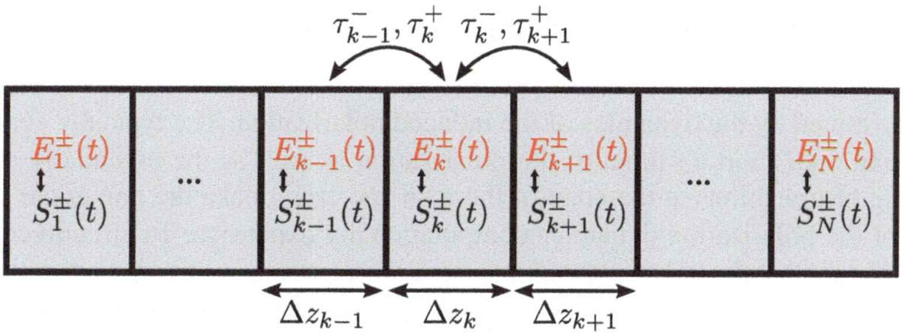

Fig. 2.2 Sketch of the spatial discretization scheme. The device is divided into N sections with length Δz_k. S_k denotes the spatially average source term in the kth section. $\tau_k^\pm$ denotes the propagation times that couple neighboring sections

describe the spatial separations $\delta_k^{\pm}$ and propagation times $\tau_k^{\pm}$ between neighboring field discretization points z_k. Note that the sign $\mp$ accounts for the directionality of the propagation. The boundary conditions, which arise at the facets of the laser cavity, apply to the section $k = 1$ for E_1^+ and to the section $k = N$ for E_N^- and must by implemented separately according to

$$E_1^+(t) \approx \sqrt{r_\mathrm{L}} E_1^-(t - \Delta z_1/v_\mathrm{g}) + \frac{\Delta z_1}{2}\left[S_1^+(t) + \sqrt{r_\mathrm{L}} S_1^-(t - \Delta z_1/v_\mathrm{g})\right] \tag{2.78}$$

$$E_N^-(t) \approx \sqrt{r_\mathrm{R}} E_N^+(t - \Delta z_N/v_\mathrm{g}) + \frac{\Delta z_N}{2}\left[S_N^-(t) + \sqrt{r_\mathrm{R}} S_N^+(t - \Delta z_N/v_\mathrm{g})\right], \tag{2.79}$$

where r_L and r_R denote the intensity reflectivity coefficients of the respective left and right cavity facets.

The complete laser model then adds equations that describe the evolution of the charge-carrier system of each section. The individual sections are delay coupled to their neighbors via the electric field with the respective propagation times τ. The resulting complete system of equations solves the traveling-wave equation equivalently to a finite difference method. However, the delay-algebraic formulation has effectively folded space into propagation time [41], which allows for independent spatial and temporal discretizations that are not subject to the Courant-Friedrichs-Lewy stability condition. This flexibility can enable much coarser spatial discretizations and can thereby massively reduce the computational costs. Hence, this approach has been successfully used to simulate optoelectronic devices from optical amplifiers [42, 43] to (mode-locked) multi-mode lasers [41, 44–47].

2.5 Integration Scheme with an Eliminated Polarization

If the exact spectral evolution of the electric field along the propagation is not required, a simplified modeling approach can be used. This sections follows the ideas of Vladimirov et al. [48] and derives a system of self-consistent equations for the electric field and charge-carrier dynamics under the assumption that the source term of the traveling-wave equation (2.61) linearly depends on the electric field itself.

Adiabatic Elimination of the Polarization

In the full Maxwell-Bloch framework, the spectral features of the active medium are determined by the dynamics of the induced polarization. The typically very fast dephasing, i.e., short T_2 time, of the polarization as well as the potentially strong detuning of transitions at the edge of the gain spectrum make the numerical simulation of the polarization dynamics computationally expensive. To circumvent this issue, the fact that the polarization dynamics are fast, can be utilized to adiabatically eliminate their dynamics. This approximation works well for semiconductors at room temperature, where typical dephasing times are shorter than 100 fs [23, 49–51], while the carrier and field dynamics rather occur on picosecond timescales [13, 52–54]. The microscopic polarization amplitudes can then be assumed to be determined by the quasi-static relation

$$\frac{\mathrm{d}}{\mathrm{d}t} p_\beta^\pm(z,t) = 0. \tag{2.80}$$

Inserting the respective dynamical equation (2.46) yields the algebraic equation

$$p_\beta^\pm(z,t) = -\frac{i\mu_\beta}{2\hbar}\left(\rho_\beta^{\mathrm{e}} + \rho_\beta^{\mathrm{h}} - 1\right) E^\pm T_{2,\beta} \frac{1 - iT_{2,\beta}\Delta\omega_\beta}{1 + \left[\Delta\omega_\beta T_{2,\beta}\right]^2} \tag{2.81}$$

for the microscopic polarization amplitude. Note that this equation represents a complex Lorentzian with respect to $\Delta\omega_\beta$, which has a FWHM of $1/T_{2,\beta}$ and corresponds to the line shape of the individual transition. With the definition of the complex gain coefficient

$$\widetilde{g}_\beta = \frac{T_{2,\beta}|\mu_\beta|^2}{2\hbar^2} \frac{1 - iT_{2,\beta}\omega_\beta}{1 + \left[\Delta\omega_\beta T_{2,\beta}\right]^2} \tag{2.82}$$

the corresponding dynamical equations for the charge-carrier occupation numbers read

$$\frac{\mathrm{d}}{\mathrm{d}t} \rho_\beta^{\mathrm{e,h}}(z,t) = -\mathrm{Re}(\widetilde{g}_\beta)\left(\rho_\beta^{\mathrm{e}} + \rho_\beta^{\mathrm{h}} - 1\right) |E^\pm(z,t)|^2. \tag{2.83}$$

The macroscopic polarization is obtained by summing over all microscopic polarization amplitudes according to Eq. (2.44). Inserting Eq. (2.81) yields

$$\begin{aligned} P^\pm(z,t) &= -\frac{i}{\hbar A_\perp^{\mathrm{act}}\mathrm{d}z} E^\pm(z,t) \sum_\beta |\mu_\beta|^2 \left(\rho_\beta^{\mathrm{e}} + \rho_\beta^{\mathrm{h}} - 1\right) T_{2,\beta} \frac{1 - i\Delta\omega_\beta}{1 + \left[\Delta\omega_\beta T_{2,\beta}\right]^2} \\ &= -\frac{i2\hbar}{A_\perp^{\mathrm{act}}\mathrm{d}z} E^\pm(z,t) \sum_\beta \widetilde{g}_\beta \left(\rho_\beta^{\mathrm{e}} + \rho_\beta^{\mathrm{h}} - 1\right), \end{aligned} \tag{2.84}$$

where the second line incorporates the complex gain coefficients $\widetilde{g}_\beta$. Note that the electric field has been factored out of the sum. This motivates the definition of the macroscopic gain function

$$g^\pm(\{\rho_\beta^{\mathrm{e}}\}, \{\rho_\beta^{\mathrm{h}}\}) = \frac{4\hbar\omega\Gamma_\perp^\pm}{\epsilon_{\mathrm{b}} v_{\mathrm{g}} A_\perp^{\mathrm{act}}\mathrm{d}z} \sum_\beta \widetilde{g}_\beta \left(\rho_\beta^{\mathrm{e}} + \rho_\beta^{\mathrm{h}} - 1\right), \tag{2.85}$$

which dynamically only depends on the set of all optically active occupation numbers $\{\rho_\beta^{\mathrm{e}}\}$ and $\{\rho_\beta^{\mathrm{h}}\}$. Using this definition, the traveling-wave equation that governs the electric field evolution can be written as

$$\left[\pm\partial_z + \frac{1}{v_{\mathrm{g}}}\partial_t\right] E^\pm(z,t) = \frac{g^\pm(\{\rho_\beta^{\mathrm{e}}\}, \{\rho_\beta^{\mathrm{h}}\})}{2} E^\pm(z,t), \tag{2.86}$$

where the factor of 2 accounts for the amplitude gain convention.

The drawback of the adiabatically eliminated polarization amplitudes, however, is that they respond to every electric field as if it were monochromatic and centered at the frequency ω. Thus, the gain spectrum has become infinitely broad and the active medium interacts with electric field components of all frequencies. Nonetheless, this approach works reasonably well for single-mode lasers [1, 13], such as DBR and micro/nano cavity lasers. For multi-mode and thus also mode-locked lasers, on the other hand, the sole use of the just derived equation is not sufficient, since the multi-mode character of the electric field itself is the object under investigation. Hence, the equations must be modified subsequently to account for the gain bandwidth of the laser and restrict the spectral evolution of the electric field. This has been done successfully for mode-locked lasers, by introducing a spectral filtering element to the laser cavity with a lumped element approach [36, 48, 55–57].

Macroscopic Charge-Carrier Equations

In many cases it is not necessary to keep track of the individual occupation numbers $\rho_\beta^{\mathrm{e,h}}$, since fast carrier-carrier and carrier-phonon interaction processes drive the charge-carrier distribution towards quasi-Fermi equilibria within the bands [1, 4, 9, 23]. In that case, the distribution functions are fully determined by the charge-carrier density and temperature. Hence, the gain function Eq. (2.85) only depends on the charge-carrier density and temperature. However, a straightforward analytical calculation of the gain function is often unfeasible due to the intricacies of the semiconductor band structures [9], especially if many-body interactions are to be included [1]. Thus, a multitude of approaches has been developed [6] to obtain effective semiconductor gain functions. At the simplest level, a linearization of the gain around the lasing threshold or any other operation point is carried out. This procedure leads to a differential gain coefficient and an amplitude-phase coupling coefficient (α-factor). To account for spectral and spatial hole burning, gain compression term can be introduced [9]. If the laser is to be operated over a larger range of charge-carrier densities, the gain function may include a logarithmic instead of a linear dependence on the density to model a gain roll-over at larger densities [9].

For the purpose of the following derivations, the gain function, however, does not need to be specified. For simplicity, it is required that it only depends on a single charge-carrier variable[2] ρ. For example, ρ may represent the charge-carrier density in a quantum well or the occupation number of a quantum-dot state. The traveling-wave equation for the forward and backward-moving slowly varying electric field amplitude can then be written as

$$\left[\pm\partial_z + \frac{1}{v_\mathrm{g}}\partial_t\right] E^{\pm}(z,t) = \left[-\frac{\alpha_\mathrm{int}}{2} + \frac{g\left(\rho(z,t)\right)}{2}\right] E^{\pm}(z,t), \tag{2.87}$$

where $g(\rho)$ is the complex valued gain function. α_int furthermore describes linear non-resonant losses. The corresponding equation for the charge-carrier dynamics is given by

[2] The derivation with a full set of occupation numbers would follow analogously.

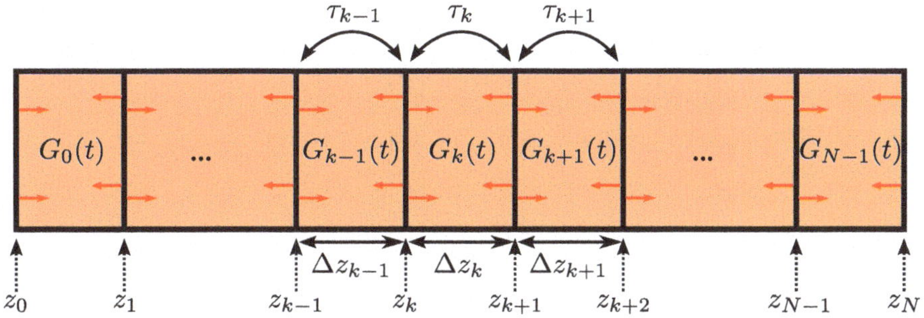

Fig. 2.3 Sketch of the discretization scheme. The device is divided into N sections with $N+1$ facets at the positions z_k. G_k denotes the integrated gain for each section and red arrows indicate the forward and backward-traveling electric field at the $N+1$ section facets. τ_k denotes the propagation time through each respective section k

$$\frac{\mathrm{d}}{\mathrm{d}t}\rho(z,t) = -\frac{\rho(z,t)}{\tau_\rho} + R - v_\mathrm{g}\mathrm{Re}\left\{g\left(\rho(z,t)\right)\right\}\left[|E^+(z,t)|^2 + |E^-(z,t)|^2\right], \tag{2.88}$$

where τ_ρ denotes the charge-carrier lifetime, R summarizes all charge-carrier scattering processes, which change the charge-carrier variable,[3] and v_g denotes the group velocity. The electric field is given in units of ρ, which is achieved by rescaling it with the factor

$$\frac{4\hbar\omega\Gamma_\perp^\pm}{\epsilon_\mathrm{b}A_\perp^\mathrm{act}\mathrm{d}z} \tag{2.89}$$

that appears as a prefactor in Eq. (2.85). All parameters may have a spatial dependence, which is not explicitly written in the equations above.

Electric Field Equations

Similar to the integration scheme presented in Sect. 2.2, the traveling-wave equation (2.87) is integrated over finite distances. Thus, before starting out with the discussion of the electric field dynamics, the spatial discretization scheme is introduced. A sketch that shows all relevant quantities is presented in Fig. 2.3. The electric field is evaluated at $N+1$ longitudinal positions z_k, which divide the device into N sections with length Δz_k. G_k denotes the integrated gain within the sections and red arrows indicate the forward and backward-traveling electric field at the $N+1$ section facets. τ_k denotes the propagation time through each respective section k, which relates to the section lengths via $\tau_k = \Delta z_k / v_\mathrm{g}$, where v_g is the group velocity.

To perform the integration of the traveling-wave equation (2.87), the method of characteristics is used to transform the partial differential equation into an ordinary differential equation

[3] For example, the influx of charge-carriers due to a pump process.

$$\frac{\mathrm{d}}{\mathrm{d}\zeta}E^{\pm}(z(\zeta), t^{\pm}(\zeta)) = \pm\left[-\frac{\alpha_{\mathrm{int}}}{2} + \frac{g\left(\rho(z(\zeta), t^{\pm}(\zeta))\right)}{2}\right]E^{\pm}(z(\zeta), t^{\pm}(\zeta)) \quad (2.90)$$

along the characteristic curve $(z(\zeta), t^{\pm}(\zeta))$. The curve is parameterized by ζ and explicitly given by

$$z(\zeta) = z' + \zeta = \zeta \quad (2.91)$$

$$t^{\pm}(\zeta) = t' \pm \frac{\zeta}{v_{\mathrm{g}}}. \quad (2.92)$$

The integration of the transformed traveling-wave equation (2.90) along a finite section $[z_k, z_{k+1}]$ of the device can be easily done via the fundamental theorem of calculus. Starting with the forward (+) direction, the integration yields

$$\int_{z_k}^{z_{k+1}} \mathrm{d}\zeta \frac{\frac{\mathrm{d}}{\mathrm{d}\zeta}E^{+}(z(\zeta), t^{+}(\zeta))}{E^{+}(z(\zeta), t^{+}(\zeta))} = \int_{z_k}^{z_{k+1}} \mathrm{d}\zeta \left[-\frac{\alpha_{\mathrm{int}}}{2} + \frac{g\left(\rho(z(\zeta), t^{+}(\zeta))\right)}{2}\right] \quad (2.93)$$

$$\ln\left(E^{+}(\zeta, t' + \frac{\zeta}{v_{\mathrm{g}}})\right)\Bigg|_{z_k}^{z_{k+1}} = -\frac{\alpha_{\mathrm{int}}}{2}\Delta z_k + \frac{1}{2}\int_{z_k}^{z_{k+1}} \mathrm{d}\zeta g\left(\rho\left(\zeta, t' + \frac{\zeta}{v_{\mathrm{g}}}\right)\right) \quad (2.94)$$

$$\Rightarrow E^{+}(z_{k+1}, t' + \frac{z_{k+1}}{v_{\mathrm{g}}}) = E^{+}(z_k, t' + \frac{z_k}{v_{\mathrm{g}}}) \times \exp\left(-\frac{\alpha_{\mathrm{int}}}{2}\Delta z_k + \frac{1}{2}\int_{z_k}^{z_{k+1}} \mathrm{d}\zeta g\left(\rho\left(\zeta, t' + \frac{\zeta}{v_{\mathrm{g}}}\right)\right)\right), \quad (2.95)$$

which corresponds to an algebraic propagation equation for the electric field that includes the yet to be determined integral over the gain function g. By defining t' via $t = t' + z_{k+1}/v_{\mathrm{g}}$ and using $z_{k+1} = z_k + \Delta z_k$, the last equation can be written as

$$E^{+}(z_{k+1}, t) = E^{+}(z_k, t - \tau_k) \times \exp\left(-\frac{\alpha_{\mathrm{int}}}{2}\Delta z_k + \frac{1}{2}\int_{z_k}^{z_{k+1}} \mathrm{d}\zeta g\left(\rho\left(\zeta, t' + \frac{\zeta}{v_{\mathrm{g}}}\right)\right)\right), \quad (2.96)$$

where $\tau_k = \Delta z_k / v_{\mathrm{g}}$ describes the propagation time from z_k to z_{k+1}. Repeating the same steps for the integration of the backward (−) direction along the section $[z_{k+1}, z_k]$ yields

$$E^{-}(z_k, t) = E^{-}(z_{k+1}, t - \tau_k) \times \exp\left(-\frac{\alpha_{\mathrm{int}}}{2}\Delta z_k + \frac{1}{2}\int_{z_k}^{z_{k+1}} \mathrm{d}\zeta g\left(\rho\left(\zeta, t' - \frac{\zeta}{v_{\mathrm{g}}}\right)\right)\right). \quad (2.97)$$

Note that the traveling-wave equation (2.90) has been integrated over the same section of the device, but for the forward direction from z_k to z_{k+1} and for the backward direction the other way around from z_{k+1} to z_k. Furthermore, both Eqs. (2.96) and (2.97) contain integrals over the gain function g on the right hand side, which differ in the time argument of the charge-carrier number. To proceed, the integrals are approximated by

$$\int_{z_k}^{z_{k+1}} \mathrm{d}\zeta\, g\left(\rho\left(\zeta, t' + \frac{\zeta}{v_g}\right)\right) = \int_{z_k}^{z_{k+1}} \mathrm{d}\zeta\, g\left(\rho\left(\zeta, t + \frac{\zeta - z_{k+1}}{v_g}\right)\right) \approx \int_{z_k}^{z_{k+1}} \mathrm{d}\zeta\, g\left(\rho\left(\zeta, t - \frac{\tau_k}{2}\right)\right) \tag{2.98}$$

for the forward (+) direction and

$$\int_{z_k}^{z_{k+1}} \mathrm{d}\zeta\, g\left(\rho\left(\zeta, t' - \frac{\zeta}{v_g}\right)\right) = \int_{z_k}^{z_{k+1}} \mathrm{d}\zeta\, g\left(\rho\left(\zeta, t - \frac{\zeta + z_k}{v_g}\right)\right) \approx \int_{z_k}^{z_{k+1}} \mathrm{d}\zeta\, g\left(\rho\left(\zeta, t - \frac{\tau_k}{2}\right)\right) \tag{2.99}$$

for the backward (−) direction. Hence, both integrals are given by integrating over the gain for a constant time $t - \tau_k/2$, which corresponds to the time where fields that start at the opposing section facets arrive in the center of the section. This motivates the definition of the integrated gain

$$G_k(t) = \int_{z_k}^{z_{k+1}} \mathrm{d}\zeta\, g\left(\rho\left(\zeta, t - \frac{\Delta t_k}{2}\right)\right) \tag{2.100}$$

within the section $[z_k, z_{k+1}]$. Using this definition, Eqs. (2.96) and (2.97) can be written as

$$E^+(z_{k+1}, t) = E^+(z_k, t - \tau_k) \exp\left(-\frac{\alpha_{\text{int}}}{2}\Delta z_k + \frac{1}{2}G_k(t)\right) \tag{2.101}$$

$$E^-(z_k, t) = E^+(z_{k+1}, t - \tau_k) \exp\left(-\frac{\alpha_{\text{int}}}{2}\Delta z_k + \frac{1}{2}G_k(t)\right), \tag{2.102}$$

which serve as the algebraic propagation equations that determine the electric field amplitudes of the forward and backward-traveling electric field at the positions z_k and z_{k+1}.

Charge-Carrier Dynamics

To obtain a closed set of equations, suitable dynamical equations for the integrated gain G_k, which appears in the electric field propagation equations, and the associated charge-carrier numbers must be found. By definition, the carrier number ρ determines the gain g and the respective dynamical equations is given by Eq. (2.88). For a sufficiently fine discretization z_k, the carrier number ρ varies only negligibly within one section $[z_k, z_{k_1}]$. The integrated gain can then be approximated by

$$G_k(t) = \int_{z_k}^{z_{k+1}} \mathrm{d}\zeta \, g\left(\rho\left(\zeta, t - \frac{\Delta t_k}{2}\right)\right) \approx \Delta z_k g\left(\langle\rho\rangle_k(t)\right), \tag{2.103}$$

where $\langle\rho\rangle_k$ is the spatially averaged carrier number within the section $[z_k, z_{k_{k+1}}]$, which is defined by

$$\langle\rho\rangle_k(t) = \frac{1}{\Delta z_k} \int_{z_k}^{z_{k+1}} \mathrm{d}z \, \rho\left(z, t - \frac{\tau_k}{2}\right). \tag{2.104}$$

To derive equations of motion for the spatially averaged carrier number, the time in Eq. (2.88) is shifted by $-\tau_k/2$ and the equation is spatially integrated over the section $[z_k, z_{k+1}]$, which yields

$$\begin{aligned}\int_{z_k}^{z_{k+1}} \frac{\mathrm{d}z}{\Delta z_k}\left(\frac{\mathrm{d}}{\mathrm{d}t}\rho\left(z, t - \frac{\tau_k}{2}\right)\right) = \int_{z_k}^{z_{k+1}} \frac{\mathrm{d}z}{\Delta z_k}\Bigg(&- \frac{\rho\left(z, t - \frac{\tau_k}{2}\right)}{\tau_\rho} + R \\ &- v_\mathrm{g}\mathrm{Re}\left\{g\left(\rho(z, t - \frac{\tau_k}{2})\right)\right\} \\ &\times \left[\left|E^+\left(z, t - \frac{\tau_k}{2}\right)\right|^2 + \left|E^-\left(z, t - \frac{\tau_k}{2}\right)\right|^2\right]\Bigg),\end{aligned} \tag{2.105}$$

where the factor $1/\Delta z_k$ ensures the proper normalization. Applying the definition of the spatially averaged carrier number Eq. (2.104) and recognizing the independent spatial and temporal evolution of the carrier numbers leads to

$$\begin{aligned}\frac{\mathrm{d}}{\mathrm{d}t}\langle\rho\rangle_k(t) = &- \frac{\langle\rho\rangle_k(t)}{\tau_\rho} + R \\ &- \frac{v_\mathrm{g}}{\Delta z_k}\int_{z_k}^{z_{k+1}} \mathrm{d}z \, \mathrm{Re}\left\{g\left(\rho(z, t - \frac{\tau_k}{2})\right)\right\}\left|E^+\left(z, t - \frac{\tau_k}{2}\right)\right|^2 \\ &- \frac{v_\mathrm{g}}{\Delta z_k}\int_{z_k}^{z_{k+1}} \mathrm{d}z \, \mathrm{Re}\left\{g\left(\rho(z, t - \frac{\tau_k}{2})\right)\right\}\left|E^-\left(z, t - \frac{\tau_k}{2}\right)\right|^2.\end{aligned} \tag{2.106}$$

To solve the remaining integrals on the right hand side, the traveling-wave equation (2.87) is multiplied by the complex conjugate $E^{\pm *}$ of the electric field amplitude to produce

$$\pm(\partial_z E^{\pm})E^{\pm *} + \frac{1}{v_g}(\partial_t E^{\pm})E^{\pm *} = \left[-\frac{\alpha_{\text{int}}}{2} + \frac{g\left(\rho(z,t)\right)}{2}\right] E^{\pm}E^{\pm *}. \tag{2.107}$$

Taking the real part of that equation yields

$$\pm\partial_z |E^{\pm}|^2 + \frac{1}{v_g}\partial_t |E^{\pm}|^2 = \left[-\alpha_{\text{int}} + \text{Re}\left\{g\left(\rho(z,t)\right)\right\}\right] |E^{\pm}|^2, \tag{2.108}$$

which represents a traveling-wave equation for the intensity $|E^{\pm}|^2$. Using the method of characteristics again, this partial differential equation is transformed into the ordinary differential equation

$$\frac{\mathrm{d}}{\mathrm{d}\zeta}|E^{\pm}(z(\zeta),t^{\pm}(\zeta))|^2 = \pm\left[-\alpha_{\text{int}} + \text{Re}\left\{g(z(\zeta),t^{\pm}(\zeta))\right\}\right] |E^{\pm}(z(\zeta),t^{\pm}(\zeta))|^2, \tag{2.109}$$

where $(z(\zeta), t^{\pm}(\zeta))$ is the previously used characteristic curve. Starting with the forward (+) direction, the first integral in Eq. (2.106), which involves the forward (+) traveling field E^+, is evaluated according to:

$$\begin{aligned}
&\int_{z_k}^{z_{k+1}} \mathrm{d}z\, \text{Re}\left\{g\left(\rho\left(z, t-\frac{\tau_k}{2}\right)\right)\right\} \left|E^+\left(z, t-\frac{\tau_k}{2}\right)\right|^2 \\
\approx &\int_{z_k}^{z_{k+1}} \mathrm{d}\zeta\, \text{Re}\left\{g\left(\rho(z(\zeta),t^+(\zeta))\right)\right\} \left|E^+(z(\zeta),t^+(\zeta))\right|^2 \\
= &\int_{z_k}^{z_{k+1}} \mathrm{d}\zeta \left[\frac{\mathrm{d}}{\mathrm{d}\zeta}|E^+(z(\zeta),t^+(\zeta))|^2 + \alpha_{\text{int}}|E^+(z(\zeta),t^+(\zeta))|^2\right] \\
\approx &\left|E^+\left(z_{k+1}, t' + \frac{z_{k+1}}{v_g}\right)\right|^2 - \left|E^+\left(z_k, t' + \frac{z_k}{v_g}\right)\right|^2 \\
&+ \frac{\alpha_{\text{int}}\Delta z_k}{2}\left[\left|E^+\left(z_k, t' + \frac{z_k}{v_g}\right)\right|^2 + \left|E^+\left(z_{k+1}, t' + \frac{z_{k+1}}{v_g}\right)\right|^2\right] \\
= &|E^+(z_{k+1},t))|^2 - |E^+(z_k, t-\tau_k))|^2 \\
&+ \frac{\alpha_{\text{int}}\Delta z_k}{2}\left[|E^+(z_k,t-\tau_k))|^2 + |E^+(z_{k+1},t))|^2\right].
\end{aligned} \tag{2.110}$$

The first approximation, i.e., from the first to the second line, changes the integration to the characteristic curve and adjusts the time argument accordingly. The integrand can then be replaced via Eq. (2.109) and the integral can be carried out. The first term is subject to the fundamental theorem of calculus and the second term is approximated using the trapezoidal rule. Using the previously derived electric field propagation equation (2.101), the last line can be written as

$$|E^+(z_k, t-\tau_k))|^2 \left\{ e^{-\alpha_{\text{int}}\Delta z_k + \text{Re}\{G_k(t)\}} - 1 + \frac{\alpha_{\text{int}}}{2}\Delta z_k \left(e^{-\alpha_{\text{int}}\Delta z_k + \text{Re}\{G_k(t)\}} + 1 \right) \right\}. \tag{2.111}$$

Assuming small non-resonant losses within the considered section, i.e., $\alpha_{\text{int}}\Delta z_k \ll 1$, allows to expand the exponentials to first order. Reorganizing the resulting terms then produces the final approximation for the forward (+) integral:

$$\begin{aligned} &\int_{z_k}^{z_{k+1}} \mathrm{d}z \, \text{Re}\left\{ g\left(\rho(z, t-\frac{\tau_k}{2})\right)\right\} \left| E^+(z, t-\frac{\tau_k}{2}) \right|^2 \\ &\approx \left| E^+(z_k, t-\tau_k)) \right|^2 e^{-\frac{\alpha_{\text{int}}}{2}\Delta z_k} \left(e^{\text{Re}\{G_k(t)\}} - 1 \right). \end{aligned} \tag{2.112}$$

Repeating the same steps for the backward (−) direction yields the similar expression

$$\begin{aligned} &\int_{z_k}^{z_{k+1}} \mathrm{d}z \, \text{Re}\left\{ g\left(\rho(z, t-\frac{\tau_k}{2})\right)\right\} \left| E^-(z, t-\frac{\tau_k}{2}) \right|^2 \\ &\approx \left| E^-(z_{k+1}, t-\tau_k)) \right|^2 e^{-\frac{\alpha_{\text{int}}}{2}\Delta z_k} \left(e^{\text{Re}\{G_k(t)\}} - 1 \right). \end{aligned} \tag{2.113}$$

Using those results, the dynamical equation of the spatially averaged carrier number $\langle\rho\rangle_k$ can be written as

$$\begin{aligned} \frac{\mathrm{d}}{\mathrm{d}t}\langle\rho\rangle_k(t) = &-\frac{\langle\rho\rangle_k(t)}{\tau_\rho} + R - \frac{v_g}{\Delta z_k} e^{-\frac{\alpha_{\text{int}}}{2}\Delta z_k} \left(e^{\text{Re}\{G_k(t)\}} - 1 \right) \\ &\times \left[\left| E^+(z_k, t-\tau_k)) \right|^2 + \left| E^-(z_{k+1}, t-\tau_k)) \right|^2 \right]. \end{aligned} \tag{2.114}$$

The spatially averaged charge-carrier number dynamics are thus governed by the integrated gain and the time-delayed forward and backward-moving electric fields.

Summary

In conclusion, Eq. (2.114) combined with the algebraic propagation equations for the electric field Eqs. (2.101) and (2.102), and the approximation for the integrated gain Eq. (2.103) self-consistently describe the electric field and the active medium dynamics within a considered small section of the laser device. Note that the presented treatment of the traveling-wave equation has lead to an exponential dependence of the light-matter interaction on the integrated gain. This already accounts for propagation effects within the individual sections unlike the general linear approximation that was presented in Sect. 2.2. Thus, one could expect an even greater tolerance towards a coarse spatial discretization with errors only appearing if the assumption of spatial charge-carrier number homogeneity breaks down. The downside of this modeling approach, however, is the infinitely broad gain bandwidth due to the adiabatically eliminated polarization dynamics. Hence, the equations must be further modified to achieve a realistic description of the gain bandwidth and thus produce a suitable model for multi-mode laser dynamics. One suitable approach is presented in Chap. 5 in order to derive a model for a mode-locked V-shaped laser.

References

1. Chow WW, Koch SW (1999) Semiconductor-laser fundamentals, 1st edn. Springer-Verlag, Berlin Heidelberg, Berlin. 978-3-662-03880-2. https://doi.org/10.1007/978-3-662-03880-2
2. Scully MO, Suhail Zubairy M (1997) Quantum optics, 1st edn. Cambridge University Press. ISBN 9780511813993 (online). https://doi.org/10.1017/cbo9780511813993
3. Loudon R (2000) The quantum theory of light. Oxford University Press, Sept 2000. ISBN 0198501765
4. Haug H, Koch SW (2004) Quantum theory of the optical and electronic properties of semiconductors, 2nd edn. World Scientific, Singapore
5. Haken H (1986) Laser light dynamics, 1st edn., vol II. North Holland. ISBN 978-0444860217
6. van Tartwijk GHM, Agrawal GP (1998) Laser instabilities: a modern perspective. Prog Quantum Electron 22(2):43–122. https://doi.org/10.1016/s0079-6727(98)00008-1
7. Agrawal GP, Dutta NK (1993) Semiconductor Lasers. Van Nostrand Reinhold, New York
8. Erneux T, Glorieux P (2010) Laser dynamics. Cambridge University Press, UK. https://doi.org/10.1017/cbo9780511776908
9. Coldren LA, Corzine SW, Mashanovitch M (2012) Diode lasers and photonic integrated circuits. Wiley series in microwave and optical enginieering, 2nd edn. Wiley
10. Haus HA, Mecozzi A (1993) Noise of mode-locked lasers. IEEE J Quantum Electron 29(3):983. https://doi.org/10.1109/3.206583
11. Haken H (1985) Light: volume PII: waves, photons, atoms. North-Holland, Amsterdam
12. Huang YZ, Pan Z, Wu RH (1996) Analysis of the optical confinement factor in semiconductor lasers. J Appl Phys 79(8):3827–3830
13. Lingnau B (2015) Nonlinear and nonequilibrium dynamics of quantum-dot optoelectronic devices. Springer International Publishing, Switzerland, Springer theses. 978-3-319-25803-4. https://doi.org/10.1007/978-3-319-25805-8
14. Schöll E (1988) Dynamic theory of picosecond optical pulse shaping by gain-switched semiconductor laser amplifiers. IEEE J Quantum Electron 24(2):435–442
15. Ohtsubo J (2013) Semiconductor lasers: stability, instability and chaos, 3rd edn. Springer, Berlin. https://doi.org/10.1007/978-3-642-30147-6
16. Rossetti M, Bardella P, Montrosset I (2011) Time-domain travelling-wave model for quantum dot passively mode-locked lasers. IEEE J Quantum Electron 47(2):139
17. Knorr A, Binder R, Lindberg M, Koch SW (1992) Theoretical study of resonant ultrashort-pulse propagation in semiconductors. Phys Rev A 46(11):7179–7186. https://doi.org/10.1103/physreva.46.7179
18. Chow WW, Indik RA, Knorr A, Koch SW, Moloney JV (1995) Time-resolved nondegenerate four-wave mixing in a semiconductor amplifier. Phys Rev A 52:2479

19. Hess O, Koch SW, Moloney JV (1995) Filamentation and beam propagation in broad-area semiconductor lasers. IEEE J Quantum Electron 31(1):35–43. ISSN 1558-1713. https://doi.org/10.1109/3.341705
20. Gehrig E, Hess O (2003) Spatio-temporal dynamics and quantum fluctuations in semiconductor lasers. Springer
21. Schneider HC, Chow WW, Koch SW (2001) Many-body effects in the gain spectra of highly excited quantum dot lasers. Phys Rev B 64:115315
22. Chow WW, Koch SW (2005) Theory of semiconductor quantum-dot laser dynamics. IEEE J Quantum Electron 41:495–505. https://doi.org/10.1109/jqe.2005.843948
23. Koch SW, Meier T, Jahnke F, Thomas P (2000) Microscopic theory of optical dephasing in semiconductors. Appl Phys A 71(5):511–517
24. Schneider HC, Chow WW, Koch SW (2004) Excitation-induced dephasing in semiconductor quantum dots. Phys Rev B 70(23):235308. https://doi.org/10.1103/physrevb.70.235308
25. Lorke M, Nielsen TR, Seebeck J, Gartner P (2006) Jahnke F (2006) Influence of carrier-carrier and carrier-phonon correlations on optical absorption and gain in quantum-dot systems. Phys. Rev. B 73:085324. https://doi.org/10.1103/physrevb.73.085324
26. Koprucki T, Wilms A, Knorr A, Bandelow U (2011) Modeling of quantum dot lasers with microscopic treatment of Coulomb effects. Opt. Quantum Electron. 42:777–783
27. Schulze A, Knorr A, Koch SW (1995) Pulse propagation and many-body effects in semiconductor four-wave mixing. Phys Rev B 51:10601–10609. https://doi.org/10.1103/physrevb.51.10601
28. Hess O, Kuhn T (1996) Maxwell-Bloch equations for spatially inhomogeneous semiconductor lasers. I. Theoretical formulation. Phys Rev A 54(4):3347–3359. https://doi.org/10.1103/physreva.54.3347
29. Knorr A, Steininger F, Hanewinkel B, Kuckenburg S, Thomas P, Koch SW (1998) Theory of ultrafast spatio-temporal dynamics in semiconductor quantum wells: electronic wavepackets and near-field optics. Phys Stat Sol (B) 206(1):139–151. https://doi.org/10.1002/(sici)1521-3951(199803)206:1<139::aid-pssb139>3.0.co;2-p
30. Helkey RJ, Morton PA, Bowers JE (1990) Partial-integration method for analysis of mode-locked semiconductor lasers. Opt Lett 15(2):112
31. Schell M, Weber A, Böttcher EH, Schöll E, Bimberg D (1991) Theory of subpicosecond pulse generation by active modelocking of a semiconductor laser amplifier in an external cavity: limits for the pulsewidth. IEEE J Quantum Electron 27:402
32. Avrutin EA, Marsh JH (2000) Portnoi EL (2000) Monolithic and multi-gigahertz mode-locked semiconductor lasers: constructions, experiments, models and applications. IEE Proc Optoelectron 147(4):251. https://doi.org/10.1049/ip-opt:20000282
33. Bandelow U, Radziunas M, Vladimirov AG, Hüttl B, Kaiser R (2006) 40 GHz mode locked semiconductor lasers: theory, simulation and experiments. Opt Quantum Electron 38:495. https://doi.org/10.1007/s11082-006-0045-2
34. Avrutin EA, Russell BM (2009) Dynamics and spectra of monolithic mode-locked laser diodes under external optical feedback. IEEE J Quantum Electron 45(11):1456
35. Javaloyes J, Balle S (2010) Mode-locking in semiconductor Fabry-Pérot lasers. IEEE J Quantum Electron 46(7):1023–1030
36. Rossetti M, Bardella P, Montrosset I (2011) Modeling passive mode-locking in quantum dot lasers: a comparison between a finite- difference traveling-wave model and a delayed differential equation approach. IEEE J Quantum Electron 47(5):569
37. Radziunas M, Vladimirov AG, Viktorov EA, Fiol G, Schmeckebier H, Bimberg D (2011) Pulse broadening in quantum-dot mode-locked semiconductor lasers: simulation, analysis, and experiments. IEEE J Quantum Electron 47(7):935–943
38. Simos H, Simos C, Mesaritakis C, Syvridis D (2012) Two-section quantum-dot mode-locked lasers under optical feedback: pulse broadening and harmonic operation. IEEE J Quantum Electron 48(7):872. ISSN 0018-9197. https://doi.org/10.1109/jqe.2012.2193387
39. Dong M, Mangan NM, Kutz JN, Cundiff ST, Winful HG (2017) Traveling wave model for frequency comb generation in single-section quantum well diode lasers. IEEE J Quantum Electron 53(6):2500311. https://doi.org/10.1109/jqe.2017.2756641

40. Bardella P, Columbo L, Gioannini M (2017) Self-generation of optical frequency comb in single section quantum dot Fabry-Perot lasers: a theoretical study. Opt Express 25(21):26234–26252
41. Javaloyes J, Balle S (2012) Multimode dynamics in bidirectional laser cavities by folding space into time delay. Opt Express 20(8):8496–8502. https://doi.org/10.1364/oe.20.008496
42. Kolarczik M, Owschimikow N, Korn J, Lingnau B, Kaptan Y, Bimberg D, Schöll E, Lüdge K, Woggon U (2013) Quantum coherence induces pulse shape modification in a semiconductor optical amplifier at room temperature. Nat Commun 4:2953. https://doi.org/10.1038/ncomms3953
43. Lingnau B, Lüdge K (2018) Rabi-oscillation-enhanced frequency conversion in quantum-dot semiconductor optical amplifiers. Opt Quantum Electron 50(2):111. ISSN 1572-817X. https://doi.org/10.1007/s11082-018-1380-9
44. Pérez-Serrano A, Javaloyes J, Balle S (2013) Multichannel wavelength conversion using four-wave mixing in semiconductor ring lasers. IEEE Photon Technol Lett 25(5):476–479
45. Pérez-Serrano A, Vilera M, Tijero JMG, Balle S, Esquivias I (2018) A voltage driven traveling-wave model for the simulation of an integrated three-section MOPA under static and modulated operation. IEEE J Quantum Electron 54(2):1–10. ISSN 0018-9197. https://doi.org/10.1109/jqe.2018.2797888
46. Meinecke S, Drzewietzki L, Weber C, Lingnau B, Breuer S, Lüdge K (2019) Ultra-short pulse generation in a three section tapered passively mode-locked quantum-dot semiconductor laser. Sci Rep 9:1783. https://doi.org/10.1038/s41598-018-38183-1
47. Lingnau B, Dillane M, O'Callaghan J, Corbett B, Kelleher B (2019) Multimode dynamics and modelling of free-running and optically injected Fabry-Pérot quantum-dot lasers. Phys Rev A 100(6):063837
48. Vladimirov AG, Pimenov AS, Rachinskii D (2009) Numerical study of dynamical regimes in a monolithic passively mode-locked semiconductor laser. IEEE J Quantum Electron 45(5):462–46. https://doi.org/10.1109/jqe.2009.2013363
49. Borri P, Schneider S, Langbein W, Woggon U, Zhukov AE, Ustinov VM, Ledentsov NN, Alferov ZI, Ouyang D, Bimberg D (2001) Ultrafast carrier dynamics and dephasing in InAs quantum-dot amplifiers emitting near 1.3-μm-wavelength at room temperature. Appl Phys Lett 79:2633. https://doi.org/10.1063/1.1411986
50. Borri P, Langbein W, Schneider S, Woggon U, Sellin RL, Ouyang D, Bimberg D (2002) Exciton relaxation and dephasing in quantum-dot amplifiers from room to cryogenic temperature. IEEE J Sel Top Quantum Electron 8(5):984–991. ISSN 1077-260X. https://doi.org/10.1109/jstqe.2002.804250
51. Vu QT, Haug H, Koch SW (2006) Relaxation and dephasing quantum kinetics for a quantum dot in an optically excited quantum well. Phys Rev B 73:205317
52. Nielsen TR, Gartner P, Jahnke F (2004) Many-body theory of carrier capture and relaxation in semiconductor quantum-dot lasers. Phys Rev B 69:235314. https://doi.org/10.1103/physrevb.69.235314
53. Majer N, Lüdge K, Schöll E (2010) Cascading enables ultrafast gain recovery dynamics of quantum dot semiconductor optical amplifiers. Phys Rev B 82:235301. https://doi.org/10.1103/physrevb.82.235301
54. Wilms A, Breddermann D, Mathe P (2012) Theory of direct capture from two- and three-dimensional reservoirs to quantum dot states. Phys Stat Sol C 9(5):1278–1280
55. Vladimirov AG, Turaev DV (2005) Model for passive mode locking in semiconductor lasers. Phys Rev A 72(3):033808. https://doi.org/10.1103/physreva.72.033808
56. Viktorov EA, Mandel P, Vladimirov AG, Bandelow U (2006) Model for mode locking of quantum dot lasers. Appl Phys Lett 88(20). https://doi.org/10.1063/1.2203937
57. Otto C, Lüdge K, Vladimirov AG, Wolfrum M, Schöll E (2012) Delay induced dynamics and jitter reduction of passively mode-locked semiconductor laser subject to optical feedback. New J Phys 14. https://doi.org/10.1088/1367-2630/14/11/113033

Chapter 3
Timing Jitter in Mode-Locked Lasers

The timing jitter of a mode-locked laser quantifies the temporal pulse train stability in the presence of continuous external forcing. In this case, temporal stability refers to the pulse train's ability to remain close to an ideal reference pulse train. This chapter reviews the established timing jitter estimation methods and introduces a new pulse-period fluctuation autocorrelation based method.[1] A special focus is placed on the application to numerically simulated time series. On that account, the practical implementation of the various methods is discussed in detail. In this context, the method's different computational demands are studied with respect to their produced timing jitter estimate fidelities. The results lead to recommendations regarding the most suitable estimation method based on the properties of the mode-locked pulse train.

This chapter is structured as follows: After introducing the basic concepts in Sect. 3.1, first the established time domain, frequency-domain and hybrid timing jitter estimation methods are presented and demonstrated in Sects. 3.2, 3.3 and 3.4, respectively. Afterwards, the pulse-period fluctuation autocorrelation method is introduced in Sect. 3.5 and finally all methods are benchmarked with respect to their required sampling data to produce reasonable timing jitter estimates in Sect. 3.6. The individual methods in this chapter are demonstrated and exemplified using simulated time series from the three-section tapered quantum-dot laser, which is presented and characterized in detail in Chap. 4.

---

[1] Parts of the chapter have been published in [1].

S. Meinecke, *Spatio-Temporal Modeling and Device Optimization of Passively Mode-Locked Semiconductor Lasers*, Springer Theses,
https://doi.org/10.1007/978-3-030-96248-7_3

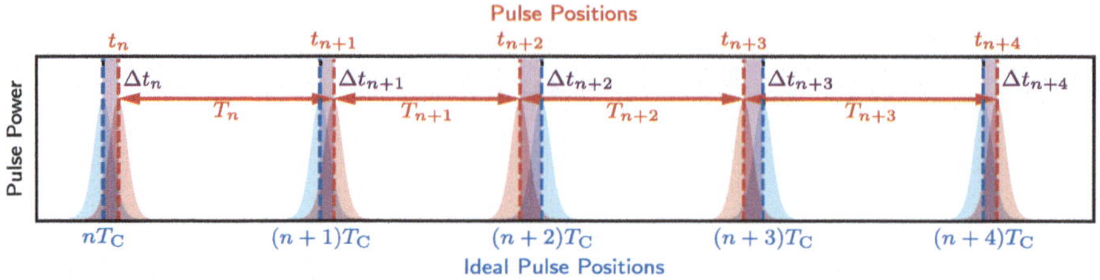

Fig. 3.1 Timing jitter sketch. Blue pulses represent an ideal noise free pulse train with a constant pulse period T_C. Red pulses represent a real pulse train, where the pulse periods T_n fluctuate. Purple shaded regions highlight the deviations Δt_n of the pulse positions t_n from the ideal pulse train nT_C

3.1 Introduction

The effect of timing jitter is illustrated by an exaggerated sketch in Fig. 3.1, where an ideal pulse train (blue) and a noisy pulse train (red) are shown. The noisy pulse train is subject to fluctuating pulse periods T_n, which lead to timing deviations Δt_n (highlighted by purple shaded regions) of the pulse positions t_n from the ideal jitter free pulse positions nT_C. T_C denotes the mean period, i.e., the inter-pulse-interval and is also referred to as the clock time. The continuous external forcing may include deterministic processes, such as shifting temperatures and mechanical vibrations, and stochastic processes, such as spontaneous emission noise and charger-carrier noise.

The analysis of a mode-locked laser's timing jitter is crucial, since an excellent temporal stability, i.e., a small timing jitter, is required for many applications [2–4]: In photonic sampling applications the timing jitter limits the sampling rate [5, 6], in optical data transmission applications the timing jitter limits the achievable bit rate [3, 7–10] and in spectroscopy applications [11–13] the timing jitter limits the spectral resolution. Hence, the timing jitter determines among other laser performance characteristics the usefulness of a given device.

The nature of the timing deviations can be furthermore separated into two fundamentally different cases: For actively mode-locked lasers, the pulses are bound to the external modulation signal, which invokes the mode-locking in the first place. The modulation acts as a restoring force, which drives the pulses back to their ideal position. The external modulation, moreover, represents a reference clock, against which timing deviations can be measured. Passively mode-locked lasers, on the other hand, lack the external modulation and are not subject to a restoring force and thus two sample pulse trains can drift arbitrarily out of synchronization.[2] Timing deviations

[2] In more mathematical terms, a passively mode-locked laser is an autonomous system, which is subject to a time shift symmetry. Any periodic solution therefore exhibits a neutral mode, which corresponds to the time shift symmetry. Perturbations in the direction of the time shift mode, will therefore remain in the system and continuous perturbations, such as spontaneous emission noise, accumulate over time.

must than be measured against an ideal, perturbation free pulse train, which can be constructed from the mean pulse repetition period.

In both cases, however, the experimental measurement is not as straightforward as one might wish, since the exact pulse positions generally can not be determined due to bandwidth limitations of the available laboratory electronics. On that account, multiple spectral methods have been developed, which estimate the timing jitter based on the power spectral density captured by a photo diode. Such spectra generally do not contain the complete pulse train information, since they are truncated by the measurement setups cut-off frequency.[3] Hence, essential to each method is a model of the noisy pulse train, such that the timing jitter can be related to the features of the (truncated) measured power spectral density. A wealth of literature has been published over the past decades on this subject regarding active mode-locking [14, 15] and passive mode-locking [16–19].

Simulated noisy pulse trains, on the other hand, provide full access to the temporal dynamics and thus permit to determine the timing jitter via both time-domain and frequency-domain methods. Hence, simulations of the timing jitter in mode-locked lasers can be used to compare time and spectral domain methods against each other and thereby probe the validity of underlying assumptions of the spectral methods. However, any conclusion derived from such a comparison only translates well into the physical reality if both the deterministic dynamics of the mode-locked lasers as well as the continuous external driving are reasonably well treated in the simulations.

Dictated by the considered three section quantum-dot laser, this section focuses on the analysis of the latter case, i.e., passive mode-locking. Nevertheless, many of the definitions and measurement techniques introduced in this section apply to both passive and active mode-locking. Moreover, the analysis is restricted to spontaneous emission noise as the timing fluctuations inducing external driving force. This, however, implies no major restriction, since it has been shown that spontaneous emission noise dominates all other contributions to the timing jitter in passively mode-locked semiconductor lasers [16, 17].

A first characterization of the timing fluctuations of the considered laser is presented in Fig. 3.2, where (a) shows a sample sequence of fluctuating pulse periods T_n and (b) shows a corresponding histogram of the observed pulse periods. The mean pulse period $\langle T \rangle = T_C = 75.173\,\text{ps}$ is indicated by red lines and is associated with the clock-time T_C, which is later used to construct an ideal noise free pulse train. The standard deviation $\langle (T - T_C)^2 \rangle = \sigma_{pp} = 16.7\,\text{fs}$ is highlighted by red shaded regions and defines the pulse-to-pulse timing jitter σ_{pp}. A detailed distinction between different time-domain timing jitter measures is presented in the next subsection.

The distribution function of the pulse periods T_n, which is presented in (b), exhibits a pronounced asymmetry due to a fat tail towards longer pulse periods. This results in the mean T_C being shifted towards larger pulse periods with respect to the mode of the distribution. Examining the sample pulse sequence in (a), almost all pulse periods larger than $T_C + \sigma_{pp}$ appear as clusters within well localized spikes. This indi-

[3] The truncated high frequencies correspond to a sampling time, which is not sufficient to resolve the individual pulses and thus prevents a time-domain estimation of the timing jitter.

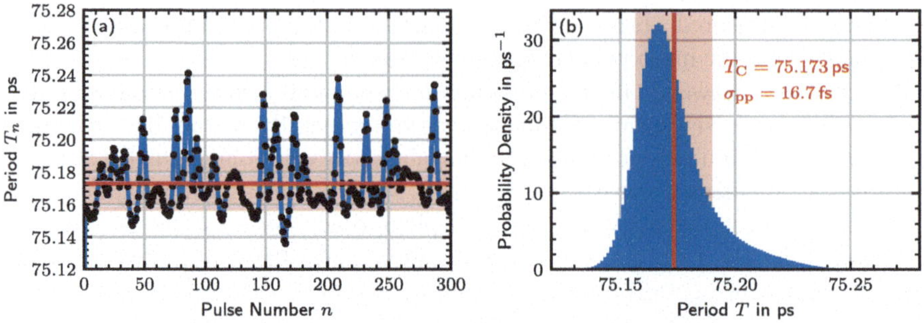

Fig. 3.2 Pulse period fluctuations. **a** Example sequence of fluctuating pulse periods (black dots) and **b** histogram of the pulse periods obtained from 10^8 pulses. The mean, denoted by the red lines, corresponds to the clock time T_C and the standard deviation, highlighted by the red shaded region, to the pulse-to-pulse timing jitter σ_{pp}. Laser parameters: $P = 0.81$ A, $U = 6$V, and $D = 10^{-6}$ cm ps$^{-1}[\eta]^{-1}$

cates strong correlations of the fluctuations due to intrinsic mode-locking dynamics. Nonetheless, the distribution quickly falls off, meaning that the fluctuations are well confined and can be thought of as small but correlated perturbations on top of the deterministic pulse train, which is governed by the mode-locking mechanics.

This simple presentation of the fluctuating pulse periods, however, does not reveal any insights into the long-term timing stability of the mode-locked laser. As mentioned above, passively mode-locked laser are not subject to a restoring force that pushes the pulses back towards an ideal jitter free position. Thus, the individual fluctuations can accumulate and induce drifts of the pulse positions. This determines the long-term stability of a pulse train and is not captured by the pulse-to-pulse jitter σ_{pp}.

On sufficiently large time scales, however, the pulse positions of passively mode-locked lasers perform a random walk with respect to an ideal jitter free pulse train [16, 17]. Therefore, the respective diffusion coefficient $D_{\Delta t}$ represents an unambiguous measure to characterize the long-term timing stability. This work in particular uses the long-term timing jitter σ_{lt} as the central timing jitter measure, which relates to the diffusion coefficient via $\sigma_{lt} = \sqrt{2D_{\Delta t}}$. The following sections present the established long-term timing jitter estimation methods and their respective application to the three-section mode-locked laser in order to illustrate their peculiarities.

3.2 Time Domain Long-Term Timing Jitter Definition

If the pulse positions can be accurately determined, as it is the case in numerical simulations, the long-term timing jitter σ_{lt} can be directly calculated using an ensemble of m pulse trains with pulse positions $\{t_n\}_m$.[4] The timing deviations relative to an

[4] The pulse positions are determined as the center of mass of their intensity distribution. This measure is more robust against pulse-shape fluctuations than the pulse maximum [20].

ideal pulse train are then calculated according to

$$\{\Delta t_n = t_n - nT_\mathrm{C}\}_m \,, \tag{3.1}$$

where T_C is the clock time, which is defined by the sample mean

$$T_\mathrm{C} = \langle T_n = t_{n+1} - t_n \rangle_{m,n} \tag{3.2}$$

of the pulse periods. Characteristic for a random walk, the ensemble variance $\mathrm{Var}_m\,(\Delta t_n)$ then grows linearly with the pulse number n for sufficiently large n, i.e., as soon as all correlations of the dynamical system have decayed. This motivates the definition of a pulse separation dependent timing jitter [20–22]

$$\sigma_{\Delta t}(n) = \sqrt{\frac{\mathrm{Var}_m(\Delta t_n)}{n}}. \tag{3.3}$$

The long-term timing jitter is then formally obtained as the limit

$$\sigma_\mathrm{lt} = \lim_{n \to \infty} \sigma_{\Delta t}(n). \tag{3.4}$$

For practical applications, the long-term timing jitter has to be evaluated at sufficiently large n. This method is referred to as the *direct time-domain method.* Note that the standard deviation of the pulse periods, which is known as the pulse-to-pulse timing jitter (sometimes also called cycle or period jitter [21]), is given by $\sigma_\mathrm{pp} = \sigma_{\Delta t}(1)$. The long-term timing jitter σ_lt is furthermore related to the diffusion constant $D_{\Delta t}$ of the random walk of the timing fluctuations for large n by

$$\sigma_\mathrm{lt} = \sqrt{2D_{\Delta t}}. \tag{3.5}$$

The above introduced quantities are illustrated in Fig. 3.3 with the three-section tapered quantum-dot laser operated at $P = 0.81$ A, $U = 6$V and $D = 10^{-6}$ cm ps$^{-1}[\eta]^{-1}$ (s. Figs. 4.5, 4.17, and 4.24). The presented results have been generated from 1000 time series, i.e., different stochastic realizations), which contain 133000 pulses each.

Figure 3.3a plots five example realizations of pulse position deviation sequences Δt_n over an interval of 10000 pulses and (b) shows the corresponding ensemble variance from the full data set. The variance increases indeed linearly with the pulse separation over the presented range, which supports the assumption of an underlying random walk.

Lastly, the pulse separation dependent timing jitter $\sigma_{\Delta t}(n)$ is plotted in Fig. 3.3c. Although $\sigma_{\Delta t}(n)$ essentially carries the same information as the ensemble variance, the subtleties introduced by finite correlations, which are not identifiable in (b), are highlighted by the division with the pulse separation n and the logarithmic scale. The timing jitter $\sigma_{\Delta t}(n)$ starts of at its pulse-to-pulse value $\sigma_\mathrm{pp} = 16.7$ fs, rises to

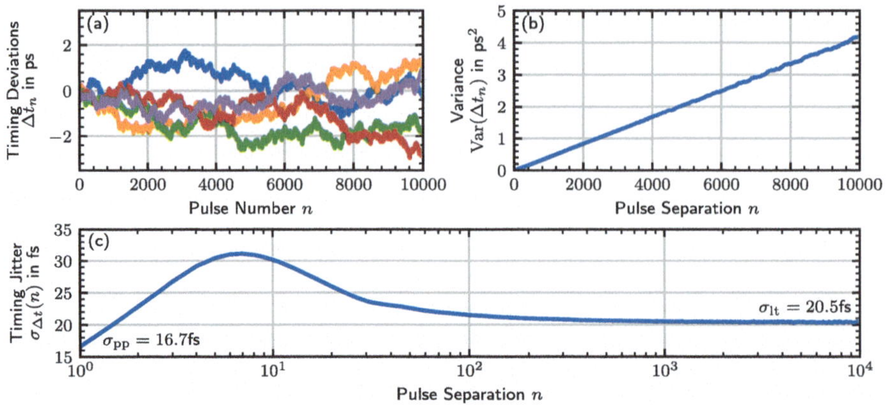

Fig. 3.3 Pulse timing deviation measures. **a** Five example realizations of pulse timing deviation sequences Δt_n, **b** ensemble variance $\mathrm{Var}_m(\Delta t_n)$ and **c** pulse separation dependent timing jitter $\sigma_{\Delta t}(n)$. **b** and **c** have been obtained from 1000 realizations with 133,000 pulses each. Laser parameters: $P = 0.81$ A, $U = 6$V, and $D = 10^{-6}$ cm ps$^{-1}[\eta]^{-1}$

≈ 32 fs at $n = 8$ and then relaxes to its long-term value $\sigma_{\mathrm{lt}} = 20.5$ fs. The shape of this curve results from finite correlations of the dynamic system, which are mostly induced by the recovery dynamics of the gain material. An ideal random walk, on the other hand, would produce a flat horizontal line.

To utilize this direct time-domain method, one must ensure that all correlations have sufficiently decayed at the pulse separation n for which the long-term timing jitter $\sigma_{\Delta t}(n)$ is to be evaluated. The example at hand Fig. 3.3c shows that the timing jitter only becomes constant for $n \gtrsim 1000$, which implies correlations of up to $\approx$ 75 ns. Examining the system equations (s. Sect. 4.2 in Chap. 4), the slowest time scale is the gain-section carrier reservoir lifetime with a value of 1 ns. Hence, the comparably long correlation is likely due to weakly damped Floquet modes[5] of the mode-locked solution. Thus, the required minimal pulse separation n must be carefully determined and can not be naively extracted from the system parameters.

3.3 Frequency Domain Timing Jitter Estimation Methods

In cases where the extraction of the pulse positions is either impossible or impractical, the long-term timing jitter can be determined from the power spectral density $S_{|E|^2}$ of the out-coupled electric field intensity. This condition almost always applies to experimental measurements. Even the fastest photo diodes and sampling oscilloscopes can not sufficiently resolve the short optical pulses generated by passively

[5] Within the framework of the Floquet theory, i.e., the linear stability analysis of periodic orbits, a periodic orbit can be expanded into Floquet modes. The corresponding eigenvalues (Floquet multipliers) determine the stability and if applicable the damping of each mode.

mode-locked lasers. This limitation will likely persist in the future as the bandwidth of electronics is fundamentally limited by the inertia of the electrons themselves. Hence, the power spectral density, which is electronically obtained from the current generated by a photo diode, must suffice for the timing jitter estimation.

In this situation, simulated stochastic pulse trains provide a crucial verification tool, since both the exact pulse positions and the power spectral densities are accessible, which allows for a direct comparison of timing jitter estimates from time and frequency-domain based methods.

The basic building block of a spectral timing jitter estimation method must be the connection between the statistics of the noisy pulse train and its power spectral density. According to the Wiener-Khinchin theorem [22, 23], the variance $\text{Var}(X)$ of a real wide-sense stationary stochastic process X with zero mean can be computed from the power spectral density $\mathcal{S}_X$ via

$$\text{Var}(X) = \Psi_X(\Delta = 0) = \int_{-\infty}^{\infty} \mathcal{S}_X(\nu)\mathrm{d}\nu = \int_{0}^{\infty} 2\mathcal{S}_X(\nu)\mathrm{d}\nu, \tag{3.6}$$

where $\Psi_X(\Delta = 0)$ is the autocorrelation function at zero lag Δ. The last equality accounts for the symmetry of the power spectral density of a real process [23]. Hence, if the power spectral density of the timing deviations $S_{\Delta t}$ is known, i.e., it has been extracted from the measured photo current, the timing jitter can be computed via the equation above.

However, every measured (simulated) power spectral density is truncated at small frequencies due to the finite measurement (integration) time. Thus, any spectral timing jitter estimation method must either make assumptions about the spectral line shape at small frequencies or suffice with the information obtained from a limited frequency bandwidth.

Furthermore, extracting the timing noise power spectral density from a real mode-locked laser turns out to be not as straightforward. Even if one limits the analysis to the case, where spontaneous emission noise dominates all other stochastic processes, both pulse timing and pulse amplitude fluctuations appear and contribute to the power spectral density of the recorded photo current. Both manifest as noise sidebands of the fundamental repetition rate and its harmonics in the power spectral density. Moreover, both kinds of fluctuations are generally not independent due to the nonlinear light-matter coupling.

3.3.1 The von der Linde Method

The separation of both parts was pioneered by von der Linde [14] by assuming a noise free ideal pulse train with added and independent timing and amplitude fluctuations.

The underlying noise free perfect pulse train[6] leads to delta peaks at the repetition rate and its harmonics. The timing and amplitude noise represent stationary stochastic processes, in this case, and manifest as sidebands of the harmonics. Von der Linde showed that the sideband contributions induced by the timing jitter grow with the square of the harmonic number h, while the amplitude jitter contributions do not depend on the harmonic number, i.e., remain constant. Thus, the two contributions can be distinguished by comparing different harmonics, which allows for the extraction of the timing jitter components. For practical purposes, the timing jitter can simply be evaluated at a sufficiently large harmonic, where the amplitude jitter contributions can be neglected compared to the timing jitter contribution.

Combining the preceding identity Eq. (3.6) and the scaling properties of the timing jitter induced side bands, the von der Linde method [14, 24] then measures the integrated root-mean-square (rms) timing fluctuations σ_{rms} within the offset-frequency band $[\nu_{\mathrm{low}}, \nu_{\mathrm{high}}]$ at the h-th harmonic:

$$\sigma_{\mathrm{rms}}(\nu_{\mathrm{low}}, \nu_{\mathrm{high}}, h) = \frac{T_{\mathrm{C}}}{2\pi h} \sqrt{\int_{\nu_{\mathrm{low}}}^{\nu_{\mathrm{high}}} 2\mathcal{S}_{\phi}(\nu_{\mathrm{off}}, h)\, \mathrm{d}\nu_{\mathrm{off}}}. \tag{3.7}$$

$\mathcal{S}_{\Phi}(\nu, h)$ is the single-sided timing phase-noise power spectral density [25, 26], which is computed according to

$$\mathcal{S}_{\Phi}(\nu_{\mathrm{off}}, h) = \frac{2\mathcal{S}_{|E|^2}(h\nu_0 + \nu_{\mathrm{off}})}{\int_{h\nu_0 \pm \nu_0/2} \mathcal{S}_{|E|^2}(\nu)\, \mathrm{d}\nu} = \frac{2\mathcal{S}_{|E|^2}(h\nu_0 + \nu_{\mathrm{off}})}{P(h)}, \tag{3.8}$$

where ν_0 denotes the fundamental repetition frequency, ν_{off} the offset frequency from the respective h-th harmonic and $P(h)$ the integrated power of the h-th harmonic. Hence, the timing phase noise spectral power density of a given harmonic h represents the measured power spectral density normalized to the integrated power of the respective harmonic.

The relation of the timing phase noise power spectral density $\mathcal{S}_{\Phi}$ to the timing deviation power spectral density $\mathcal{S}_{\Delta t}$ follows from the definition of the timing phase deviations

$$\phi_n = \frac{2\pi}{T_{\mathrm{C}}} \Delta t_n \Rightarrow \mathcal{S}_{\Delta t} = \left(\frac{T_{\mathrm{C}}}{2\pi}\right)^2 \mathcal{S}_{\phi}, \tag{3.9}$$

which leads to the factor $2\pi/T_{\mathrm{C}}$ in Eq. (3.7). The factor two in the numerator in Eq. (3.8) is commonly not found in the literature, but is crucial for the correct esti-

[6] This construction of the pulse train implicitly assumes an actively more-locked laser with a fixed clock time dictated by an external modulation. The added independent timing noise to each ideal pulse position is neither correlated nor accumulative.

mation of the long-term jitter,[7] since it reflects the noise contributions in the negative frequency components of $\mathcal{S}_{|E|^2}$.[8]

The power in the h-th harmonic $P(h)$ is often approximated by

$$P(h) = \int_{h\nu_0-\nu_0/2}^{h\nu_0+\nu_0/2} \mathcal{S}_{|E|^2}(\nu)d\nu \approx \mathcal{S}_{|E|^2}(h\nu_0)R_{\mathrm{BW}} \tag{3.10}$$

where R_{BW} is the resolution bandwidth of the power spectral density. This approximation is valid as long as the spectral power contained within the resolution bandwidth is much larger than the power contained in the remainder of the sidebands. This requirement is fulfilled as long as the FWHM of the line shape is much smaller than the resolution bandwidth R_{BW}.

By the construction of the pulse train, the von der Linde method is implicitly intended to be used with actively mode-locked lasers. Subsequent works [16–18] pointed out that the naive application of the von der Linde method to passively mode-locked lasers can lead to a severe underestimation of the timing jitter.

The absence of the external modulation, i.e., a restoring force that drives the pulses towards their ideal position and the finite measurement (integration) time, lead to a repetition-rate line shape in the timing phase noise spectral density, which can be approximated by a Lorentzian [16, 17]. The characteristic FWHM of the Lorentzian line shape then produces a corner frequency, which indicates the transition from the finite repetition rate peak to the $1/\nu^2$ decay predicted by the von der Linde theory. Hence, choosing an offset frequency band $[\nu_{\mathrm{low}}, \nu_{\mathrm{high}}]$, which integrates frequencies smaller than the corner frequencies results in an underestimation of the timing jitter. Moreover, the FWHM of the Lorentzian line shape grows with the square of the harmonic number h [17]. This requires to not only specify the integration offset frequency band, but also the respective harmonic h in order to compare different measurements. Hence, the von der Linde method can be used for passive mode-locking [16, 27–29], but only when carefully applied.

To avoid the ambiguities induced by the somewhat arbitrary frequency band and the corner frequency in passively mode-locked lasers, it is desirable to translate the integrated rms timing jitter σ_{rms} into the long-term timing jitter σ_{lt}. The long-term timing jitter, as explained in the introducing paragraphs of this chapter, represents a more intuitive measure of the long-term diffusive behavior of the timing deviations Δt_n.

A conversion formula between the two measures was derived by Otto [22, 29], by assuming a pure random walk of the timing deviations Δt_n, i.e., uncorrelated

[7] Without the factor two, the integrated rms timing jitter σ_{rms} is nonetheless a suitable measure of the timing fluctuations. The factor two only becomes relevant when σ_{rms} is converted into the long-term jitter and compared to other long-term jitter estimation methods.

[8] The power spectral density $\mathcal{S}_{|E|^2}$ is symmetric with respect to the origin, i.e., $\mathcal{S}_{|E|^2}(-\nu) = \mathcal{S}_{|E|^2}(\nu)$, since $|E|^2$ itself is real. Thus, the power of the noise components at the absolute frequency $|\nu|$ is equally shared between the frequencies ν and $-\nu$.

pulse-period fluctuations T_k that accumulate over time. The long-term timing jitter is than computed from the integrated rms jitter according to

$$\sigma_{\mathrm{lt}}(\nu_{\mathrm{low}}, \nu_{\mathrm{high}}, h) = \sigma_{\mathrm{rms}}(\nu_{\mathrm{low}}, \nu_{\mathrm{high}}, h)\pi \left[\frac{1}{T_{\mathrm{C}}}\left(\frac{1}{\nu_{\mathrm{low}}} - \frac{1}{\nu_{\mathrm{high}}}\right)\right]^{-0.5} . \tag{3.11}$$

In practice, the assumption of uncorrelated timing fluctuations requires the integration domain of the von der Linde method to be beyond the corner frequency, i.e., to only cover the part of the spectrum, which falls of as $1/\nu^2$. Note that assuming an uncorrelated random walk does not lead to the pulse-to-pulse, but to the long-term timing jitter, which includes the correlations that exist in a real passively mode-locked laser.

The combination of the von der Linde method and the Otto conversion formula then represents a completely independent approach to estimate the long-term timing jitter σ_{lt} and will be referred to as the *von der Linde/Otto method*. In the case of simulated time series, the validity of this method can be easily checked by comparing the obtained estimate to the result produced by the direct time-domain method. This immediately reveals both an improper normalization[9] of the timing phase noise power spectral density S_Φ and a badly chosen offset frequency band $[\nu_{\mathrm{low}}, \nu_{\mathrm{high}}]$ or harmonic number h for that matter.

3.3.2 The Kéfélian Method

Mode-locked lasers with high repetition rates only permit to electronically record the first few harmonics due to the bandwidth limitations of the measurement electronics. This restriction often applies to monolithically integrated lasers [8, 9, 30–33]. To overcome this constraint, Kéfélian et al. developed a method [19], which estimates the long-term timing jitter σ_{lt} based on the fundamental repetition-rate linewidth.

In high repetition-rate passively mode-locked laser, the timing jitter is dominated by spontaneous emission noise [16], which manifest as a Lorentzian shaped repetition-rate linewidth broadening [16, 17]. The linewidth parameter $\Delta\nu$ extracted from a Lorentzian fit then uniquely describes the single-sided timing phase noise power spectral density [34]

$$S_\Phi(\nu_{\mathrm{off}}) = \frac{\Delta\nu}{\pi\,\nu_{\mathrm{off}}^2}, \tag{3.12}$$

---

[9] Approximating the power $P(h)$ contained within a harmonic becomes worse at higher harmonics, since the characteristic linewidth grows quadratically. The constant factor of two in Eq. (3.8) becomes especially relevant if the long-term timing jitter is to be estimated and compared against other methods.

which follows the familiar $1/\nu^2$ decay. Utilizing the stochastic Parseval theorem [18], this simple analytic timing phase noise spectral power density yields an explicit equation for the long-term timing jitter [19]

$$\sigma_{\mathrm{lt}} = T_{\mathrm{C}}\sqrt{\frac{\Delta\nu T_{\mathrm{C}}}{2\pi}}. \tag{3.13}$$

Note that this quantity is sometimes (including the original publication) referred to as the pulse-to-pulse jitter [19, 31, 32]. However, in this work the pulse-to-pulse timing jitter describes the standard deviation of the pulse-period fluctuations.

The quadratic scaling of the repetition rate linewidth with the harmonic number [17, 35], i.e.,

$$\Delta\nu_h = h^2\Delta\nu \tag{3.14}$$

allows for a straightforward extension of the this method to higher harmonics. The corresponding equation then reads

$$\sigma_{\mathrm{lt}}(h) = T_{\mathrm{C}}\sqrt{\frac{\Delta\nu_h T_{\mathrm{C}}}{2\pi h^2}}, \tag{3.15}$$

where the factor $1/h^2$ compensates the increasing linewidth $\Delta\nu_h$ of the h-th harmonic. The quadratically increasing repetition rate linewidth $\Delta\nu_h$ corresponds to the quadratically increasing timing jitter contribution to the timing phase noise spectral density in the von der Linde theory. Hence, the use of higher harmonics can be similarly used to distinguish between the amplitude and timing jitter contributions.

For the characterization of passively mode-locked lasers, the Kéfélian method has the distinct advantage of directly producing an unambiguous figure of merit, namely the long-term timing jitter, which is well suited for the description of the long-term diffusive drift of the pulse position deviations. It, furthermore, does not require the specification of a somewhat arbitrary frequency bandwidth like the von der Linde/Otto method. On that account, the Kéfélian method is also much easier to implement, since only an adequate curve-fitting procedure has to be applied to the repetition-rate line shape in order to extract the linewidth parameter.

3.3.3 Numerical Application to the Three-Section Laser

Both the von der Linde/Otto and the Kéfélian method are illustrated in Fig. 3.4. The three-section tapered quantum-dot laser (s. Chap. 4) is operated at the same parameters as in Fig. 3.3, i.e., at $P = 0.81$ A, $U = 6$V and $D = 10^{-6}\,\mathrm{cm\,ps^{-1}}[\eta]^{-1}$ (s. Figs. 4.5, 4.17, and 4.24). The presented results have been generated from 1000 time series (stochastic realizations), each 10 μs long, utilizing Bartlett's method [36] to average the power spectral densities. The integration time of 10 μs corresponds to a 100 kHz resolution of the phase noise spectra.

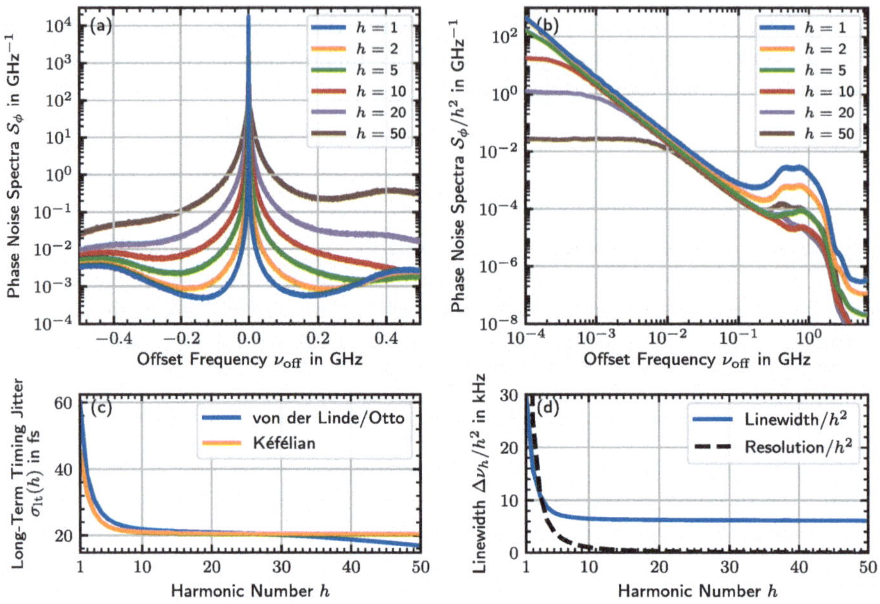

Fig. 3.4 Timing phase noise spectra and extracted long-term timing jitter. **a** Timing phase noise spectra for the harmonic numbers $h \in \{1, 2, 5, 10, 20, 50\}$. **b** Single-sided timing phase noise spectra normalized to the squared harmonic numbers. **c** Long-term timing jitter extracted via the von der Linde/Otto and Kéfélian method as a function of the harmonic number. **d** Extracted Lorentzian linewidth parameter $\Delta\nu_h$ and spectral resolution scaled by the squared harmonic number h. Power spectra are calculated from 1000 realizations, each 10 μs long, corresponding to a 100 kHz spectral resolution. rms timing jitter integration range from 5 MHz to 6 GHz. Laser parameters: $P = 0.81$ A, $U = 6$V, and $D = 10^{-6}\,\mathrm{cm\,ps^{-1}}[\eta]^{-1}$

To provide a first glance at the timing phase noise spectral densities, Fig. 3.4a plots the repetition-rate line shapes for the harmonic numbers $h \in \{1, 2, 5, 10, 20, 50\}$ as a function of the offset frequency ν_{off}. The respective carrier frequencies are centered at $\nu_{\mathrm{off}} = 0$ GHz. The different harmonics can be identified by their color as specified in the legend. $h = 1$ is also referred to as the fundamental repetition-rate. The maximum of each line shape is determined by the normalization, i.e., the integrated power within the respective harmonic. For an increasing harmonic number h, i.e., from blue to brown, the power in the noise side bands grows. Thus, the fundamental repetition-rate line shape exhibits the largest peak among all harmonics.

However, the qualitative and quantitative features of the timing-jitter induced sidebands are better presented in double logarithmic plot, which is shown in Fig. 3.4b. In this subfigure, the phase noise spectra have been scaled by the factor $1/h^2$, to counter their respective increase as predicted by von der Linde [14]. The plotted phase noise spectra confirm this scaling, since all of them approximately show the same $1/\nu^2$ falloff, i.e., produce the same timing phase noise spectral density, within the offset frequency range from ≈ 10 to ≈ 100 MHz.

At higher frequencies, all phase noise spectra exhibit a side peak at $\nu_{\text{off}} \approx 0.5$ GHz, which is associated with the relaxation oscillations of the quantum-dot active medium [13, 22, 37, 38]. Due to this origin, the respective side peaks do not scale with the harmonic number as the timing jitter induced sidebands. To a first-order approximation, the peaks of lower harmonics are therefore observed at higher spectral power densities. At higher harmonics, however, nonlinear frequency mixing effects [14, 39, 40] result in asymmetries among the left- and right-sided side bands. In the case at hand, the respective right-sided sidebands of the 20th and 50th harmonic, as displayed in Fig. 3.4a, exhibit relaxation oscillations peaks much larger than their left-sided counterpart. This results in their pronounced appearance in the double logarithmic plot Fig. 3.4b compared to the 10th harmonic.

At smaller frequencies ν_{off}, the corner frequency, as induced by the Lorentzian line shape, manifests depending on the considered harmonic h. For the larger considered harmonics, the corner frequency can be read of Fig. 3.4b: For $h = 50$ the corner frequency is observed at ≈ 6 MHz, for $h = 20$ at ≈ 1 MHz and for $h = 10$ at $\approx$ 250 kHz. The evolution of the corner frequency as a function of the harmonic number thus follows the quadratic increase of the Lorentzian linewidth parameter as predicted by [16, 17]. For smaller harmonics ($h \lessapprox 10$), the corner frequency is obscured by the 100 kHz spectral resolution.

In the subsequent step, the long-term timing jitter σ_{lt} is extracted from all harmonics between the 1st and 50th, using both the von der Linde/Otto and the Kéfélian method. In the case of the von der Linde/Otto method, the frequency integration range is chosen to be from 5 MHz to 6 GHz. The offset frequency band considered for the Lorentzian fit in the Kéfélian method spans from $\nu_{\text{off}} = -100$ MHz to $\nu_{\text{off}} = 100$ MHz to maximize the data, but also to avoid fitting the relaxation oscillation peak centered around 500 MHz. The results are shown in Fig. 3.4c, where the timing jitter produced by the von der Linde/Otto method is plotted in blue and the timing jitter produced by the Kéfélian method is plotted in orange. If of interest, the integrated rms jitter σ_{rms} can be calculated via Eq. (3.11).

Using the long-term timing jitter $\sigma_{\text{lt}} = 20.5$ fs, which computed via the direct time-domain method, as a benchmark, both the von der Linde/Otto and the Kéfélian method overestimate the long-term jitter for harmonics $h \lesssim 10$. The von der Linde/Otto method starts off with $\sigma_{\text{lt}}(h = 1) \approx 60$ fs, then approximately converges within the first ten harmonics to $\sigma_{\text{lt}} = 20$ fs, but starts drifting towards smaller timing jitter values for harmonic $h \gtrsim 35$. The Kéfélian method, on the other hand, starts of at $\sigma_{\text{lt}}(h = 1) \approx 46$ fs and then converges to $\sigma_{\text{lt}} = 20.5$ fs within the first ten harmonics, but contrary to the von der Linde/Otto method remains stable at this value for higher harmonics.

The observed drop-off of the long-term timing jitter values, which are produced by the von der Linde/Otto method, at higher harmonics is caused by an inappropriate choice of integrated offset frequencies. For increasing harmonics, the lower integration boundary $\nu_{\text{low}} = 5$ MHz shifts into the vicinity of the corner frequencies, which causes a deviation from the $1/\nu^2$ spectral decay and thereby causes the underestimation of the long-term timing jitter as observed in Fig. 3.4c. This issue could be circumvented by adjusting the lower integration boundary ν_{low} accordingly. However, this measure was not taken, to highlight the potential issues that arise when using the von der Linde/Otto method.

The overestimation of the timing jitter at low harmonics is a common feature of both spectral methods. Naively, the overestimation could be attributed to strong amplitude jitter contributions to the noise induced sidebands, which become attenuated be the relative scaling with the harmonic number [14]. It, however, turns out that the dominant source of the overestimation is an insufficient spectral resolution and spectral leakage introduced by the DFT in the calculation of the power spectra [36]. While the latter is analyzed in detail in the following paragraphs and in Fig. 3.5, the former, i.e., the lacking resolution, is shown in Fig. 3.4d: The relationship between the extracted Lorentzian linewidth parameter $\Delta\nu_h$ and the relative spectral resolution is illustrated in dependence on the harmonic number h by plotting both quantities scaled by the inverse squared harmonic number. The scaled resolution (black dashed line) is then simply given by $100\,\mathrm{kHz/h^2}$. The scaled linewidth (blue solid line) starts of at $\approx 30\,\mathrm{kHz}$ and then converges within the first ten harmonic to $\Delta\nu_h/\mathrm{h}^2 \approx 6.07\,\mathrm{kHz}$. Both lines cross at $h \approx 3$, which implies that for $h \lessapprox 3$ the peak of the respective harmonic, i.e., the spectral components contained within the linewidth, is covered by no more than one data point. This greatly magnifies the spectral leakage issue of the DFT and handicaps the Lorentzian fit. As the harmonic number h increases, this issue is mitigated by an increasing linewidth parameter, which leads to a convergence of $\Delta\nu_h/h^2$ for regions, where $\Delta\nu_h \gg 100\,\mathrm{kHz}$ holds. These results can be used as a rule of thumb requirement for the application of both the von der Linde/Otto and the Kéfélian method for simulated time series.

The convergence of the long-term timing jitter estimate, which is produced by the two considered spectral methods with respect to the spectral resolution, is further investigated by simulating time series of lengths between 1 and 100 μs.[10] The corresponding spectral resolutions are than given by the inverse integration times, i.e., from 1 MHz to 10 kHz. The results are displayed in Fig. 3.5, where (a) and (b) plot the single-sided timing phase noise power spectral densities $\mathcal{S}_\Phi$, which are computed from various simulation times, and are evaluated at the fundamental and the 20th harmonics, respectively.

In the case of the fundamental harmonics, the two shortest considered integration times 1 and 2 μs produce phase noise spectra (blue and orange lines), which visibly exhibit much larger spectral power densities than the spectra produced by the longer integration times. Moreover, the overestimation at short integration times extends to frequencies much larger than the respective spectral resolution and thereby conserves the characteristic $1/\nu^2$ falloff of the timing jitter induced noise bands.

The 20th harmonics, on the other hand, exhibit no visible differences in the timing phase noise spectra, which are produced by the different integration times. This contrast to the fundamental harmonic is attributed to the 400-fold increase of the linewidth of the 20th harmonic, which amounts to $\Delta\nu_{20} \approx 2.5\,\mathrm{MHz}$ according to the

[10] The respective power spectra are computed from 50 realizations for integration times from 1 to 20 μs and from 20 realizations for integration times from 50 to 100 μs.

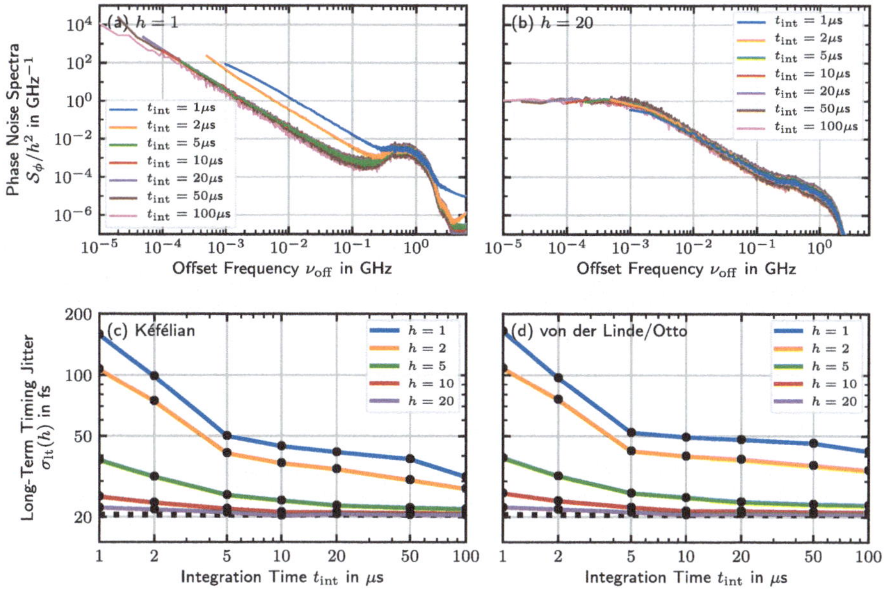

Fig. 3.5 Spectral timing jitter estimation methods in dependence on the integration time. Single-sided phase noise spectra at the 1st (**a**) and 20th (**b**) harmonic, respectively. Long-term timing jitter obtained from the Kéfélian (**c**) and the von der Linde/Otto (**d**) method, respectively, evaluated at the 1st, 2nd, 5th, 10th and 20th harmonic as a function of the integration time. The black dotted lines indicate the long-term jitter $\sigma_{\mathrm{lt}} \approx 20.5$ fs as obtained from the direct time-domain method. Spectra computed from 50 realizations for integration times from 1 to 20 μs and from 20 realizations for integration times from 50 to 100 μs. rms timing jitter integration range from 5 MHz to 6 GHz. Laser parameters: $P = 0.81$ A, $U = 6$V, and $D = 10^{-6}\,\mathrm{cm\,ps^{-1}}[\eta]^{-1}$

data presented in Fig. 3.4d. Thus, the linewidth exceeds the spectral resolution for all integration times.

The quantitative consequences of the different integration times in terms of the resulting long-term timing jitter σ_{lt} are presented in Fig. 3.5c, d for the Kéfélian and von der Linde/Otto method, respectively. Both subfigures plot the estimated timing jitter as functions of the integration time for the 1st, 2nd, 5th, 10th and 20th harmonic. The black dotted lines indicate the long-term jitter $\sigma_{\mathrm{lt}} \approx 20.5$ fs, which has been obtained from the direct time-domain method. Similar to the results presented in Fig. 3.4, the rms integration range is chosen to span from 5 MHz to 6 GHz and the offset frequency band considered for the Lorentzian fit to span from $\nu_{\mathrm{off}} = -100$ MHz to $\nu_{\mathrm{off}} = 100$ MHz.

Both spectral methods produce quantitative and qualitative similar results, since all considered harmonics benefit from an increased integration time. The quantitative dependence of the chosen harmonic on the integration time decreases for an increasing harmonic number. Examining the fundamental harmonic (blue lines), the extracted long-term jitter reduces from ≈ 160 to ≈ 40 fs, i.e., the overestimation factor reduces from eight to two. However, no convergence within the considered simulation times to the benchmark value $\sigma_{\mathrm{lt}} \approx 20.5$ fs is achieved. The smallest considered harmonic that accomplishes convergence to this benchmark value is the 10th harmonic (red lines) at the longest integration time of $t_{\mathrm{int}} = 100\ \mu$s. Finally, the 20th harmonic (purple lines) provides reasonable approximations of the long-term timing jitter for integration times $t_{\mathrm{int}} \gtrsim 10\ \mu$s. Based on this, the use of the 20th harmonic and an integration time of $t_{\mathrm{int}} = 10\ \mu$s can be recommended for evaluation of the long-term timing jitter in the presented case.

Concluding based on the presented results, both the von der Linde/Otto and the Kéfélian method are suitable for the characterization of the long-term timing jitter of simulated passively mode-locked lasers. For an accurate estimation of the timing jitter, a sufficient spectral resolution, i.e., simulation time, is required. While sufficient spectral resolution is generally not an issue in experimental characterizations, it may lead to excessively high demands in both time and memory requirements for numerical characterizations. This issue can be somewhat mitigated by evaluating higher harmonics and exploiting the quadratic scaling of the repetition-rate linewidth with the harmonic number. It is, however, important to note that even though numerically simulated power spectral densities are in principle not limited at large frequencies, high harmonics with a sufficient signal-to-noise ratios are generally only available for short optical pulses, i.e., broad frequency combs.

Compared to the direct time-domain method, both presented spectral methods do not require pulse detection and pulse position extraction routines, which makes them easier to implement in that respect. Both spectral methods, however, require a careful choice of the integration time and the considered frequency bands to produce accurate timing jitter estimates. The presented results further show that the validity of a timing jitter estimate can be easily checked and maintained by ensuring an appropriate ratio between the repetition-rate linewidth parameter and the spectral resolution.

3.4 Hybrid Long-Term Timing Jitter Estimation Methods

The timing phase noise power spectral density $\mathcal{S}_{\Phi}$ can also be computed from sets of timing deviations $\{\Delta t_n\}$ [20, 22, 41] via

$$\mathcal{S}_{\Phi}^{\Delta t}(\nu) = \left(\frac{2\pi}{T_{\mathrm{C}}}\right)^2 \mathcal{S}_{\Delta t}(\nu), \tag{3.16}$$

which follows from the definition of the timing phase Eq. (3.9). $\mathcal{S}_{\Delta t}$ represents the power spectral density of the timing deviations $\{\Delta t_n\}$. Note that the inherently discrete nature of the timing deviations $\{\Delta t_n\}$ limits the maximal frequency of $S_{\Delta t}(\nu)$ to the Nyquist frequency $1/(2T_{\mathrm{C}}) = \nu_0/2$, where T_{C} is the mean period (clock time) and ν_0 the repetition rate of the mode-locked laser. The frequency ν is then measured in discrete steps of $1/(2NT_{\mathrm{C}}) = \nu_0/(2N)$, where N is the length of the sequences $\{\Delta t_n\}$.

This way, timing phase noise power spectral densities can be computed with models that only produce discrete pulse positions [32, 42] and no complete time traces of the electric field intensity. Consequently, the von der Linde method can be applied to such cases in order to evaluate the integrated rms timing jitter σ_{rms} via

$$\sigma_{\text{rms}}(\nu_{\text{low}}, \nu_{\text{high}}) = \frac{T_{\text{C}}}{2\pi}\sqrt{\int_{\nu_{\text{low}}}^{\nu_{\text{high}}} 2\mathcal{S}_{\phi}^{\Delta t}(\nu)\text{d}\nu} = \sqrt{\int_{\nu_{\text{low}}}^{\nu_{\text{high}}} 2\mathcal{S}_{\Delta t}(\nu)\text{d}\nu}. \tag{3.17}$$

By construction, the considered power spectral densities $\mathcal{S}_{\Phi}^{\Delta t}$ and $\mathcal{S}_{\Delta t}$ do not contain any other contributions except those from the timing deviations $\{\Delta t_n\}$. Hence, the spectra $\mathcal{S}_{\Phi}^{\Delta t}$ and $\mathcal{S}_{\Delta t}$ do not follow the Lorentzian line shape, but the $1/\nu^2$ divergence at small frequencies, which is characteristic for a random walk. Thus, the lower integration bound ν_{low} for the rms jitter σ_{rms} is not restricted by a corner frequency as in the conventional von der Linde method.

To obtain the long-term timing jitter σ_{lt}, Eq. (3.11) can be used exactly in the same way as in the von der Linde/Otto method. Due to the reliance on the time-domain pulse positions, this approach is referred to as the *hybrid von der Linde/Otto method* within this work.

The Kéfélian method can also be adapted to work with the power spectrum $\mathcal{S}_{\Phi}^{\Delta t}$, which is generated from the timing deviations $\{\Delta t_n\}$. To extract the linewidth parameter $\Delta\nu$ the spectrum must not be fitted with a Lorentzian, but directly with the $1/\nu^2$ decay

$$\mathcal{S}_{\Phi}^{\text{fit}}(\nu) = \frac{\Delta\nu}{\pi\nu^2}. \tag{3.18}$$

Identical to the Kéfélian method, the long-term timing jitter σ_{lt} can then be computed according to Eq. (3.15). This approach is referred to as the *hybrid Kéfélian method* within this work.

Figure 3.6 illustrates the timing phase noise power spectral density $\mathcal{S}_{\Phi}^{\Delta t}$, which is obtained from the timing deviations $\{\Delta t_n\}$ (blue solid line) and compares it to the timing phase noise spectra, which are computed from the 1st (orange dashed line) and 20th harmonic (green dashed line) of the power spectral density $\mathcal{S}_{|E|^2}$ of the electric field. The three-section tapered quantum-dot laser (s. Chap. 4) is operated at the same parameters as in Fig. 3.3, i.e., at $P = 0.81$ A, $U = 6$V and $D = 10^{-6}\,\text{cm ps}^{-1}[\eta]^{-1}$ (s. Figs. 4.5, 4.17, and 4.24). The presented results have been generated from 1000 (realizations) each 10 μs long. This integration time corresponds to ≈ 133000 pulses in each sequence $\{\Delta t_n\}$ and a spectral resolution of 100 kHz.

The power spectrum computed from the timing deviations $\mathcal{S}_{\Phi}^{\Delta t}$ produces the same power densities as the power spectrum $\mathcal{S}_{\Phi}(h = 20)$ obtained from the 20th harmonic of $\mathcal{S}_{|E|^2}$ within the regime, where both exhibit the $1/\nu^2$ decay. However, contrary to $\mathcal{S}_{\Phi}(h = 20)$, $\mathcal{S}_{\Phi}^{\Delta t}$ does not exhibit a corner frequency that limits the timing jitter evaluation at small frequencies. At larger frequencies around $\nu \approx 0.5$ GHz, $\mathcal{S}_{\Phi}^{\Delta t}$ pro-

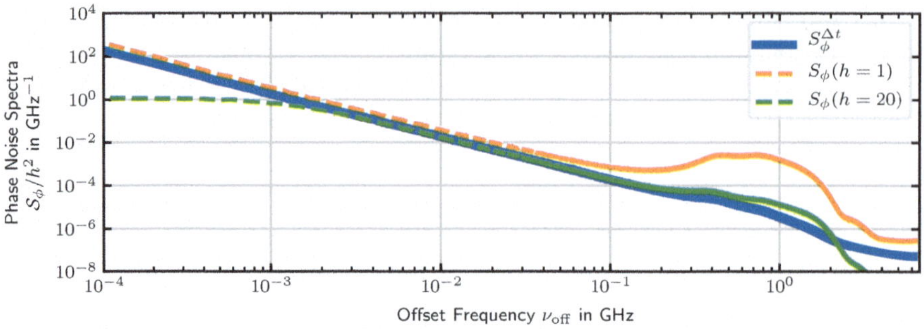

Fig. 3.6 Single-sided timing phase noise power spectral density obtained from the timing deviations $\{\Delta t_n\}$ (blue solid line). Phase noise spectra obtained from the 1st (orange dashed line) and 20th (green dashed line) harmonic of power spectrum $\mathcal{S}_{|E|^2}$ are shown for comparison. All spectra are computed from 1000 realizations, each 10 μs long. The rms timing jitter integration domain ranges from 5 MHz to 6 GHz. Laser parameters: $P = 0.81$ A, $U = 6$V, and $D = 10^{-6}$ cm ps$^{-1}[\eta]^{-1}$

duces the previously observed signatures of relaxation oscillations, which manifest in timing deviations due to correlations between amplitude and timing fluctuations. The power spectral density $\mathcal{S}_\Phi(h = 1)$ obtained from the 1st harmonic of $S_{|E|^2}$ is plotted for reference and suffers from insufficient resolution and spectral leakage, as discussed in the previous section, but does not display a corner frequency within the considered frequency range.

Applying both the hybrid von der Linde/Otto and the hybrid Kéfélian method to the timing phase noise power spectral density $\mathcal{S}_\Phi^{\Delta t}$ yields a long-term timing jitter of $\sigma_{\text{lt}} \approx 20.5$ fs, which agrees with the results, which have been obtained with the direct time-domain method and the von der Linde/Otto and Kéfélian method applied to $\mathcal{S}_\Phi(h = 20)$. Hence, the two hybrid approaches represent another set of valid long-term timing jitter estimation methods.

Compared to the two spectral methods, the hybrid methods do not suffer from potentially insufficient spectral resolution and can therefore be used with lasers that exhibit arbitrarily small timing jitter, i.e., arbitrarily small repetition rate linewidths. This advantage, however, comes at the cost of the additional time series processing, which is required to extract the pulse positions $\{\Delta t_n\}$.

3.5 The Pulse-Period Fluctuation Autocorrelation Method

The previous subsection 3.2 presented a pulse separation depended timing jitter $\sigma_{\Delta t}(n)$, which converged to the long-term timing jitter σ_{lt} for sufficiently large n. Variations of $\sigma_{\Delta t}(n)$ were attributed to correlations between pulses, which are introduced by the charge-carrier system of the mode-locked laser. This section quantitatively constructs the pulse separation depended timing jitter $\sigma_{\Delta t}(n)$ and consequently also the long-term timing jitter σ_{lt} from exactly those correlations.

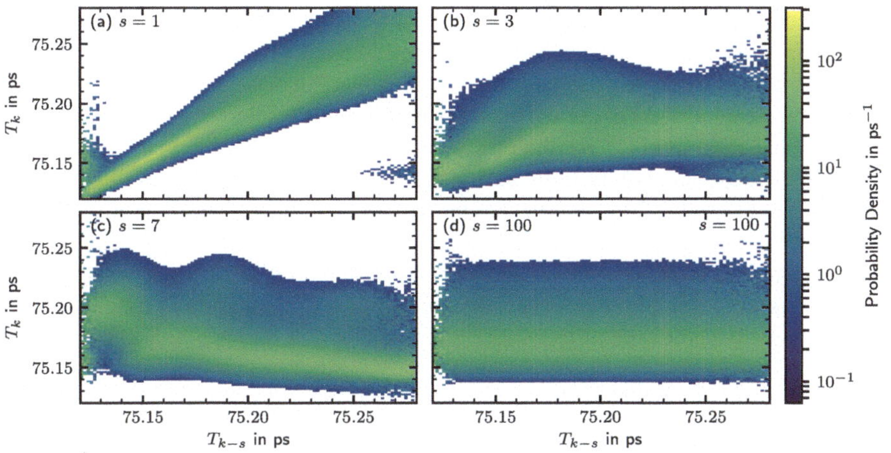

Fig. 3.7 Pulse-period correlations. The columns in each subfigure show the color-coded conditional probability densities of observing the pulse period T_k (indicated by the y-axis) after having observed a pulse period T_{k-s} (indicated by the x-axis) s pulses before. Subfigures **a–d** show increasing pulse separation $s \in \{1, 3, 7, 100\}$. The histograms have been computed from 1.33×10^8 pulses. Laser parameters: $P = 0.81$ A, $U = 6$V, and $D = 10^{-6}\,\text{cm}\,\text{ps}^{-1}[\eta]^{-1}$

This construction is largely motivated by finding a method that requires fewer computational resources for the estimation of the long-term timing jitter than the direct time-domain method for cases, where the generation of the pulse sequences $\{\Delta t_n\}$ is the computationally limiting factor. Requiring fewer resources translates to requiring fewer simulated pulses in this case.

However, before developing a quantitative theory, selected consequences of the pulse-period correlations are illustrated in Fig. 3.7, where (a) to (d) show conditional probability densities. For all subfigures, each column plots the color-coded probability densities of observing the pulse period T_k (indicated by the y-axis) after having observed a pulse period T_{k-s} (indicated by the x-axis) s pulses before. Subfigures (a) to (d) present increasing pulse separations $s \in \{1, 3, 7, 100\}$. The three-section tapered quantum-dot laser (s. Chap. 4) is operated at the same parameters as in Fig. 3.3, i.e., at $P = 0.81$ A, $U = 6$V, and $D = 10^{-6}\,\text{cm}\,\text{ps}^{-1}[\eta]^{-1}$ (s. Figs. 4.5, 4.17, and 4.24). The histograms have been generated from 1.33×10^8 pulses.

The pulse-to-pulse relations ($s = 1$) are presented in Fig. 3.7a and exhibit a very strong correlation, i.e., the maximum of the probability densities follows the diagonal $T_k = T_{k-1}$. Thus, observing a short (long) pulse period T_k is most likely after having just observed a short (long) pulse period T_{k-1}. Increasing the pulse separation to $s = 3$ in Fig. 3.7b already drastically changes the respective probability densities. Observing a short period T_k is most likely after having observed a short period T_{k-3} three pulses before. Observing a long period T_k, however, is most likely after having observed the mean period T_{k-3} three pulses before. Further increasing the pulse separation to $s = 7$ (Fig. 3.7c), which roughly corresponds to the maximum of the pulse separation dependent timing jitter $\sigma_{\Delta t}(n)$ in Fig. 3.3c, again drastically changes the respective probability densities. Observing a short period T_k is now most likely after having observed a long period T_{k-7} seven pulses before. Observing a long period T_k, on the other hand, is now equally likely after having observed either a short period T_{k-7} or the mean period T_{k-7} seven pulses before. Lastly, for the long pulse separation $s = 100$ presented in Fig. 3.7d, the probability densities of observing a pulse period T_k become independent of the pulse period T_{k-100} observed 100 pulses before. Hence, every column in Fig. 3.7d is identical to the one-dimensional histogram presented in Fig. 3.2.

In a complete picture, the probability of observing a pulse period T_k depends on all previous pulses s and their respective periods T_{k-s}. Hence, this probability can be written as a conditional probability

$$P(T_k) = P(T_k | T_{k-1}, T_{k-2}, \ldots, T_{k-n}, \ldots), \tag{3.19}$$

where the T_{k-n} denote the previous pulse periods. Finite correlation times imply that only a finite number of pulse periods T_{k-n} has to be considered. In the case of delta correlated pulse periods, i.e., a uncorrelated random walk, no previous periods have to be considered. The conditional probability densities shown in Fig. 3.7 are obtained by integrating Eq. (3.19) over all T_{k-n} except for T_{k-s}. In a Bayesian picture, knowledge of previous pulse periods T_{k-n} improves the accuracy of the prediction for the next pulse period. If one starts with no observations, the probability density approximated from the relative frequencies of the pulse-periods, as shown in Fig. 3.2b, has to be used as the best guess for the next pulse. Each additional observation can then be used to update the probability density for the next pulse period.

While the conditional probability density Eq. (3.19) provides a somewhat intuitive formalization of the pulse-period correlations, it is cumbersome to obtain and to work with. Hence, the quantitative framework derived in the following omits its utilization. To construct the timing jitter from the pulse-period fluctuations, the long-term timing jitter definition is recalled

$$\sigma_{\Delta t}(n) = \sqrt{\frac{\mathrm{Var}_m(\Delta t_n)}{n}}, \tag{3.20}$$

where $\mathrm{Var}_{\mathrm{m}}$ denotes the ensemble variance of m sequences of pulse timing deviations $\{\Delta t_n\}$. The timing deviations $\{\Delta t_n\}$ can be described by a cumulative stochastic process

$$\Delta t_n = \sum_{k=1}^{n} \delta T_k, \tag{3.21}$$

where the increments δT_k are the pulse-period fluctuations, which are defined by

$$\{\delta T_k = T_k - T_C = t_{k+1} - t_k - T_C\}\,. \tag{3.22}$$

While the timing deviations $\{\Delta t_n\}$ form a weak-sense stationary process, the fluctuations $\{\delta T_k\}$ are bounded by the laser dynamics and thus form a stationary process[11] that has zero mean by definition. This also implies a zero mean for $\{\Delta t_n\}$. Hence, the variance of the timing deviation reads

$$\mathrm{Var}(\Delta \mathrm{t_n}) = \left\langle \Delta \mathrm{t_n^2} \right\rangle = \left\langle \sum_{\mathrm{k,l=1}}^{\mathrm{n}} \delta \mathrm{T_k} \delta \mathrm{T_l} \right\rangle, \tag{3.23}$$

where $\delta T_k \delta T_l$ can be identified as the correlation function of the pulse-period fluctuations. Since δT_k is described by a stationary process with constant variance $\langle \delta T^2 \rangle = \langle \delta T_k^2 \rangle$, the correlation function only depends on the distance $|k - l|$ between two pulses and the variance can be written as

$$\mathrm{Var}(\Delta \mathrm{t_n}) = \sum_{\mathrm{k,l=1}}^{\mathrm{n}} \langle \delta \mathrm{T_k} \delta \mathrm{T_l} \rangle = \langle \delta \mathrm{T}^2 \rangle \sum_{\mathrm{k,l=1}}^{\mathrm{n}} \Psi_{\delta \mathrm{T}} \left(|\mathrm{k} - \mathrm{l}| \right), \tag{3.24}$$

where $\Psi_{\delta T}$ is the normalized autocorrelation function of the pulse-period fluctuations. Rewriting the sum in terms of the distance between pulses $d = |k - l|$, the timing jitter $\sigma_{\Delta t}$ can be expressed as

$$\sigma_{\Delta t}^2(n) = \langle \delta T^2 \rangle \left[1 + 2 \sum_{d=1}^{n-1} \frac{n-d}{n} \Psi_{\delta T}(d) \right], \tag{3.25}$$

where the normalization $\Psi_{\delta T}(0) = 1$ yields the first term in the square brackets. Note that delta correlated timing fluctuations, i.e., $\Psi_{\delta T}(m) = \delta_{m,0}$, lead to a pulse separation independent timing jitter. This case corresponds to a true random walk with $\langle \delta T^2 \rangle = \sigma_{\mathrm{pp}}^2 = \sigma_{\mathrm{lt}}^2$. While the authors of [17] use a very similar framework to analytically analyze the consequences of a simple exponentially decaying autocorrelation function, the approach presented in the following intends to utilize the autocorrelation function, which is computed from sets of simulated $\{\delta T_k\}$.

Computing $\sigma_{\Delta t}(n)$ according to Eq. (3.25) up to a pulse separation of n requires a good estimate of the autocorrelation function up to n, which can be as computationally demanding as directly computing the long-term timing jitter. Nonetheless, a significant improvement of the computational cost can be achieved by fitting the autocorrelation function with a suitable model function using only small n. This procedure is referred to as the *pulse-period fluctuation autocorrelation method*.

---

[11] The ensemble expectation value $\langle \delta T_k^2 \rangle_m$ is stationary, i.e., does not depend on k. This condition is fulfilled as long as the mode-locking state is stable. The opposite would imply the breakup of the mode-locking state and thereby also render the mean pulse period T_C ill-defined. The timing deviations $\{\Delta t_n\}$ form a weak-sense stationary process, since they accumulate the fluctuations $\{\delta T_k\}$. The respective variance therefore depends on the number of accumulated increments δT_k and thus on the time.

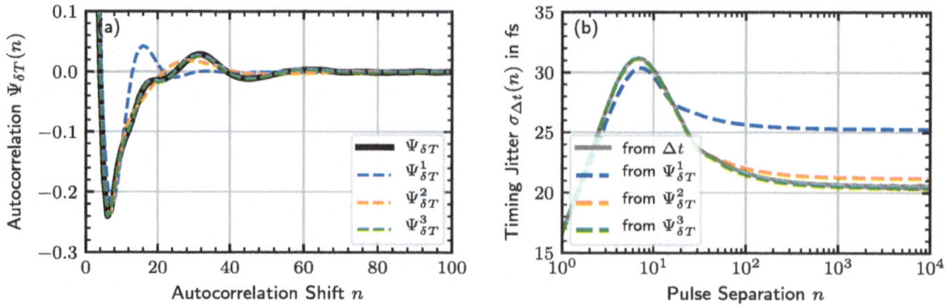

Fig. 3.8 Pulse-period fluctuation autocorrelation method. **a** Auto-correlation function of the pulse-period fluctuations calculated from $\{\delta T_k\}$ (solid gray line) and fitted model functions $\Psi^j_{\delta T}$ for $j = 1, 2, 3$ (blue, orange, and green dashed lines). **b** Timing jitter as a function of the pulse separation calculated from $\{\Delta t_n\}$ and the fitted model functions. Laser parameters: $P = 0.81$ A, $U = 6$V, and $D = 10^{-6}$ cm ps$^{-1}[\eta]^{-1}$. Adapted from [1], with the permission of AIP Publishing

The basic dynamic response of semiconductor lasers to perturbations are relaxation oscillations between the field intensity and the carrier populations of the active medium. While propagating within the resonator, the pulse interacts with spatially varying carrier populations due to different biasing and bleaching conditions. Hence, the proposed model function is chosen to be a linear combination of j exponentially damped harmonic oscillations

$$\Psi^j_{\delta T}(n) = \sum_{k=1}^{j} a_k \exp(-\Gamma_k n) \cos(\omega_k n + \phi_k) \tag{3.26}$$

with the free parameters a_k, Γ_k, ω_k and ϕ_k, which represent the amplitude, damping, frequency and phase offset of the kth term. A successful fit then has to optimize $4j$ free parameters to reproduce the autocorrelation function $\Psi_{\delta T}$.

The pulse-period fluctuation autocorrelation method is illustrated in Fig. 3.8, where (a) shows the computed autocorrelation function $\Psi_{\delta T}$ (solid gray line) and fitted model functions for $j = 1, 2, 3$ (blue, orange, and green dashed lines). The three-section tapered quantum-dot laser (s. Chap. 4) is operated at the same parameters as in Fig. 3.3, i.e., at $P = 0.81$ A, $U = 6$V, and $D = 10^{-6}$ cm ps$^{-1}[\eta]^{-1}$ (s. Figs. 4.5, 4.17, and 4.24). The model functions $\Psi^{1,2,3}_{\delta T}$ were fitted using autocorrelation shifts $n \in [0, 100]$.

A fast drop of the autocorrelation function (not completely shown) with a dip below zero is observed, which is reproduced well by all model functions. The following relaxation to zero is characterized by further modulations, which are only captured well by the model function with $j = 3$. The resulting pulse separation dependent timing jitter is shown in Fig. 3.8b. Starting with a pulse-to-pulse jitter of $\sigma_{\text{pp}} \approx 16.7$ fs, the directly computed timing jitter (solid gray line) first rises to ≈ 32 fs and then relaxes to its long-term value $\sigma_{\text{lt}} \approx 20.5$ fs. While the qualitative behavior is captured by all model functions, the maximum is only correctly described by

$j = 2, 3$ and the long-term value only by $j = 3$. Based on this, it can be concluded that the proposed model function Eq. (3.26) works well for the considered case if j is chosen properly.

The ratio of the long-term timing jitter and the pulse-to-pulse timing jitter $\sigma_{\mathrm{lt}}/\sigma_{\mathrm{pp}}$ can be interpreted as the contributions of finite time correlations to the long-term timing jitter. The correlations are quantitatively described by autocorrelation function values $\Psi_{\delta T}(n)$ for $n > 0$ and contribute to the ratio via the second term in the square brackets in Eq. (3.25). In a different picture, the long-term timing jitter can by thought of as the pulse-to-pulse timing jitter multiplied with the correlation contribution ratio. In the presented case, the correlation contribution ratio amounts to $\sigma_{\mathrm{lt}}/\sigma_{\mathrm{pp}} \approx 1.23$.

Equation (3.25) can be furthermore used to approximate the required pulse separation n for which the timing jitter $\sigma_{\Delta t}(n)$ converges to the long-term timing jitter. In the respective second term

$$2\sum_{d=1}^{n-1} \frac{n-d}{n} \Psi_{\delta T}(d), \tag{3.27}$$

not only the autocorrelation function $\Psi_{\delta T}$ must have decayed to zero, but also the factor $(n-d)/n$ must be approximately constant for all non-zero autocorrelation shifts d. The latter is true to first-order for $n \gtrsim 10 d_0$, where d_0 denotes the largest autocorrelation shift with a non-zero autocorrelation function $\Psi_{\delta T}$. Revisiting Fig. 3.8a, b, this approximation checks out, since the autocorrelation has approximately decayed to zero at $d \approx 80$ and the long-term timing jitter has converged at $n \approx 1000$.[12]

The factor $(n-d)/n$, moreover, is critical for the complete construction of pulse separation dependent timing jitter $\sigma_{\Delta t}(n)$ from the pulse-period autocorrelation function. From Eq. (3.25) it is evident that for a selected n, positive correlations $\Psi_{\delta T}(d)$ increase the ratio $\sigma_{\mathrm{lt}}/\sigma_{\mathrm{pp}}$, while negative correlations decrease it. The factor $(n-d)/n$ represents the weight of the autocorrelation function $\Psi_{\delta T}(d)$, which changes with the considered pulse separation. As a result, the timing jitter $\sigma_{\Delta t}(n)$ increases until it reaches its maximum at $n = 7$, even though the autocorrelation function $\Psi_{\delta T}(d)$ becomes negative for $d > 3$, since the relative weights for the positive autocorrelation values ($d \leqq 3$) increase and thereby dominate the negatively correlated components $\Psi_{\delta T}(d)$, which only exhibit relatively small weights yet. With a similar argument, the convergence to the long-term timing jitter for pulse separations $80 < n < 1000$ is only generated by changes of the factor $(n-d)/n$, since the autocorrelation function has already decayed to zero.

---

[12] With this knowledge, an educated guess can also be performed based on the laser parameters if one assumes that all correlations have sufficiently decayed within five lifetimes of the slowest quantity ($e^{-5} \approx 0.007 < 1\%$). For the considered laser, the gain section carrier reservoir with a lifetime on 1 ns represents the slowest quantity. This corresponds to correlation time of ≈ 5 ns $\approx 66 T_C$ and yields an approximate minimum pulse separation $n = 660$, which is of the correct order of magnitude.

3.5.1 Guidelines to Calculate the Long-Term Timing Jitter

One disadvantage of the proposed pulse-period fluctuation autocorrelation method is the rather involved implementation and evaluation. Hence, the following itemization presents a recipe, which guides with practical recommendations through the individual steps that are to be carried out, in order to calculate the long-term timing jitter σ_{lt}.

- Simulate m realizations of timing fluctuations $\{\delta T_k\}$ with the length n or one realization with the length $n \times m$. The parameter n corresponds to the maximum autocorrelation shift and should be long enough to capture all relevant features of the autocorrelation function.
- Calculate the pulse-to-pulse timing jitter σ_{pp} and the autocorrelation function $\Psi_{\delta T}$. The parameter m has to be chosen such that the autocorrelation function has sufficiently converged up to the pulse separation n. The convergence of σ_{pp} is most likely fast compared to $\Psi_{\delta T}$ and thus represents no practical limitation.
- Optimize the free parameters of the model function $\Psi^j_{\delta T}$, given by Eq. (3.26), to best replicate the calculated autocorrelation function $\Psi_{\delta T}$. Both reasonable parameter bounds and initial parameter guesses may be required to achieve a converging fit.

  - Increase the order j until the fit works well: The *goodness of the fit* can either be evaluated 'by eye' or automatized via the Pearson correlation coefficient $r_j = r(\Psi_{\delta T}, \Psi^j_{\delta T})$ between the calculated and the fitted autocorrelation functions. In the latter case, a correlation coefficient $r^{\text{th}}_j \leq 1 - 10^{-4}$ (based on $n \in [0, 100]$ for the fit) has been proven empirically to be a suitable threshold for increasing the order j (s. Sect. 4.6.1 for an example).
  - All investigated cases within this work either required $j = 2$ or $j = 3$. Using $j = 4$ never lead to a better timing jitter estimation but rather to fit convergence and over-fitting issues.
  - An insufficiently converged autocorrelation function $\Psi_{\delta T}$ produces relatively large Pearson correlation coefficients independent of the order j.
  - The parameter n has to be chosen sufficiently large, such that the fit has converged with respect to increasing n.

- Calculate the long-term timing jitter σ_{lt} using the fitted model function $\Psi^j_{\delta T}$ according to Eq. (3.25).

The premise of choosing multiple parameters correctly in order to achieve a sufficient convergence and thus a reliable long-term timing jitter estimations demands a delicate application of the pulse-period fluctuation autocorrelation method. Hence, a spot check of the timing jitter estimates generated by the means of the pulse-period fluctuation autocorrelation method against, e.g., the direct time-domain method is recommended to ensure their validity.

3.6 Computational Timing Jitter Estimation Efficiency

In the process of understanding, optimizing, and designing passively mode-locked lasers, models and simulations are vital tools [43–52]. However, the direct estimation of the timing jitter from either simulated spectra or sequences of pulse positions [20, 21, 48] often poses the problem of excessively large computational costs, as large numbers of simulated pulses are required. In the case of weak noise sources, analytical approaches have been developed for the Haus master equation [16, 18] and semi-analytical approaches for delay-differential equation descriptions [53, 54].

Many cases, however, e.g., the three-section tapered quantum-dot laser, which is considered Chap. 4, do not permit the use of either of those techniques. Since the simulation of noisy time series generally dominates the overall computational cost, the goal is to obtain a good timing jitter estimate from a sample set, which is as small as possible. For this reason, this section compares the standard errors of the previously introduced long-term timing jitter estimation methods with respect to the sample-set size.

As the long-term timing jitter is a statistic measure of the stochastic properties of the pulse positions, its standard error depends on the considered sample-set size, i.e., the number of pulses in the computed time series realizations. This is equally true for the spectral and autocorrelation based methods, where increased sample-set sizes generate better estimates of the power spectral density and autocorrelation function. Hence, the sample-set size depended standard error of five methods is analyzed: The direct time-domain method based on $\{\Delta t_n\}$ with $n = 10000$, the hybrid von der Linde/Otto method performed on the spectrum $\mathcal{S}_{\Delta t}$, the Kéfélian method performed on the 20th harmonic of $\mathcal{S}_{|E|^2}$, the pulse-period fluctuation autocorrelation method, and finally the von der Linde/Otto method applied to 20th harmonic the spectrum $\mathcal{S}_{|E|^2}$.

The three-section tapered quantum-dot laser (s. Chap. 4) is operated at the same parameters as in Fig. 3.3, i.e., at $P = 0.81$ A, $U = 6$V, and $D = 10^{-6}$ cm ps$^{-1}[\eta]^{-1}$ (s. Figs. 4.5, 4.17, and 4.24). The full sample set contains 1000 time series of 10 μs simulated time, which corresponds to $\approx 133,000$ pulses each. The standard error analysis is then performed by using a subsampling method with subset sizes between one and 100. The sample-set size, which is used for a given timing jitter estimate, is measured by the total number of pulses in the respective subset. The standard error is determined based on the fluctuating timing jitter estimates originating from the different subsets.

The power spectra $\mathcal{S}_{|E|^2}$ are directly calculated from the full time series using a FFT algorithm. For subsets containing multiple time series, Bartlett's method [36] is used to average the individual power spectra. The time series are further processed by extracting the pulse positions t_n, which are then subdivided into sets of 10000 for the further calculation of the timing deviations $\{\Delta t_n\}$. The power spectra $\mathcal{S}_{\Delta t}$ are calculated from the sets $\{\Delta t_n\}$ with a FFT algorithm and if applicable Bartlett's method for averaging. Finally, the autocorrelation functions $\Psi_{\delta T}$ are computed from the pulse-period fluctuations $\{\delta T_k = t_{k+1} - t_k - T_c\}$ via the Wiener Khinchin theorem, also using Bartlett's method if applicable.

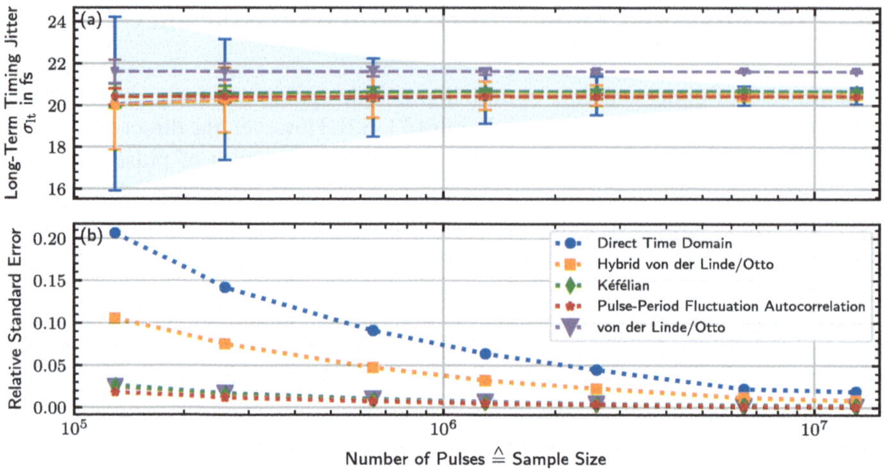

Fig. 3.9 Performance of the long-term timing jitter estimation methods. **a** Long-term timing jitter and respective standard errors as a function of the underlying number of pulses. The curves are computed from the direct time-domain method (blue), the hybrid von der Linde/Otto method applied to $\mathcal{S}_{\Delta t}$ (orange), the Kéfélian method (green), the pulse-period fluctuation autocorrelation method (red), and the von der Linde/Otto method applied to $\mathcal{S}_{|E|^2}$ (purple). **b** Relative standard errors. Laser parameters: $P = 0.81$ A, $U = 6$V, and $D = 10^{-6}$ cm ps$^{-1}[\eta]^{-1}$. Adapted from [1], with the permission of AIP Publishing

In all cases, an increased sample-set size improves the fidelity of the estimated quantity, i.e., the timing deviation variance $\mathrm{Var}(\Delta t_n)$, the power spectral densities $\mathcal{S}_{|E|^2}$ and $\mathcal{S}_{\Delta t}$, and the autocorrelation function $\Psi_{\delta T}$.

The results are presented in Fig. 3.9, where (a) shows the computed long-term jitter as a function of the sample-set size, with error bars indicating the respective standard errors. Since the direct time-domain method (blue) makes no assumptions about the underlying system, no systematic deviations are expected and the resulting error range (highlighted in light blue) is used as a benchmark for the validity of the other methods. The hybrid von der Linde/Otto, Kéfélian and pulse-period fluctuation autocorrelation method all lie within that range and thus exhibit no evidence of a systematic error. The von der Linde/Otto method applied to the 20th harmonic (purple), on the other hand, converges to a systematic overestimation the long-term timing jitter, which is attributed to spectral leakage, i.e., an insufficient spectral resolution as discussed in Sect. 3.3.3.

The respective relative standard errors of all five methods are plotted additionally in Fig. 3.9b. A $1/\sqrt{\text{sample size}}$ dependence, i.e., the law of large numbers, is observed for all curves, but with vastly different absolute values. While the Kéfélian and the pulse-period fluctuation autocorrelation methods produce very similar relative standard errors, the hybrid von der Linde/Otto method performs worse by a factor of ≈ 5, and the direct time-domain method performs worse by a factor of ≈ 10. In terms of computational cost (sample-set size), this translates into a factor of $\approx$

25 for the hybrid von der Linde/Otto method and a factor of ≈ 100 for the direct time-domain method. While proven inaccurate by a systematic error, the von der Linde/Otto method applied to the 20th harmonic of $\mathcal{S}_{|E|^2}$ produces standard errors as low as the autocorrelation and Kéfélian method.

The difference between the methods can be explained by their underlying assumptions about the stochastic properties and their usage of the calculated time series: The direct method assumes a random walk on large time scales and only uses the last pulse position of each realization.[13] The von der Linde/Otto method assumes a $1/\nu^2$ decay of the phase noise spectrum, which corresponds to a random walk, but integrates over a large frequency band and thus utilizes a larger number of data points. The Kéfélian methods assumes a specific line shape of the harmonics, namely a Lorentzian, and evaluates the linewidth parameter from a large number of spectral data points, which are calculated not only from the pulse positions, but from the full time-traces. Finally, the autocorrelation based method assumes an exponentially decaying oscillating pulse period autocorrelation function, where a model function has to be only fitted to small n for which the estimated $\Psi_{\delta T}(n)$ is of much higher fidelity.

Based on those results, the use of either the Kéfélian or the pulse-period fluctuation autocorrelation method is advantageous and therefore highly recommended in order to achieve computationally efficient timing jitter estimates. However, both methods have their specific advantages but also caveats, which can lead to systematic errors in the timing jitter estimate: The Kéfélian method requires a sufficient spectral resolution, i.e., simulated time, of $\mathcal{S}_{|E|^2}$ to successfully extract the linewidth parameter $\Delta\nu$ as discussed in Sect. 3.3.3. This problem can be somewhat mitigated by analyzing higher harmonics, but nonetheless prohibits an efficient analysis of lasers with low timing jitter.[14] The autocorrelation method requires a good model function to reproduce all features of the autocorrelation function, whose existence can not by guaranteed. However, it is not limited by low timing jitter values and can furthermore also be applied to models that only generate pulse positions and no complete time traces [32, 42]. In cases where both fail, the hybrid von der Linde/Otto or hybrid Kéfélian method should be used. Finally, the direct time-domain method can only be recommended for the purpose of validating the other methods.

---

[13] The direct time-domain estimate of the long-term timing jitter could in principle be improved by using multiple data points and fitting the final relaxation to its equilibrium long-term value with the exponential decay $\sigma_{\Delta t}(n) = A\exp(-\gamma_n n) + \sigma_{\mathrm{lt}}$. In practice, however, the utilized curve-fitting routines (SciPy 1.10) produced uncertainties in the fit parameters, especially in the constant offset σ_{lt}, which outweighed the potentially gained advantages. Note that this issue does not arise in the pulse-period fluctuation autocorrelation method, since the autocorrelation functions are guaranteed to decay to zero.

[14] Low timing jitter values produce small repetition-rate linewidths, which are computationally costly to properly resolve.

3.7 Chapter Conclusions

This chapter has presented a review of the established long-term timing jitter estimation methods in the context of their application to numerically simulated time series. Time-domain, frequency-domain, and hybrid methods have been considered. Special attention has been paid to the respective estimates' convergence with respect to the time series length. In the case of the spectral methods, it has been found that the repetition-rate linewidth must by sufficiently resolved in order to produce reliable timing jitter estimates.

If the long-term timing jitter is to be computed via the von der Linde/Otto method, this chapter has further highlighted a factor of two in the timing phase noise spectral power density normalization, which is missing in the existing literature. This factor is required in order to reconcile the timing phase noise power spectral densities obtained from the power spectral densities of the out-coupled intensity and the power spectral density of the extracted pulse timing deviations.

Moreover, this chapter has proposed a new time-domain based method, which extracts the long-term timing jitter from the pulse-to-pulse timing jitter and the pulse-period fluctuation autocorrelation function. If supplied with a suitable model function of the said autocorrelation function, this method is able to produce long-term timing jitter estimates with relatively few stochastic pulse train realizations.

On that account, the standard error convergence of the various timing jitter estimation methods has been benchmarked with respect to the supplied stochastic time series sample-set size. While all methods follow the law of large numbers, the differences in the relative standard errors for a given sample-set size is up to two orders of magnitude. Based on that, recommendations on the choice of timing jitter estimation method have been formulated.

Those results are summarized in the flowchart presented in Fig. 3.10: Due to the relatively simple implementation, the Kéfélian method is recommended as the first choice as long as the repetition-rate linewidth of any harmonic can be sufficiently resolved. If that is not the case, the pulse-period fluctuation auto-correlation method performs equally well, but the implementation is a bit more involved. Lastly, if no suitable model function for the autocorrelation function can be found, the hybrid von der Linde/Otto method is recommended to be employed. The direct time-domain method is not recommended in any case unless for the validation of the results obtained from the other methods.

Lastly, the relative standard error of all methods can be estimated by a subsampling approach: The long-term timing jitter is evaluated for different subsamples and different subsampling sizes. The resulting relative standard errors can then be extrapolated via the law of larger numbers to produce the required sample-set size, i.e., the number of stochastic realization, which yields the desired relative standard error of the long-term timing jitter estimate.

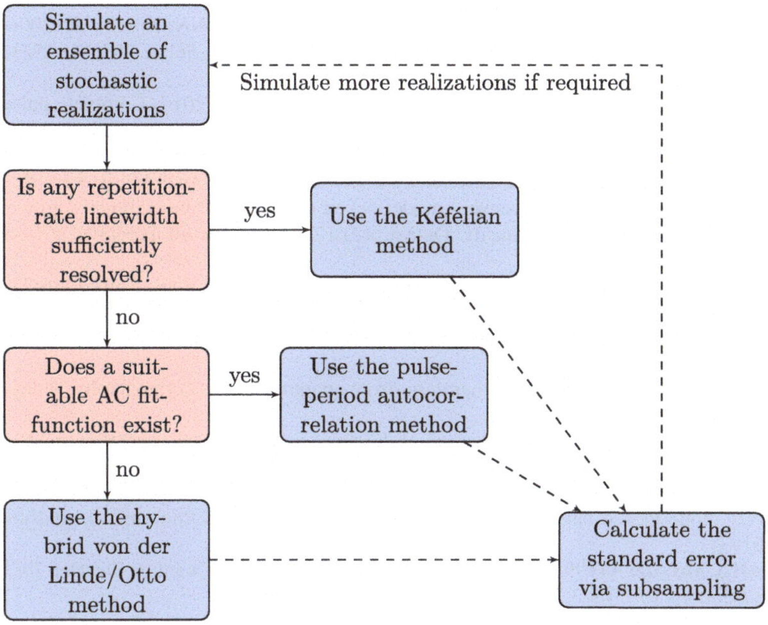

Fig. 3.10 Decision flowchart. Depending on the properties of the simulated mode-locked laser and the obtained stochastic time series realizations, the most suitable long-term timing jitter estimation method is recommended

References

1. Meinecke S, Lüdge K (2021) Efficient timing jitter simulation for passively mode-locked semiconductor lasers. App Phys Lett 118(1):011104. https://doi.org/10.1063/5.0036928
2. Jiang LA, Grein ME, Ippen EP, McNeilage C, Searls J, Yokoyama H (2002) Quantum-limited noise performance of a mode-locked laser diode. Opt Lett 27(1):49–51. https://doi.org/10.1364/ol.27.000049
3. Kuntz M, Fiol G, Lämmlin M, Meuer C, Bimberg D (2007) High-speed mode-locked quantum-dot lasers and optical amplifiers. Proc IEEE 95(9):1767–1778. ISSN 0018-9219. https://doi.org/10.1109/jproc.2007.900949
4. Lelarge F, Dagens B, Renaudier J, Brenot R, Accard A, van Dijk F, Make D, Gouezigou OL, Provost JG, Poingt F, Landreau J, Drisse O, Derouin E, Rousseau B, Pommereau F, Duan GH (2007) Recent advances on InAs/InP quantum dash based semiconductor lasers and optical amplifiers operating at 1.55 μm. IEEE J Sel Top Quantum Electron 13(1):111. ISSN 1077-260X. https://doi.org/10.1109/jstqe.2006.887154
5. Valley GC (2007) Photonic analog-to-digital converters. Opt Express 15(5):1955–1982. https://doi.org/10.1364/oe.15.001955
6. Bajek D, Cataluna MA (2021) Fast optical sampling by electronic repetition-rate tuning using a single mode-locked laser diode. Opt Express 29(5):6890–6902. https://doi.org/10.1364/oe.413045
7. Jiang LA, Ippen EP, Yokoyama H (2005) Semiconductor mode-locked lasers as pulse sources for high bit rate data transmission. J Opt Fiber Comm Rep 2:21–51

8. Bimberg D, Fiol G, Kuntz M, Meuer C, Lämmlin M, Ledentsov NN, Kovsh AR (2006) High speed nanophotonic devices based on quantum dots. Phys Stat Sol A 203(14):3523–3532. https://doi.org/10.1002/pssa.200622488
9. Schmeckebier H, Fiol G, Meuer C, Arsenijević D, Bimberg D (2010) Complete pulse characterization of quantum dot mode-locked lasers suitable for optical communication up to 160 Gbit/s. Opt Express 18(4):3415. https://doi.org/10.1364/oe.18.003415
10. Liu S, Wu X, Jung D, Norman JC, Kennedy MJ, Tsang HK, Gossard AC, Bowers JE (2019) High-channel-count 20GHz passively mode-locked quantum dot laser directly grown on Si with 4.1Tbit/s transmission capacity. Optica 6(2):128–134. https://doi.org/10.1364/optica.6.000128
11. Udem T, Holzwarth R, Hänsch TW (2002) Optical frequency metrology. Nature 416:233–237. https://doi.org/10.1038/416233a
12. Keller U (2003) Recent developments in compact ultrafast lasers. Nature 424:831–838. https://doi.org/10.1038/nature01938
13. Lüdge K (2012) Nonlinear laser dynamics—from quantum dots to cryptography. Wiley-VCH, Weinheim. 978-3-527-41100-9
14. von der Linde D (1986) Characterization of the noise in continuously operating mode-locked lasers. Appl Phys B 39(4):201. https://doi.org/10.1007/bf00697487
15. Eliyahu D, Salvatore RA, Yariv A (1996) Noise characterization of a pulse train generated by actively mode-locked lasers. J Opt Soc Am B 13(7):1619. https://doi.org/10.1364/josab.13.001619
16. Haus HA, Mecozzi A (1993) Noise of mode-locked lasers. IEEE J Quantum Electron 29(3):983. https://doi.org/10.1109/3.206583
17. Eliyahu D, Salvatore RA (1997) Yariv A (1997) Effect of noise on the power spectrum of passively mode-locked lasers. J Opt Soc Am B 14(1):167. https://doi.org/10.1364/josab.14.000167
18. Jiang LA, Grein ME, Haus HA, Ippen EP (2001) Noise of mode-locked semiconductor lasers. IEEE J Sel Top Quantum Electron 7(2):159. https://doi.org/10.1109/2944.954125
19. Kefelian F, O'Donoghue S, Todaro MT, McInerney JG, Huyet G (2008) RF linewidth in monolithic passively mode-locked semiconductor laser. IEEE Photon Technol Lett 20(16):1405. ISSN 1041-1135. https://doi.org/10.1109/lpt.2008.926834
20. Mulet J, Mørk J (2006) Analysis of timing jitter in external-cavity mode-locked semiconductor lasers. IEEE J Quantum Electron 42(3):249. https://doi.org/10.1109/jqe.2006.869808
21. Lee DC (2002) Analysis of jitter in phase-locked loops. IEEE Trans Circuits Syst II 49(11):704. ISSN 1057-7130. https://doi.org/10.1109/tcsii.2002.807265
22. Otto C (2014) Dynamics of quantum dot lasers—effects of optical feedback and external optical injection. Springer Theses. Springer, Heidelberg, 2014. ISBN 978-3-319-03785-1. https://doi.org/10.1007/978-3-319-03786-8
23. Gardiner CW (2002) Handbook of stochastic methods for physics, chemistry and the natural sciences. Springer, Berlin
24. Kolner B, Bloom D (1986) Electrooptic sampling in GaAs integrated circuits. IEEE J Quantum Electron 22(1):79. ISSN 0018-9197. https://doi.org/10.1109/jqe.1986.1072877
25. Demir A, Mehrotra A, Roychowdhury J (2000) Phase noise in oscillators: a unifying theory and numerical methods for characterization. IEEE Trans Circuits Syst I Regul Pap 47(5):655–674
26. Demir A (2006) Computing timing jitter from phase noise spectra for oscillators and phase-locked loops with white and $1/f$ noise. IEEE Trans Circuits Syst I Regul Pap 53(9):1869–1884
27. Paschotta R (2004a) Noise of mode-locked lasers (part II): timing jitter and other fluctuations. Appl Phys B 79(2):163–173. ISSN 0946-2171. https://doi.org/10.1007/s00340-004-1548-9
28. Mangold M, Link SM, Klenner A, Zaugg CA, Golling M, Tilma BW, Keller U (2014) Amplitude noise and timing jitter characterization of a high-power mode-locked integrated external-cavity surface emitting laser. IEEE Photon J 6(1):1500309. ISSN 1943-0655. https://doi.org/10.1109/jphot.2013.2295464
29. Otto C, Jaurigue LC, Schöll E, Lüdge K (2014) Optimization of timing jitter reduction by optical feedback for a passively mode-locked laser. IEEE Photon J 6(5):1501814. https://doi.org/10.1109/jphot.2014.2352934

30. Breuer S, Elsäßer W, McInerney JG, Yvind K, Pozo J, Bente EAJM, Yousefi M, Villafranca A, Vogiatzis N, Rorison J (2010) Investigations of repetition rate stability of a mode-locked quantum dot semiconductor laser in an auxiliary optical fiber cavity. IEEE J Quantum Electron 46(2):150. ISSN 0018-9197. https://doi.org/10.1109/jqe.2009.2033255
31. Lin CY, Grillot F, Li Y (2011) Microwave characterization and stabilization of timing jitter in a quantum dot passively mode-locked laser via external optical feedback. IEEE J Sel Top Quantum Electron 17(5):1311. https://doi.org/10.1109/jstqe.2011.2118745
32. Drzewietzki L, Breuer S, Elsäßer W (2013) Timing jitter reduction of passively mode-locked semiconductor lasers by self- and external-injection: numerical description and experiments. Opt Express 21(13):16142–16161. https://doi.org/10.1364/oe.21.016142
33. Auth D, Liu S, Norman J, Bowers JE, Breuer S (2019) Passively mode-locked semiconductor quantum dot on silicon laser with 400 Hz rf line width. Opt Express 27(19):27256–27266. https://doi.org/10.1364/oe.27.027256
34. Yamamoto Y (1983) AM and FM quantum noise in semiconductor lasers—part I: theoretical analysis. IEEE J Quantum Electron 19(1):34–46. ISSN 0018-9197. https://doi.org/10.1109/jqe.1983.1071726
35. Haug H, Koch SW (1993) Quantum theory of the optical and electronic properties of semiconductors, 3rd edn. World Scientific, Singapore. 9810218648 (pbk)
36. Press WH, Flannery BP, Teukolsky SA, Vettering WT (2007) Numerical recipes, 3rd edn. Cambridge University Press, Cambridge
37. Lüdge K, Schöll E (2009) Quantum-dot lasers–desynchronized nonlinear dynamics of electrons and holes. IEEE J Quantum Electron 45(11):1396–1403. https://doi.org/10.1109/jqe.2009.2028159
38. Lingnau B, Chow WW, Lüdge K (2014) Amplitude-phase coupling and chirp in quantum-dot lasers: influence of charge carrier scattering dynamics. Opt Express 22(5):4867–4879. https://doi.org/10.1364/oe.22.004867
39. Zajnulina M, Lingnau B, Lüdge K (2017) Four-wave mixing in quantum dot semiconductor optical amplifiers: a detailed analysis of the nonlinear effects. IEEE J Sel Top Quantum Electron 23(6):3000112. ISSN 1077-260X. https://doi.org/10.1109/jstqe.2017.2681803
40. Lingnau B, Lüdge K (2018) Rabi-oscillation-enhanced frequency conversion in quantum-dot semiconductor optical amplifiers. Opt Quantum Electron 50(2):111. ISSN 1572-817X. https://doi.org/10.1007/s11082-018-1380-9
41. Paschotta R (2004) Noise of mode-locked lasers (part I): numerical model. Appl Phys B 79:153–162. ISSN 0946-2171. https://doi.org/10.1007/s00340-004-1547-x
42. Schelte C, Javaloyes J, Gurevich SV (2018) Functional mapping for passively mode-locked semiconductor lasers. Opt Lett 43(11):2535–2538. https://doi.org/10.1364/ol.43.002535
43. Vladimirov AG, Turaev DV (2005) Model for passive mode locking in semiconductor lasers. Phys Rev A 72(3):033808. https://doi.org/10.1103/physreva.72.033808
44. Bandelow U, Radziunas M, Vladimirov AG, Hüttl B, Kaiser R (2006) 40 GHz mode locked semiconductor lasers: theory, simulation and experiments. Opt Quantum Electron 38:495. https://doi.org/10.1007/s11082-006-0045-2
45. Radziunas M, Vladimirov AG, Viktorov EA, Fiol G, Schmeckebier H, Bimberg D (2011) Pulse broadening in quantum-dot mode-locked semiconductor lasers: simulation, analysis, and experiments. IEEE J Quantum Electron 47(7):935–943
46. Javaloyes J, Balle S (2011) Anticolliding design for monolithic passively mode-locked semiconductor lasers. Opt Lett 36(22):4407–4409
47. Rossetti M, Tianhong X, Bardella P, Montrosset I (2011) Impact of gain saturation on passive mode locking regimes in quantum dot lasers with straight and tapered waveguides. IEEE J Quantum Electron 47(11):1404. https://doi.org/10.1109/jqe.2011.2167131
48. Otto C, Lingnau B, Schöll E, Lüdge K (2014) Manipulating coherence resonance in a quantum dot semiconductor laser via electrical pumping. Opt Express 22:13288. https://doi.org/10.1364/oe.22.013288
49. Jaurigue LC, Nikiforov O, Schöll E, Breuer S, Lüdge K (2016) Dynamics of a passively mode-locked semiconductor laser subject to dual-cavity optical feedback. Phys Rev E 93(2):022205. https://doi.org/10.1103/physreve.93.022205

50. Bardella P, Drzewietzki L, Krakowski M, Krestnikov I, Breuer S (2018) Mode locking in a tapered two-section quantum dot laser: design and experiment. Opt Lett 43(12):2827–2830. https://doi.org/10.1364/ol.43.002827
51. Meinecke S, Drzewietzki L, Weber C, Lingnau B, Breuer S, Lüdge K (2019) Ultra-short pulse generation in a three section tapered passively mode-locked quantum-dot semiconductor laser. Sci Rep 9:1783. https://doi.org/10.1038/s41598-018-38183-1
52. Hausen J, Meinecke S, Lingnau B, Lüdge K (2019) Pulse cluster dynamics in passively mode-locked semiconductor vertical-external-cavity surface-emitting lasers. Phys Rev Appl 11:044055. https://doi.org/10.1103/physrevapplied.11.044055
53. Pimenov AS, Habruseva T, Rachinskii D, Hegarty SP, Huyet G, Vladimirov AG (2014) The effect of dynamical instability on timing jitter in passively mode-locked quantum-dot lasers. Opt Lett 39(24):6815. https://doi.org/10.1364/ol.39.006815
54. Jaurigue LC, Pimenov AS, Rachinskii D, Schöll E, Lüdge K, Vladimirov AG (2015) Timing jitter of passively mode-locked semiconductor lasers subject to optical feedback; a semi-analytic approach. Phys Rev A 92(5):053807. https://doi.org/10.1103/physreva.92.053807

Chapter 4
Three-Section Tapered Quantum-Dot Laser

4.1 Introduction

Monolithically integrated edge-emitting mode-locked semiconductor lasers are favorable for future photonic integration [1–3], because they have the advantage of straightforward growth and processing, while keeping a relatively small footprint. However, compared to surface-emitting mode-locked semiconductor disk lasers [4–7], monolithically integrated edge-emitting lasers mostly show inferior pulse performance in terms of the pulse peak power and the pulse width.

Some of the shortcomings of edge-emitting mode-locked lasers can be mitigated by the post emission amplification and compression of the pulses. This, however, introduces additional active and passive optical elements and thereby increases the complexity and cost of a given photonic setup. Therefore, it is highly desirable to optimize monolithically integrated device designs, in order for them to already meet the performance requirements of a considered application. Optimization approaches may be separated into two categories: The first category focuses on the device geometry and the cavity design. Previous works have demonstrated that the precise tuning of the saturable absorber length [8–12] and the resonator facet reflectivities [10, 13, 14] can lead to shorter pulses and an increased pulse train stability. The positioning of the absorber section within the laser cavity has been furthermore shown to affect the mode-locking performance and thus to yield improvements if properly tuned [15–17]. Moreover, tapered gain sections have been demonstrated to produce an additional pulse shortening and a substantial increase of the output power, while maintaining high-quality output beams [8, 11, 18–23]. The second category addresses the electronic properties of the active medium, where one option is to employ semiconductor quantum dots as an active medium. Devices based on them exhibit advantages such as high differential gain, ultra-fast recovery, broad gain spectra, small chirp, and low-temperature sensitivity due to the atom-like discrete energy levels [24–28]. These properties have been shown to generate very stable mode-locked pulse trains with sub-picosecond pulses at high repetition rates [8, 12, 29–33].

S. Meinecke, *Spatio-Temporal Modeling and Device Optimization of Passively Mode-Locked Semiconductor Lasers*, Springer Theses,
https://doi.org/10.1007/978-3-030-96248-7_4

Relevant to passively mode-locked semiconductor lasers, strong spontaneous emission noise and the absence of an external reference clock lead to relatively pronounced timing and amplitude fluctuations [34–38]. Both are limiting factors for many applications [39–41]. Techniques such as hybrid mode-locking [42–44], optical injection [45–49], opto-electronic self-feedback [38, 50], and optical self-feedback [37–39, 47, 51–55] allow to improve the timing stability considerably, but again come at the cost of additional electronics and optics, which need to be properly calibrated and controlled. To avoid this, it is therefore preferable to also optimize the laser design towards an excellent pulse train stability that renders any additional control schemes unnecessary.

This chapter focuses on a device that combines both approaches from the aforementioned categories. A detailed experimental and numerical analysis of a three-section tapered semiconductor quantum-dot laser with a saturable absorber section, which is positioned at approximately one-third of the cavity length, is performed.[1] The results represent the foundation for further numerical explorations.[2]

4.1.1 Investigated Device

The investigated laser contains an optically active region that consists of ten layers of InAs quantum dots, which are grown on a GaAs substrate using molecular beam epitaxy. The cavity length amounts to 3.0 mm, which leads to a round-trip time of $T \approx 75.5$ ps and corresponds to a mode spacing of $\nu_c \approx 13.24$ GHz. The laser has been processed into a 0.7 mm straight section, a 0.7 mm long absorber section, and a 1.6 mm long tapered section with a full taper angle that amounts to $\theta = 2°$. The absorber to gain section length ratio thus yields $\approx$1:4.28. The straight section active region width is 14 μm. The confinement of the optical field in the lateral direction is achieved by gain guiding. The tapered output facet on the right is anti-reflection coated, resulting in a reflectivity of $r_R = 0.03$, while the facet on the left is high-reflection coated, resulting in a reflectivity of $r_L = 0.95$. The two forward biased gain sections are contacted in parallel and are therefore always subject to the same current density. The absorber section is driven by a reverse bias. A sketch of the device geometry is shown in Fig. 4.1a, where G and A denote the gain and saturable absorber sections, and z the longitudinal coordinate. Figure 4.1b shows the top view picture of the laser, which has been taken with a light microscope. The white lines indicate the optically active region.

To put the investigated device into perspective, other representative mode-locked tapered quantum-dot lasers that have been studied and published [8, 11, 19–23] exhibit absorber to gain section length ratios that range from 1:4 to 1:7, total lengths from 2.4 to 3.0 mm, taper angles from 0.6° to 3.6°, and five to ten quantum-dot

[1] The experimental characterization, which is presented in this chapter, has been carried out by Lukas Drzewietzki, Christoph Weber and Stefan Breuer at the Technische Universität Darmstadt.

[2] Parts of this chapter have been published in [56–58].

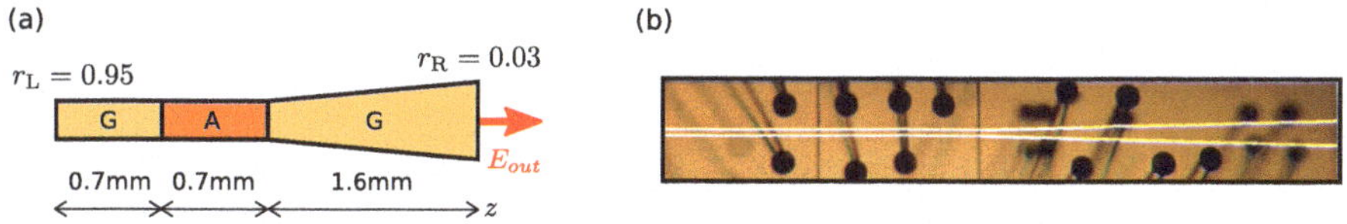

Fig. 4.1 **a** Sketch of the three-section quantum-dot semiconductor laser. The total length of 3.0 mm corresponds to a 13.24 GHz fundamental repetition rate. G denotes the two respective gain sections and A denotes the reverse biased absorber section. The left facet is high-reflection coated and the right output facet is anti-reflection coated with the reflectivities $r_L = 0.95$ and $r_R = 0.03$, respectively. z denotes the longitudinal coordinate of the cavity. **b** Corresponding light microscope picture of the device, where the white lines indicate the laser's active region. Light microscope picture by Stefan Breuer. Adapted from [56]

layers. Hence, the device, which is considered in this work, is in comparison rather long, exhibits a small absorber to gain ratio, an intermediate taper angle, and a larger number of quantum-dot layers. The defining feature, however, that makes this work's laser stand out from all other devices, is the three-section configuration with the absorber section positioned in the middle.

4.1.2 Chapter Outline

This chapter is organized as follows: Subsequent to this introduction, the derivation of the numerical model is performed in Sect. 4.2. The presentation and discussion of the results is then divided into multiple parts: First, the basic emission states are introduced and thoroughly characterized by means of the numerical model in Sect. 4.3. The succeeding Sect. 4.4 then juxtaposes experimental and simulated results. Building onto that, an in-depth study of the emission states and the pulse performance in the operation parameter space is shown in Sect. 4.5. The complementary timing jitter analysis is presented in Sect. 4.6. With the goal of further optimizing the device performance, the influence of the taper angle and the saturable absorber position are studied in Sects. 4.7 and 4.8, respectively. Finally, the chapter conclusions are drawn in Sect. 4.9.

4.2 Three-Section Tapered Quantum-Dot Laser Model

The goal of numerically simulating the three-section tapered quantum-dot laser is to not only reproduce the experimentally observed emission dynamics and perform operation parameter studies, but to also gain an in-depth understanding of the pulse-shaping mechanism and study the implications of the device design.

Hence, a model that is capable of describing the device-specific spatio-temporal evolution of the electric field within the quantum-dot active medium is required.

However, the simulations also need to be numerically efficient enough to calculate sufficiently long time series for the evaluation of timing and amplitude jitter as well as to perform multi-parameter studies.

The established delay-differential equation modeling frameworks [37, 59–64] are computationally very efficient, but require limiting assumptions about the device geometry and spectral field evolution.[3] Hence, the model derived in this sections follows the ideas of [9, 14, 65–69] and couples a traveling-wave equation for the propagation of the electric field via effective Maxwell-Bloch equations to optically active quantum-dot populations. The latter are modeled by microscopically motivated rate equations, which describe the electronic degrees of freedom of the active medium.

This section presents the detailed derivation based on Chap. 2 of the proposed model, supplies the utilized laser parameters, and discusses the numerical implementation.

Quantum-Dot Active Medium

The stochastic Stranski-Krastanov growth process produces quantum dots of varying sizes and shapes. Resulting from that, the respective optical transition frequencies ω_j, which depend on the quantum-dot geometries, are subject to an inhomogeneous broadening [26, 70, 71]. The probability of finding a certain quantum-dot transition frequency follows the central limit theorem, i.e., a Gaussian distribution. Hence, the complete gain spectrum is constructed from the characteristic Lorentzian line shapes of the individual quantum dots, whose transition frequencies follow the inhomogeneously broadened Gaussian distribution.

While the homogeneous linewidth of individual quantum dots typically amounts to $\approx$10 meV [72, 73], the FWHM of inhomogeneously broadened quantum-dot gain spectra easily exceeds more than 30 meV [28, 74–76]. The broad gain spectra make quantum dots suitable for mode-locked lasers, where short pulses are desired [8, 12, 30, 31]. A sketch of an inhomogeneously broadened quantum-dot gain spectrum is shown in Fig. 4.2a, where a Gaussian envelope (black dashed line) encloses the homogeneously broadened QD transitions (blue lines).

The sketch in Fig. 4.2a only plots a small number of homogeneously broadened transitions, which is representative for practical numerical implementations [28, 69, 76, 77]: In order to reduce the degrees of freedom, the inhomogeneously broadened spectrum is averaged over finite transition energy ranges with their respective mean transition energy $\langle\omega_j\rangle$. Each quantum-dot subgroup then contains n_j quantum dots and can be thought of as a single optical transition that is n_j-fold degenerate.

The model proposed in this chapter takes this procedure to its limit and averages over all quantum dots, such that only one optical transition remains. To still account for the proper gain bandwidth, the homogeneous linewidth of the remaining transition is set to approximate the inhomogeneously broadened gain bandwidth. This process

[3] Unidirectional propagation in ring geometry is assumed. Hence, saturation effects that counter-propagating pulses experience can not be described. The spectral evolution is modeled by a lumped-element Lorentzian filter.

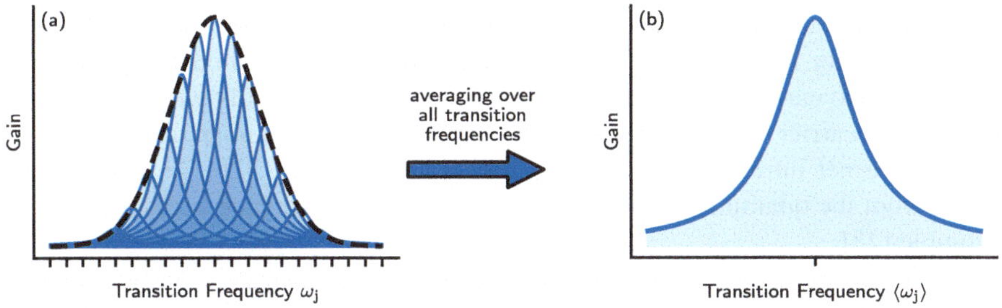

Fig. 4.2 Sketch of the inhomogeneously broadened quantum-dot ensemble. **a** The stochastic growth of the quantum dots produces a Gaussian envelope (black dashed line) in the gain spectrum that encloses the homogeneously broadened quantum-dot transitions (blue lines). **b** The model averages over all transitions and thus approximates the ensemble by a single homogeneously broadened transition with a corresponding effective lifetime

and the resulting gain spectrum are sketched in Fig. 4.2b. Note that the resulting gain spectrum has more pronounced tails than the original inhomogeneous broadened gain spectrum due to the $1/\omega^2$ fall-off of the Lorentzian at larger frequencies.

The effective homogeneous linewidth of the averaged quantum-dot ensemble, furthermore, implies a shorter effective lifetime T_2^{eff} of the medium polarization. Hence, the proposed approximation is only valid if the system dynamics are much slower than the actual polarization dynamics of the individual quantum dots. As long as this condition holds, the quantum-dot polarization dynamics can be assumed to be in a quasi-equilibrium with no dynamical effect on the overall system dynamics. If this condition breaks, one must resort to sufficiently resolving the inhomogeneous broadening via quantum-dot subgroups with their true characteristic lifetime.

Charger-Carrier Scattering Dynamics

The charge-carrier model includes excitonic occupation numbers of the quantum-dot ground state ρ^{GS}, quantum dot excited state ρ^{ES}, and the charge-carrier density n in the surrounding quantum well. Those quantities are spatially resolved along the optical axis of the laser. Figure 4.3b shows a sketch of the considered levels and their interaction via charge-carrier scattering processes as well as their coupling to the electric field. The quantum-dot excited state is assumed to be two-fold degenerate with respect to the quantum-dot ground state. The dynamics at each spatial coordinate are described by a set of coupled rate equations [28]

$$\frac{\mathrm{d}}{\mathrm{d}t}n(z,t) = -\frac{n}{\tau^n} + J - 4N^{\mathrm{QD}}R^{\mathrm{ES}}_{\mathrm{cap}}(n,\rho^{\mathrm{ES}}) \tag{4.1}$$

$$\frac{\mathrm{d}}{\mathrm{d}t}\rho^{\mathrm{ES}}(z,t) = -\frac{\rho^{\mathrm{ES}}}{\tau^{\mathrm{ES}}} + R^{\mathrm{ES}}_{\mathrm{cap}}(n,\rho^{\mathrm{ES}}) - \frac{1}{2}R_{\mathrm{rel}}(\rho^{\mathrm{GS}},\rho^{\mathrm{ES}}) \tag{4.2}$$

$$\frac{\mathrm{d}}{\mathrm{d}t}\rho^{\mathrm{GS}}(z,t) = -\frac{\rho^{\mathrm{GS}}}{\tau^{\mathrm{GS}}} + R_{\mathrm{rel}}(\rho^{\mathrm{GS}},\rho^{\mathrm{ES}}) - \partial_t\rho^{\mathrm{GS}}|_{\mathrm{stim}} \tag{4.3}$$

with the pump current density J, the characteristic carrier lifetimes τ^{GS}, τ^{ES}, and τ^{n} and the factor four for the spin and quantum-dot excited state degeneracy. The parameters also carry a spatial dependence, which is not explicitly written. Furthermore, the net carrier capture from the quantum-well layer is summarized by the term $R^{\mathrm{ES}}_{\mathrm{cap}}$ and the net intra-dot carrier relaxation by the term R_{rel} [26, 78, 79]. Direct capture from the quantum well into the quantum-dot ground state is assumed to be negligible [28].

The carrier scattering processes are mediated by the carrier-carrier and the carrier-phonon interaction [26, 76, 79]. Within the proposed model, those processes are treated in a relaxation rate approximation [28]. The net intra-dot carrier-relaxation reads

$$R_{\mathrm{rel}}(\rho^{\mathrm{GS}}, \rho^{\mathrm{ES}}) = \tilde{R}_{\mathrm{rel}} \left[(1 - \rho^{\mathrm{GS}})\rho^{\mathrm{ES}} - \rho^{\mathrm{GS}}(1 - \rho^{\mathrm{ES}}) e^{-\frac{\Delta\varepsilon^{\mathrm{ESGS}}}{k_{\mathrm{B}}T}} \right] \tag{4.4}$$

and includes Pauli-blocking terms and a Boltzmann-factor. The energy difference $\Delta\varepsilon^{\mathrm{ESGS}}$ between the quantum-dot excited and ground state and the effective temperature T account for the proper detailed balance between the in and out-scattering processes. Relaxation towards the quasi-equilibrium is thereby ensured. The net carrier-capture rate is given by

$$\begin{aligned} R_{\mathrm{cap}}(n, \rho^{\mathrm{ES}}) &= \tilde{R}_{\mathrm{cap}} \left[F(\varepsilon^{\mathrm{QWES}}, \mu, T) - \rho^{\mathrm{ES}} \right] \\ &= \tilde{R}_{\mathrm{cap}} \left[\left(1 + \frac{e^{\Delta\varepsilon^{\mathrm{QWES}}/k_{\mathrm{B}}T}}{e^{n/\mathcal{D}^{\mathrm{2D}}k_{\mathrm{B}}T} - 1} \right)^{-1} - \rho^{\mathrm{ES}} \right], \end{aligned} \tag{4.5}$$

where $F(\varepsilon, \mu, T)$ denotes the quasi-Fermi function with the chemical potential μ and the quantum-well band edge to quantum-dot excited state energy difference $\Delta\varepsilon^{\mathrm{QWES}}$. The last equality expands the quasi-Fermi function and the chemical potential, which itself is a function of the charge-carrier density n. $\mathcal{D}^{\mathrm{2D}}$ denotes the two-dimensional density of states D^{2D} [76, 80], with the proper effective mass.

Lastly, the coupling to the electric field is described by the term $\partial_t \rho^{\mathrm{GS}}|_{\mathrm{stim}}$, which is derived in the following subsections. Dynamic carrier gratings and spatial hole burning, which are induced by standing-wave patterns of the electric field [9], as well as carrier diffusion, are neglected since the strong nonlinear interaction with the optical pulses and the fast carrier relaxation are assumed to dominate. Those effects only become relevant for single section devices that are not subject to the strong modulation induced by a saturable absorber [81–83].

Electric Field Dynamics

The propagation of the electric field is described in the slowly-varying envelope and rotating wave approximation, which yield first-order partial differential equations for the forward (+) and backward (−) moving envelope functions $E^{\pm}(z, t)$. The details have been presented in Chap. 2 and the resulting traveling-wave equation reads

$$\left(\pm\partial_z + \frac{1}{v_g}\partial_t\right) E^{\pm}(z,t) = -\frac{\alpha_{\text{int}}}{2} E^{\pm}(z,t) + \frac{i\omega\Gamma_{\perp}}{2\epsilon_b v_g} P^{\pm}(z,t), \tag{4.6}$$

where $P^{\pm}$ is the active medium macroscopic polarization, ω the optical center frequency, $\Gamma_{\perp}$ the geometrical transverse confinement factor, v_g the group velocity, and ϵ_b the background permittivity. Compared to the traveling-wave equation (2.22) in Chap. 2, linear waveguide losses with the coefficient α_{int} have been added phenomenologically. Applying the integration and discretization technique presented in Chap. 2, Sect. 2.2, i.e., integrating the traveling-wave equation along its characteristic and approximating the right-hand side with the trapezoid rule, yields the delay-algebraic field propagation scheme

$$E_k^{\pm}(t) = \frac{4 - \delta_k^{\pm}\alpha_{int}}{4 + \delta_k^{\pm}\alpha_{int}} E_{k\mp 1}^{\pm}(t - \tau_k^{\pm}) + \frac{\delta_k^{\pm}}{2 + \delta_k^{\pm}\alpha_{int}} \left[S_k^{\pm}(t) + S_{k\mp 1}^{\pm}(t - \tau_k^{\pm})\right], \tag{4.7}$$

where the macroscopic polarization along with the corresponding parameters have been summarized to a field source term $S_k^{\pm}$ and the linear losses have been treated separately. Within that scheme the longitudinal dimension z has been discretized into N sections with the length Δz_k. The electric field $E_k^{\pm}$, the source term $S_k^{\pm}$, and the active medium states ρ_k^m are averaged over those lengths and are considered to be localized in the middle of each section. The sections are coupled via the propagating electric fields to their respective neighbors via the spatial separations $\delta_k^{\pm}$ and propagation times $\tau_k^{\pm}$, which are defined by

$$\delta_k^{\pm} = \frac{\Delta z_k + \Delta z_{k\mp 1}}{2} \quad \text{and} \quad \tau_k^{\pm} = \frac{\Delta z_k + \Delta z_{k\mp 1}}{2v_g}. \tag{4.8}$$

The propagating scheme and spatial discretization is furthermore sketched in Fig. 4.3a. Note that, e.g., the discretization points z_k and z_{k+1} are connected by both the propagation time τ_{k+1}^{+} and τ_k^{-}, which correspond to the forward (+) and backward (−) moving fields. The boundary conditions are given by the reflections at the left and right facet with the respective intensity reflectivity coefficients r_L and r_R. As discussed in Chap. 2, Sect. 2.2, this approach allows for a much coarser spatial discretization than commonly used finite difference methods, while maintaining a high temporal accuracy [77, 83–86].

Coherent Field-Matter Coupling

The macroscopic active medium polarization $P^{\pm}(z,t)$ is calculated semi-classically and is determined by the sum of all microscopic polarization amplitudes $p_{\alpha}^{\pm}$, where α denotes a suitable set of quantum numbers. Reproducing Eq. (2.44) from the derivation of the spatially inhomogeneous Maxwell-Bloch equations presented in Chap. 2, the macroscopic polarization reads

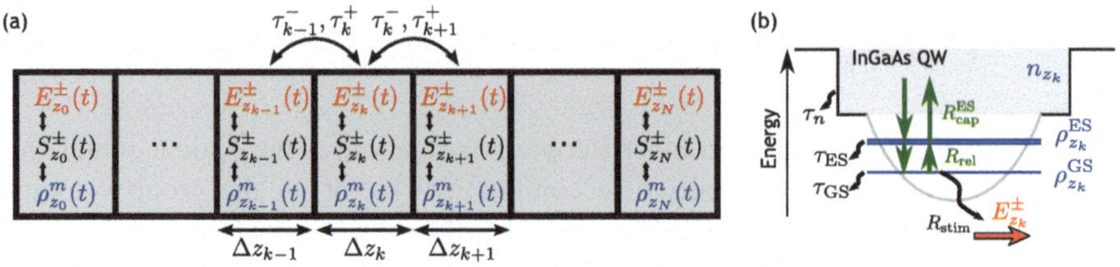

Fig. 4.3 **a** Field propagation scheme. The left (−) and right (+) traveling waves $E_k^{\pm}$ in one section k are coupled to the respective carrier population ρ_k^m via the field source term $S_k^{\pm}$ and to neighboring sections via the propagation time $\tau_k^{\pm}$. **b** Sketched quantum-dot conduction band structure with carrier interaction processes. Only the quantum-dot ground state couples to the in-phase gain of the electric field. Adapted from [56]

$$P^{\pm}(z,t) = \frac{2}{A_{\perp}^{\mathrm{act}}\mathrm{d}z}\sum_{\beta}\mu_{\beta}^{*}p_{\beta}^{\pm}(z,t), \tag{4.9}$$

where the index β represents a set of quantum numbers, which describe all excitonic quantum-dot states localized at the longitudinal coordinate z and μ_β their respective dipole moments. $A_{\perp}^{\mathrm{act}}\mathrm{d}z$ represents the normalization volume, which is constructed from the transverse cross-section of the active region $A_{\perp}^{\mathrm{act}}$ and an infinitesimal section $\mathrm{d}z$ along z.

Averaging over all inhomogeneously broadened transition frequencies, as previously described, and assuming that only the quantum-dot ground (GS) and first excited state (ES) transitions contribute to the optical gain, the sum over β reduces to a simple multiplication with the quantum-dot sheet density N^{QD}

$$P^{\pm}(z,t) = \frac{4a_{\mathrm{L}}N^{\mathrm{QD}}}{h^{\mathrm{QW}}}\left(\nu_{\mathrm{GS}}\mu_{\mathrm{GS}}^{*}p_{\mathrm{GS}}^{\pm}(z,t) + \nu_{\mathrm{ES}}\mu_{\mathrm{ES}}^{*}p_{\mathrm{ES}}^{\pm}(z,t)\right). \tag{4.10}$$

Here, h^{QW} denotes the height of the surrounding quantum-well reservoir, a_{L} the number of QD layers, $\mu_{\mathrm{GS,ES}}$ the respective dipole moments, and $\nu_{\mathrm{GS,ES}}$ the relative degeneracy of the quantum-dot ground and excited state. The factor four contains the factor two from Eq. (4.9) and another factor two that accounts for the spin degeneracy.

The dynamics of the microscopic polarization amplitudes are determined by the Maxwell-Bloch equations, which have been derived within the free-carrier theory in Chap. 2, Sect. 2.3. Adapted to the quantum-dot ground and excited state ($m \in \{\mathrm{GS}, \mathrm{ES}\}$), these equations read

$$\frac{\mathrm{d}}{\mathrm{d}t}p_{\mathrm{m}}^{\pm}(z,t) = -\left[i\Delta\omega_{\mathrm{m}} + \frac{1}{T_2^{\mathrm{eff}}}\right]p_{\mathrm{m}}^{\pm}(z,t) - i\frac{\mu_{\mathrm{m}}}{2\hbar}E^{\pm}(z,t)\left(2\rho^{\mathrm{m}}(z,t) - 1\right), \tag{4.11}$$

where $\Delta\omega_{\mathrm{m}}$ denotes the detuning from the optical center frequency and T_2^{eff} the effective polarization dephasing time. $(2\rho^m - 1)$ denotes the inversion produced by the excitonic quantum-dot occupation probability ρ^m. Note that the effective dephasing time T_2^{eff} reflects the full gain bandwidth of the quantum-dot GS(ES) ensemble and not the homogeneous linewidth of an individual optical transition.

The dynamical equation for the microscopic polarization equation (4.11) can be formally integrated to yield the expression

$$p_{\text{m}}^{\pm}(t) = -i\frac{\mu_{\text{m}}}{2\hbar}\int_{\infty}^{t} \mathrm{d}\tau\, E^{\pm}(\tau)\,(2\rho^{\text{m}}(\tau) - 1)\exp\left(-\left[i\Delta\omega_{\text{m}} + T_2^{\text{eff}^{-1}}\right](t-\tau)\right), \tag{4.12}$$

which corresponds to a convolution integral that describes memory effects, which decay with T_2^{eff}. Under the assumption that the quantum-dot occupation numbers ρ^{m} evolve slowly compared to the effective polarization dephasing time T_2^{eff} [69], they can be pulled out of the integral. The microscopic polarization then reads

$$\begin{aligned} p_{\text{m}}^{\pm}(t) = &-i\frac{\mu_{\text{m}} T_2^{\text{eff}}}{2\hbar}(2\rho^{\text{m}}(t) - 1) \\ &\times \underbrace{\int_{\infty}^{t} \mathrm{d}\tau \frac{1}{T_2^{\text{eff}}} E^{\pm}(\tau)\exp\left(-\left[i\Delta\omega_{\text{m}} + T_2^{\text{eff}^{-1}}\right](t-\tau)\right)}_{=G^{\pm}(t)}, \end{aligned} \tag{4.13}$$

where the underbraced integral defines the new variable $G_{\text{m}}^{\pm}$, which describes the fast polarization dynamics. The time evolution of $G_{\text{m}}^{\pm}$ can be obtained by taking the total derivative of the definition, i.e., the convolution integral, which produces the dynamical equation

$$\frac{\mathrm{d}}{\mathrm{d}t}G_{\text{m}}^{\pm} = \frac{1}{T_2^{\text{eff}}}(E^{\pm} - G_{\text{m}}^{\pm}) - i\Delta\omega_{\text{m}} G_{\text{m}}^{\pm}. \tag{4.14}$$

Transforming this equation to frequency space yields the Lorentzian line shape centered at the frequency $\Delta\omega_{\text{m}}$ and with the broadening given by the inverse effective lifetime $1/T_2^{\text{eff}}$. Alternatively, the impact of this equation can be interpreted as a first-order low pass filter relative to the frequency $\Delta\omega_{\text{m}}$ and the characteristic time T_2^{eff}. Thus, the $G_m^{\pm}$ can either be seen as an effective polarization or a filtered electric field.

Assuming that the electric field frequency ω is centered at the quantum-dot ground state transition, the dynamical equation for $G_{\text{ES}}^{\pm}$ is adiabatically eliminated and yields

$$G_{\text{ES}}^{\pm} = (1 + iT_2^{\text{eff}}\Delta\omega_{\text{ES}})^{-1}E^{\pm}, \tag{4.15}$$

which has a vanishing real part compared to the magnitude of the imaginary part for the considered quantum dots. Hence, the quantum-dot excited state only contributes to the amplitude-phase coupling of the laser. This approximation avoids the computationally demanding integration of the fast oscillation that occur in the dynamic equations for $G^{\pm}_{\mathrm{ES}}$ due to the frequency detuning $\Delta\omega_{\mathrm{ES}}$. The validity of such an approximation has been shown for various quantum-dot laser models [63, 69, 87, 88] via comparisons to models that do not eliminate the excited state polarization dynamics.

Collecting all the terms and performing the sum over all quantum-dot states, the electric field source term reads

$$S^{\pm}(z,t) = \frac{i\omega\Gamma_{\perp}}{2\epsilon_{\mathrm{b}} v_{\mathrm{g}}} P^{\pm}(z,t) = \frac{g_{\mathrm{GS}}}{2}(2\rho^{\mathrm{GS}} - 1)G^{\pm}_{\mathrm{GS}} - i\frac{\delta\omega^{\mathrm{ES}}}{2}\left(2\rho^{\mathrm{ES}} - 1\right)E^{\pm}, \tag{4.16}$$

where the shorthands

$$g_{\mathrm{m}} = \frac{2\omega\Gamma T_2^{\mathrm{eff}} a_{\mathrm{L}} N^{\mathrm{QD}} \nu_{\mathrm{m}} |\mu_{\mathrm{m}}|^2}{\epsilon_{\mathrm{b}} \hbar v_{\mathrm{g}} h^{\mathrm{QW}}} \quad \text{and} \quad \delta\omega^{\mathrm{ES}} = g_{\mathrm{ES}} \frac{T_2^{\mathrm{eff}} \Delta\omega_{\mathrm{ES}}}{1 + (T_2^{\mathrm{eff}} \Delta\omega_{\mathrm{ES}})^2} \tag{4.17}$$

denote the electric field linear gain coefficient g_{m} and the amplitude-phase coupling coefficient $\delta\omega^{\mathrm{ES}}$, which is induced by the quantum-dot excited state transitions. Note that the effective dephasing time T_2^{eff} not only determines the gain bandwidth but also the gain coefficients g_{m}. However, the optical gain, which can be calculated from the full semiconductor Bloch-equations, depends upon the microscopic dephasing time of the individual quantum-dot transitions [26, 73, 77, 79]. This effect is not explicitly included in the proposed model. Hence, the gain coefficients g_{m} that are used for the simulations are treated as fit parameters, which are adjusted to best match the experiment. In conjunction with the effective polarization lifetime T_2^{eff} both the magnitude and width of the gain spectrum can be adjusted.

Lastly, the coherent coupling term in the dynamical equations for the quantum-dot occupation probabilities, which accounts for stimulated emission and absorption, has to be computed. Adapting the corresponding Maxwell-Bloch equation (2.45) from Chap. 2 yields

$$\begin{aligned} \partial_t \rho^{\mathrm{GS}}|_{\mathrm{stim}} &= \frac{\mu_{\mathrm{GS}}}{\hbar} \mathrm{Im}\left(E^{+} p^{+^*}_{\mathrm{GS}} + E^{-} p^{-^*}_{\mathrm{GS}}\right) \\ &= g^{\mathrm{GS}} \eta (2\rho^{\mathrm{GS}} - 1) \mathrm{Re}\left(G^{+}_{\mathrm{GS}} E^{+^*} + G^{-}_{\mathrm{GS}} E^{-^*}\right), \end{aligned} \tag{4.18}$$

where $\eta = \varepsilon_b v_{\mathrm{g}} h^{\mathrm{QW}} / (4\hbar\Gamma a_{\mathrm{L}} N^{\mathrm{QD}})$ describes the photon-to-field conversion factor. Note that η does not affect the dynamics of the laser, but rather defines the units of the electric field and effective polarization. Thus, η can be used to suitably rescale the electric field and effective polarization variables.

With that step completed, a set of equations has been derived that self-consistently describe the spatially inhomogeneous evolution of the electric field and the considered charge-carrier variables. The specifics of an investigated laser must be implemented via the (spatially dependent) laser parameters.

Implementing Spontaneous Emission

Within the semi-classical framework, spontaneous emission of photons must be added phenomenologically [89–91]. Hence, the electric field source term in Eq. (4.16) is modified according to

$$S^{\pm}(z,t) \rightarrow S^{\pm}(z,t) + S^{\pm}_{\text{sp}}(z,t), \tag{4.19}$$

where the noise amplitude $S^{\pm}_{\text{sp}}$ describes spontaneously emitted photons that enter the lasing modes. In order to correctly account for the spectral properties of the spontaneous emission, a stochastic time-domain approach that follows the ideas of [69, 76, 92] is chosen. In particular, the mean frequency and the spectral width of the spontaneous emission must be properly described. Hence, the spontaneous emission is modeled by a two-dimensional Ornstein-Uhlenbeck process

$$\frac{\mathrm{d}}{\mathrm{d}t} S^{\pm}_{\text{sp}}(z,t) = -\left[\gamma_{\text{sp}} - i\Delta\omega_{\text{GS}}\right] S^{\pm}_{\text{sp}} + \sqrt{\tilde{D}(z,t)}\xi(z,t), \tag{4.20}$$

which is centered at the frequency $\Delta\omega_{\text{GS}}$ and has the correlation time γ_{sp}^{-1}. The noise amplitude $S^{\pm}_{\text{sp}}$ is driven by the δ-correlated (in space and time) complex Gaussian white noise $\xi(t)$ with the noise strength $\tilde{D}$. For a slowly varying spontaneous emission noise strength $\tilde{D}$, the correlation function reads [76, 93]

$$\left\langle S^{\pm^*}_{\text{sp}}(z,t) S^{\pm}_{\text{sp}}(z',t+\tau)\right\rangle \approx \frac{\tilde{D}(z,t)}{\gamma_{\text{sp}}} e^{-(\gamma_{\text{sp}}+i\Delta\omega_{\text{GS}})\tau}\delta(z-z'). \tag{4.21}$$

Using the Wiener-Khinchin theorem, the spontaneous emission power spectral density $\mathcal{S}_{S_{\text{sp}}}$ can by calculated and yields

$$\mathcal{S}_{S_{\text{sp}}} = \frac{\tilde{D}(z,t)}{\pi} \frac{1}{(\omega - \Delta\omega^{\text{GS}})^2 + \gamma_{\text{sp}}^2}, \tag{4.22}$$

which represents a Lorentzian centered at $\Delta\omega^{\text{GS}}$ with a width (FWHM) of $2\gamma_{\text{sp}}$. Thus, the correlation time γ_{sp}^{-1} can be identified as the effective polarization lifetime T_2^{eff}. Further acknowledging that the spontaneous emission noise strength should be directly proportional to the quantum-dot occupation probability, the dynamical equation for the spontaneous emission noise reads

$$\frac{\mathrm{d}}{\mathrm{d}t} S^{\pm}_{\text{sp}}(z,t) = -\left[\frac{1}{T_2^{\text{eff}}} + i\Delta\omega_{\text{GS}}\right] S^{\pm}_{\text{sp}} + \sqrt{D(z)\rho^{\text{GS}}(z,t)}\xi(z,t). \tag{4.23}$$

This stochastic differential equation must be solved alongside the other equations for each section to provide the proper spontaneous emission contribution to the electric field. Lastly, for convenience in the numerical implementations, the spontaneous emission term can be integrated into the dynamical equation for the effective polarization

$$\frac{d}{dt}G^{\pm}_{GS} = \frac{1}{T^{eff}_2}(E^{\pm} - G^{\pm}_{GS}) - i\Delta\omega_{GS}G^{\pm}_{m} + \sqrt{D(z)\rho^{GS}(z,t)}\xi(z,t). \tag{4.24}$$

The noise strength $D(z)$ itself depends on various laser parameters as well as the β-factor, which describes the probability that a spontaneously emitted photon ends up in the considered lasing modes. Thus, the spontaneous emission strength is commonly specified in terms of the β-factor, when comparing different lasers. This work, however, treats the noise strength D itself as a fit parameter, in order to reproduce the timing jitter of the fundamentally mode-locked emission state.

Modeling the Absorber Reverse Bias

The absorber section is subject to a reverse bias U, whose effect is twofold: Firstly, the static transverse electric field reduces the barrier height, which leads to enhanced thermionic carrier escape rates [94, 95]. Following [96], this is implemented by an effective quantum-dot excited state lifetime $\tau^{ES}_{abs}(U)$, which exponentially depends on the absorber bias U. The quantum-well carrier reservoir is assumed to only carry vanishing charge-carrier densities and its explicit modeling is therefore neglected within the absorber section. The corresponding equation for the effective excited state lifetime reads

$$\tau^{ES}_{abs}(U) = \tau^{ES}_{abs,0} \exp\left(U/U_0^{\tau^{ES}_{abs}}\right), \tag{4.25}$$

where the parameters $\tau^{ES}_{abs,0}$ and $U_0^{\tau^{ES}_{abs}}$ are chosen based on Refs. [94, 96].

Secondly, the optical transitions are slightly redshifted due to the quantum confined stark effect [94, 97], which is implemented via the detuning parameter $\Delta\omega_{GS}$ in the dynamical equation (4.14) for the effective polarization $P^{\pm}$. The redshift is assumed to scale linearly with the applied reverse bias [94]. The constituting equation is then given by

$$\Delta\omega^{abs}_{GS}(U) = \Delta\omega^{abs}_{GS,0}U, \tag{4.26}$$

where the parameter $\Delta\omega^{abs}_{GS,0}$ is chosen similarly to [8, 94, 96, 97].

The different operating conditions, compared to the forward biased gain sections, are furthermore accounted for by choosing larger differential gain coefficients in the absorber section [8, 67]. The relationship between the gain coefficients $g^{abs}_{GS} > g^{gain}_{GS}$ is motivated by the decreased carrier density in the surrounding quantum well, which leads to reduced Coulomb scattering and thereby to an increased microscopic dephasing time of the optical transitions [98]. This, as previously explained, then produces a larger differential gain coefficient [26, 76, 79].

Accounting for the Tapered Gain Section

To complete the model of the three-section laser, the effect of the tapered gain structure has to be included. On that account, the transverse electric field is assumed to follow the profile of the active region by adiabatically expanding and reducing in

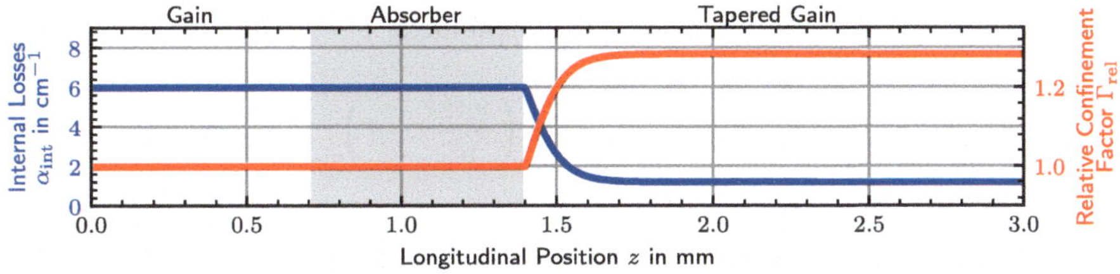

Fig. 4.4 Effects of the tapered gain structure on the waveguide parameters. The blue line plots the non-resonant waveguide losses as a function of the longitudinal position z and the red line the relative optical confinement factor Γ_{rel}. Taper angle $\Theta = 2.0°$

its spatial extend [11, 23, 99]. In the proposed model, this is described by rescaling the stimulated emission term in Eq. (4.3) with the relative change of the ridge width $w(z)$

$$\partial_t \rho^{GS}(z)|_{stim}^{taper} = \frac{w_0}{w(z)} \partial_t \rho^{GS}(z)|_{stim}, \tag{4.27}$$

where w_0 is the width of the straight section active region. Since the transverse electric field profile is distributed over a larger active area in the tapered region, the quantum dots see only a reduced field strength, depending on the local waveguide width $w(z)$. This results in a reduced local stimulated recombination rate and therefore a higher saturation energy of the tapered gain medium. With that modeling approach, the electric field strength itself is not rescaled, but only the interaction with the active medium.

Similarly, the spontaneous emission, which is proportional to the number of quantum dots within one section, increases with the ridge width $w(z)$ according to

$$\sqrt{D(z)\rho^{GS}(z,t)}\xi(z,t) \rightarrow \sqrt{\frac{w(z)}{w_0} D\rho^{GS}(z,t)}\xi(z,t), \tag{4.28}$$

where the spatial dependence of the noise strength $D(z)$ has been identified with the relative ridge width $w(z)/w_0$. Since the spontaneous emission enters the electric field equation (or effective polarization equation), the inverse relative ridge width ratio has to be used, when compared to the modification of the stimulated emission term, which enters the quantum-dot occupation probability equation.

Moreover, increasing the ridge width is assumed to lead to a better overlap between the electric field and the active medium and thereby improves the transverse confinement factor $\Gamma_\perp(z)$. This effect is described by decomposing the transverse confinement factor $\Gamma_\perp(z)$ according to

$$\Gamma_\perp(z) = \Gamma_\perp^{rel}(z)\Gamma_\perp^0 \tag{4.29}$$

and modeling the relative transverse confinement factor $\Gamma_{\perp}^{\mathrm{rel}}(z)$ with the fit function

$$\Gamma_{\perp}^{\mathrm{rel}}(z) = 1 + \Gamma_{\mathrm{T}} \tanh\left(\frac{w(z)}{w_0} - 1\right). \tag{4.30}$$

The parameter Γ_{T} and the fit function Eq. (4.30) itself are chosen to mimic results from beam-propagation calculations as presented in Refs. [11, 21, 99].

The improved overlap, furthermore, leads to a reduction of the internal non-resonant waveguide losses $\alpha_{\mathrm{int}}(z)$ [11, 21]. This effect is modeled via the fit function

$$\alpha_{\mathrm{int}}(z) = \alpha_{\mathrm{int}}^{0} - \alpha_{\mathrm{int}}^{\mathrm{T}} \tanh\left(\frac{w(z)}{w_0} - 1\right). \tag{4.31}$$

The parameter $\alpha_{\mathrm{int}}^{\mathrm{T}}$ and the function Eq. (4.31) are chosen to match the behavior of the aforementioned beam-propagation calculations [11, 21] and to reproduce waveguide characterization measurements [56].

The impact of both the reduced waveguide losses and the improved confinement factor is presented in Fig. 4.4 for a taper angle $\Theta = 2.0°$, where the blue line plots the non-resonant waveguide losses α_{int} as a function of the longitudinal position z and the red line plots the relative optical confinement factor $\Gamma_{\perp}^{\mathrm{rel}}$. Both quantities quickly change within a small region between $z = 1.4$ mm and $z \approx 1.7$ mm. While the waveguide losses drop from $\alpha_{\mathrm{int}} = 6\,\mathrm{cm}^{-1}$ to $\alpha_{\mathrm{int}} \approx 1.2\,\mathrm{cm}^{-1}$, the relative confinement factor increases from $\Gamma_{\mathrm{rel}} = 1$ to $\Gamma_{\mathrm{rel}} \approx 1.28$.

Specifying the Pump Current and the Output Power

Lastly, the applied pump current P can be related to the pump current density J that enters the quantum-well carrier density Eq. (4.1) via the relation

$$P = J A_{\mathrm{G}} a_{\mathrm{L}} e_0 \varrho^{-1}, \tag{4.32}$$

where A_{G} is the area of the active region, a_{L} the number of QD layers, e_0 the electron charge, and ϱ the injection efficiency. The latter is related to the internal quantum efficiency and is fitted to match experimental data. The out-coupled power at the tapered low-reflectivity facet to the right is calculated according to

$$\mathcal{P}(t) = 2\hbar\omega a_{\mathrm{L}} w_0 N^{\mathrm{QD}} \eta \left|E_{\mathrm{out}}(t)\right|^2, \tag{4.33}$$

where $\hbar\omega$ denotes the mean photon energy, w_0 the untapered ridge width, N^{QD} the quantum-dot area density, and η the previously introduced photon-to-field conversion factor. The out-coupled electric field E_{out} at the right facet of the tapered section is given by

$$E_{\mathrm{out}}(t) = \sqrt{1 - r_{\mathrm{R}}}\, E^{+}(z = l, t). \tag{4.34}$$

Due to the chosen modeling of the tapered gain section, the electric field is scaled as if it were confined to the untapered ridge width w_0, which therefore appears in the output power. As discussed previously, η can be used to define the units of the electric field and thus effectively rescale it. However, any transformation $\eta \to \tilde{\eta}$ leaves the out-coupled power untouched, since η appears in conjunction with $|E_{\text{out}}(t)|^2$ in Eq. (4.33). Hence, if one is only interested in the out-coupled power and not the electric field strength itself, an arbitrary η can be chosen. Therefore, the quantum-well height h^{QW} and the confinement factor $\Gamma_{\perp}^{0}$, which appear in η, do not need to be specified in order to calculate the out-coupled power.

Laser Parameters

To most accurately describe the lasing operation of the three-section tapered quantum-dot laser within the proposed model framework, the laser parameters must be properly adjusted. Some parameters, like the facet reflectivities, are directly defined by the device design specifications, while others, like the internal losses, must be determined by separate characterization measurements. The majority of the parameters, which mostly concern properties of the active medium dynamics, however, are fitted to reproduce the experimentally observed emission dynamics. Parameters from other works, which have investigated similar semiconductor quantum-dot active media, are used as initial educated guesses as well as to define reasonable magnitudes. Especially the latter is important, since the high-dimensional parameter space and the limited experimental data set do not guarantee the uniqueness of a given parameter set that reproduces the experimental results.

A list of all model parameters and their values is presented in Table 4.1. The last column shortly defines their meaning and indicates the origin of the respective parameter value in parenthesis via the keywords 'specified', 'measured', 'fitted', and 'adapted' and, if applicable, provides the respective references.

Numerical Implementation

The integration of the ordinary differential equation is implemented via a standard equidistant fourth-order Runge-Kutta method [100]. The electric field is recorded at identical time steps and the delay-algebraic propagation equations are evaluated via a third-order polynomial interpolation routine, which is implemented via Neville's algorithm [100]. The complex Gaussian white noise that drives the spontaneous emission is produced with the Box-Muller method [100]. The corresponding stochastic differential equations are solved by a simple Euler method.

The numerical integration has converged with a time-step $\Delta t = 31.25$ fs and a spatial discretization along the optical axis of the device into 60 sections. Thus, the mean spatial discretization length amounts to $\langle \Delta z_k \rangle = 50\,\mu$m, which corresponds to a mean propagation time $\langle \tau_k \rangle = 625$ fs.

The discretization into 60 sections produces 180 real valued ordinary differential equations that describe the evolution of the charge carrier dynamics, 120 complex valued ordinary differential equations that describe the evolution of the effective polarization, and 120 complex valued delay-algebraic electric field propagation equations. All equations can be evaluated in parallel. On a state-of-the-art desktop CPU

Table 4.1 Parameters used in the three-section tapered quantum-dot laser model unless otherwise specified

Symbol	Value	Meaning (origin)
$\hbar\omega$	0.99 eV	Photon energy (specified)
l	3 mm	Device length (specified)
w_0	4 μm	Effective active region width (specified[a])
$\alpha_{\mathrm{int}}^{0}$	$6\,\mathrm{cm}^{-1}$	Internal losses (measured)
κ_{L}	0.95	Left facet reflectivity (specified)
κ_{R}	0.03	Right facet reflectivity (specified)
v_{g}	$0.008\,\mathrm{cm\,ps}^{-1}$	Group velocity (calculated)
a_{L}	10	Number of QD layers (specified)
N^{QD}	$0.2 \times 10^{11}\,\mathrm{cm}^{-2}$	QD area density (measured)
ϱ	0.77	Internal pump efficiency (fitted)
$D^{2\mathrm{D}}\hbar^2$	6.96×10^{-32} kg	QW 2D density of states (adapted from [28])
τ^{n}	1 ns	QW carrier lifetime (adapted from [28])
τ^{ES}	1 ns	QD ES carrier lifetime (adapted from [28])
τ^{GS}	1 ns	QD GS carrier lifetime (adapted from [28])
$\tilde{R}_{\mathrm{rel}}$	$5\,\mathrm{ps}^{-1}$	QD GS-ES relaxation rate (adapted from [28])
$\tilde{R}_{\mathrm{cap}}$	$0.1\,\mathrm{ps}^{-1}$	QD ES capture rate (adapted from [28])
$\Delta\varepsilon^{\mathrm{ESGS}}$	64 meV	GS-ES energy separation (adapted from [28])
$\Delta\varepsilon^{\mathrm{QWES}}$	24 meV	ES confinement (adapted from [28])
T	300 K	Effective charge-carrier temperature (specified)
$\nu_{\mathrm{GS}}, \nu_{\mathrm{ES}}$	1, 2	Relative QD GS, ES state degeneracy
$2\hbar/T_2^{\mathrm{gain}}$	50 meV	Effective gain bandwidth (fitted, based on [73])
$2\hbar/T_2^{\mathrm{abs}}$	50 meV	Effective gain bandwidth (fitted, based on [73])
$g_{\mathrm{GS}}^{\mathrm{gain}}$	$30\,\mathrm{cm}^{-1}$	Differential gain (fitted, based on [8, 73])
$g_{\mathrm{GS}}^{\mathrm{abs}}$	$60\,\mathrm{cm}^{-1}$	Differential gain (fitted, based on [8, 73])
$\delta\omega_{\mathrm{gain}}^{\mathrm{ES}}$	$17.4\,\mathrm{cm}^{-1}$	Amplitude-phase coupling coefficient (computed)
$\delta\omega_{\mathrm{abs}}^{\mathrm{ES}}$	$34.8\,\mathrm{cm}^{-1}$	Amplitude-phase coupling coefficient (computed)
$\Delta\omega_{\mathrm{GS},0}^{\mathrm{abs}}$	$0.367\,\mathrm{THz\,V}^{-1}$	Absorber redshift coefficient (adapted from [94])
$\tau_{\mathrm{abs},0}^{\mathrm{ES}}$	20 ps	Absorber base ES lifetime (adapted from [96])
$U_0^{\tau^{\mathrm{ES}}}$	2.0 V	Characteristic reverse bias (adapted from [96])
$\alpha_{\mathrm{int}}^{\mathrm{T}}$	$4.8\,\mathrm{cm}^{-1}$	Taper model coefficient (fitted to measurements)
Γ_{T}	0.28	Taper model coefficient (adapted from [11])

The last column shortly presents the meaning and indicates the origin of the respective parameters in parenthesis

[a]An effective active region width of $w_0 = 4\,\mu\mathrm{m}$ is assumed to approximate the effects of the gain-guided structure with a width of 14 μm

(Intel Core i7-8770, 2021, using a single core), a 75 μs time series (around 1000 round-trips) takes roughly two minutes to compute and already produces reasonable pulse-train statistics.

Model Summary

In summary, a system of coupled delay-algebraic differential equations has been derived, which describes the spatio-temporal evolution of the electric field and the considered charge-carrier populations of the three-section tapered quantum-dot laser. The electric field propagation is modeled in the traveling-wave equation framework and is numerically solved via delay-algebraic propagation equations. The active medium is described by an excitonic multi-population model, where charge-carrier interactions are treated by relaxation-rate approximations that drive the carriers towards a quasi-Fermi equilibrium. The light-matter interaction is treated semi-classically via an effective polarization equation and allows for both spatially varying gain bandwidths and frequency detunings.

The specifics of the laser configuration, i.e., the position and size of the absorber section and the tapered gain section, have been accounted for by spatially modified equations and laser parameters. Particularly, the effects induced by the tapered gain section, which is a transverse property, have been captured by relatively simple modifications in a one-dimensional longitudinal laser model.

The direct integration of these equations produces time series of all dynamical variables, which can be used for analysis and characterization. Hence, not only the emitted electric field, but also the internal laser dynamics can be studied.

The key approximations that allow for the computational efficiency of the model have been the averaging over the inhomogeneously broadened quantum-dot ensemble and the use of an effective polarization equation. If effects related to both are of interest, the implementation of equations, which describe the full polarization dynamics of the corresponding quantum-dot subgroups, is straightforward by reversing the respective approximations.

4.3 Basic Emission States

The three-section tapered quantum-dot laser analysis starts off by discussing the qualitatively different fundamental emission states that can be observed for various driving conditions. While the respective stability boundaries of those states are discussed in Sect. 4.5, this section presents and characterizes representative examples of the observed mode-locked emission states. Complementing the mode-locking states, the laser can also be found in the off state, i.e., without coherent light emission, and in the continuous-wave (cw) state, i.e., with constant-intensity light emission. Those states, however, are not discussed in the following due to their simple structure.

4.3.1 *Illustration of the Mode-Locked Emission States*

The attainable mode-locking states are illustrated in the rows of Fig. 4.5. The columns from left to right represent: Pseudo space-time plots, where the individual time-traces have been sliced into sections of the cold cavity round-trip time, power spectral densities of the out-coupled intensity $\mathcal{S}_{|E_{\text{out}}|^2}$, which are normalized to the zero frequency component, and autocorrelation functions of the out-coupled intensity $|E_{\text{out}}|^2$, which are normalized to the zero time-shift value.

The first row ((a1)–(a3)) presents Q-switched mode-locking (QSML) at $P = 0.75\,\text{A}$ and $U = 6.0\,\text{V}$, which is composed of sets of broad pulses with inter-pulse spacings of about 25 ps as shown in the space-time plot (a1). Pulse emission ceases in between Q-switched bursts. The dominant frequency in the power spectrum (a2) appears at $\nu \approx 40\,\text{GHz}$, which corresponds to the third harmonic. While the fundamental and second harmonic are still visible, they are attenuated by about two orders of magnitude. The slow envelope, which is seen in the pseudo space-time plot, has a long period of 5–10 μs and leads to the low-frequency components ν of about 100 to 200 MHz in the power spectrum (a2). The autocorrelation (a3) exhibits characteristic peaks at $\Delta t \approx 25\,\text{ps}$, $\Delta t \approx 50\,\text{ps}$, and $\Delta t \approx 75\,\text{ps}$, which represent the pulse spacing. While the third peak corresponds to the round-trip time and almost shows no attenuation, the first and second peak are smaller than the zeroth peak by a factor of ≈ 2. This indicates some dissimilarity between the pulses circulating within the resonator.

The second row ((b1)–(b3)) shows fundamental mode-locking (FML) at $P = 0.81\,\text{A}$ and $U = 6.0\,\text{V}$. Its representation in the space-time plot (b1) is a narrow line that tilts to the right as the pulse period is slightly longer than the cold-cavity round-trip time. The resulting power spectrum (b2) exhibits well-defined and sharp peaks at the fundamental frequency $\nu \approx 13.3\,\text{GHz}$ and the respective harmonics, which sit on pedestals generated by relaxation oscillations. The peaks are separated from the pedestals by about four orders of magnitude. The signature of the relaxation oscillations can also be seen in the low-frequency range of the spectrum, where the strongest component is located at $\nu \approx 1.0\,\text{GHz}$. The autocorrelation function of the FML state only contains one sharp peak at $\Delta t \approx 75.16\,\text{ps}$, which is generated by the pulse period.

Increasing the pump current leads to a loss of stability of the FML pulse train and noise-induced perturbations create competing pulse trains that periodically end up taking the gain from the previous pulse trains. This pulse-switching instability (FML_{PS}) of the FML state at $P = 0.95\,\text{A}$ and $U = 6.0\,\text{V}$ is displayed in the third row ((c1)–(c3)). The switching between pulse trains is shown in the space-time plot (c1). It occurs at the round-trip numbers ≈ 20 and ≈ 80 and takes about 10 round-trips to complete. The power spectrum (c2) shows strong peaks at the fundamental and respective harmonic frequencies, which are, however, heavily smeared out when compared to the FML state (b2). Moreover, the period of pulse-switching instability induced pulse-train switching results in a characteristic slow-frequency component ν of approximately 150 MHz in the power spectrum. When compared to the FML state,

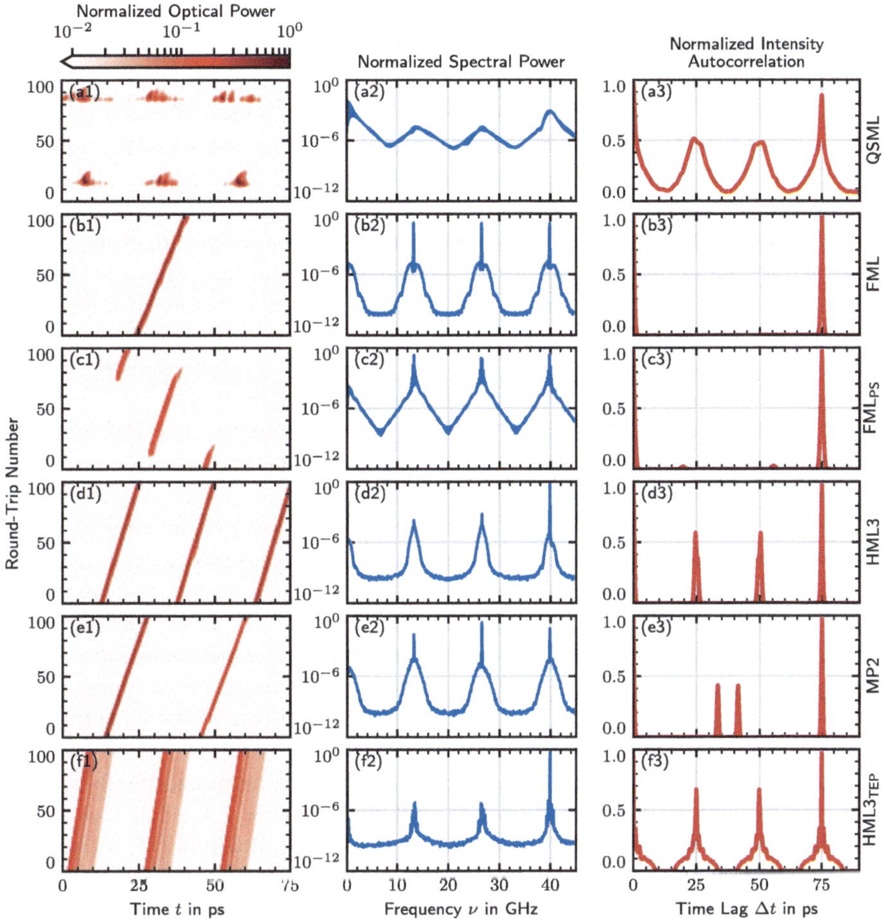

Fig. 4.5 Illustration of the observed mode-locking states. From left to right: Pseudo space-time plots normalized to the maximum power, normalized power spectral densities, and normalized autocorrelation functions. From top to bottom: Q-switched mode-locking (QSML, $P = 0.75\,\text{A}$, $U = 6.0\,\text{V}$), fundamental mode-locking (FML, $P = 0.81\,\text{A}$, $U = 6.0\,\text{V}$), fundamental mode-locking with a pulse-switching instability (FML_{PS}, $P = 0.95\,\text{A}$, $U = 6.0\,\text{V}$), third-order harmonic mode-locking (HML3, $P = 1.08\,\text{A}$, $U = 6.0\,\text{V}$), two non-identical pulses (MP2, $P = 1.02\,\text{A}$, $U = 8.0\,\text{V}$), and third-order harmonic mode-locking with a trailing-edge plateau (HML3_{TEP}, $P = 1.5\,\text{A}$, $U = 4.0\,\text{V}$). Subfigures a1, b1, c1, and d1 adapted from [56]

the autocorrelation function (c3) has gained small bumps at the time lags $\Delta t \approx 20$ ps and $\Delta t \approx 55$ ps, which indicate that the competing new pulse train is generated $\approx$20 ps 'in front of' the dying pulse train.

The fourth row ((d1)–(d3)) shows third-order harmonic mode-locking (HML3) at $P = 1.08$ A and $U = 6.0$ V. The space-time plot (d1) resembles FML, with the exception that three pulse trains are observed within the laser cavity. The corresponding power spectrum (d2) therefore displays a strong peak at the third harmonic frequency at $\nu \approx 40$ GHz. The fundamental and second harmonic are observed but attenuated by about three to four orders of magnitude with respect to the peak of the third harmonic. The autocorrelation function (d3) shows characteristic peaks at $\Delta t \approx 25$ ps, $\Delta t \approx 50$ ps, and $\Delta t \approx 75$ ps. The first and second peak are not as pronounced as the third peak, which is likely due to a dissimilarity between the pulses and suggests an imperfect third-order harmonic mode-locking.

The fifth row ((e1)–(e3)) presents a multi-pulse state with two non-identical and non-equidistant pulses (MP2) at $P = 1.02$ A and $U = 8.0$ V. The space-time plot (e1) shows the two pulses with the inter-pulse separations of $\approx$33 ps and $\approx$42 ps, respectively. The peak power of the pulse to the left is larger by a factor of about $\approx$1.5 when compared to the pulse on the right, since the left pulse has access to a larger gain recovery time. The power spectrum (e2) has its strongest non-zero component at $\nu \approx 26.6$ GHz, corresponding to the second harmonic. The mode-locking state, however, is periodic with roughly the cavity round-trip time, which results in a strong fundamental component at $\nu \approx 13.3$ GHz as well. The autocorrelation function (e3) resembles the non-equidistant pulse spacing with characteristic peaks at lag times of $\Delta t \approx 33$ ps and $\Delta t \approx 42$ ps.

Lastly, the sixth row ((f1)–(f3)) depicts a third-order harmonic mode-locking state with a strong trailing-edge plateau (HML3$_{\mathrm{TEP}}$) at $P = 1.5$ A and $U = 4.0$ V. The trailing-edge plateau strikingly appears in the space-time plot (f1) as a broad area behind the initial pulse that extends for about $\approx$8 ps. Otherwise identical to the HML3 state (d2), the power spectrum (f2) is noisier, since the trailing-edge plateau is somewhat modulated by spontaneous emission noise. The autocorrelation function (f3) provides further evidence for the trailing-edge plateau, as the characteristic peaks at $\Delta t \approx 25$ ps, $\Delta t \approx 50$ ps, and $\Delta t \approx 75$ ps now sit on broad pedestals.

It can be concluded that each state can be uniquely identified by the features of its power spectral density and intensity autocorrelation function. This insight is crucial for the experimental investigation of this lasers, since sufficiently high-resolution time traces are not accessible, but power spectra and autocorrelation functions can be measured by relatively simple means and thus allow for the characterization of the emission dynamics.

4.3.2 Net-Gain Analysis

The mode-locking dynamics are further studied in terms of the net gain $\mathcal{G}$, which was introduced by New [101], to assess the stability of a mode-locked state against perturbations. The frequency dependent net gain is defined by

$$\mathcal{G}(t, \omega) = \ln(\mathcal{A}(t, \omega)), \tag{4.35}$$

where $\mathcal{A}(t, \omega)$ describes the total amplification that a small perturbation, which travels at the group velocity at an optical frequency ω, experiences within one complete round trip in the laser cavity. Hence, positive values of $\mathcal{G}$ correspond to an amplification and negative values to an attenuation of such a perturbation. Mode-locking is labeled as stable if a window of positive net gain only exists for the desired mode-locked pulses.

However, since the net gain is a small-signal, i.e., a linear, measure, mode-locking can still be stable despite positive net-gain windows in front of or behind the mode-locked pulses [59, 62]. In those cases, the group velocity dispersion causes the mode-locked pulses to travel either faster or slower than the small perturbation, which leads to a merging of both and thereby protects the mode-locked state.

In the case of the three-section tapered quantum-dot laser model, the net gain can be computed by integrating the amplitude gain along one complete round-trip within the co-moving frame

$$\mathcal{G}(\omega, t) = \int_0^l \mathrm{d}z \left\{ A^-(z, t + \frac{l-z}{v_\mathrm{g}}, \omega) + A^+(z, t + \frac{l+z}{v_\mathrm{g}}, \omega) - \frac{\alpha_\mathrm{int}(z)}{2} \right\} \\ + \ln\left(\sqrt{r_\mathrm{L} r_\mathrm{R}}\right), \tag{4.36}$$

where $A^\pm(z, t, \omega)$ represents the local resonant gain, $\alpha_\mathrm{int}(z)$ the non-resonant waveguide losses, and r_L and r_R the facet losses. l denotes the length of the laser cavity and the time arguments in $A^\pm$ result from the co-moving frame. The local gain is calculated by adiabatically eliminating the equation of the effective polarization $G^\pm$ and the result reads

$$A^\pm(z, t, \omega) = \frac{g_\mathrm{GS}(z)}{2} \left[2\rho_\mathrm{GS}(z, t) - 1\right] \frac{T_2^{\mathrm{eff}^2}}{(\Delta\omega_\mathrm{GS}(z) - \omega)^2 + T_2^{\mathrm{eff}^2}}. \tag{4.37}$$

Both the time and the frequency dependence of the net gain are contained in this expression via the time dependence of $\rho_\mathrm{GS}(z, t)$ and the detuning $\Delta\omega_\mathrm{GS}(z)$ of the Lorentzian line shape that results from the elimination of $G^\pm$.

Fundamental Mode-Locking

The results obtained from applying the above technique to the fundamentally mode-locked state at $P = 0.81\,\mathrm{A}$, $U = 6.0\,\mathrm{V}$ are shown in Fig. 4.6, where (a) shows the time trace of the out-coupled power $\mathcal{P}$ (red line) and the respective instantaneous frequency shift $\Delta\nu$ along the pulse (blue line), (b) the frequency-dependent net gain, and (c) the optical spectrum calculated from the out-coupled electric field.

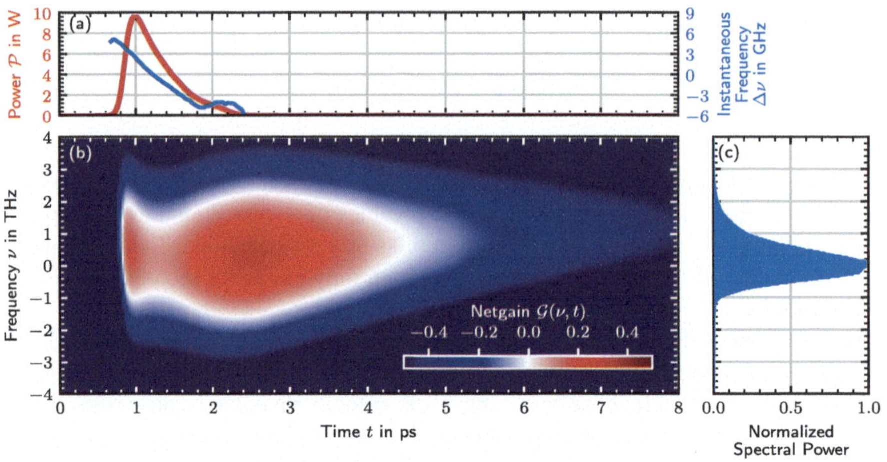

Fig. 4.6 Fundamental mode-locking pulse shape and net gain at $P = 0.81$ A and $U = 6.0$ V. **a** Time series slice portraying the pulse shape (red line) and instantaneous frequency shift along the pulse (blue line, only for optical powers $\gtrsim$100 mA). **b** Frequency-resolved net gain calculated from the carrier distribution. Frequencies ν are relative to the mean quantum-dot ground state emission frequency of the gain sections. **c** Normalized optical spectrum calculated from the out-coupled field

At those operation parameters, the out-coupled pulse reaches almost $\mathcal{P}^{\max} = 10$ W peak power at $t \approx 1.0$ ps with a time-domain pulse width of $\Delta_{\text{pw}} \approx 600$ fs. The pulse shape itself is strongly asymmetric with a steep leading edge and a comparatively flat trailing-edge.

The position of the pulse maximum is accompanied by a local maximum of the net gain, which is slightly blueshifted due to the reverse-bias-induced redshift of the absorber section. Thus, the redshifted absorption spectrum results in an overall blueshift of the positive net-gain spectrum. The net-gain window is opened by the saturation of the absorption by the incident pulse, which is followed by the saturation of the gain. The latter leads to a decrease of the net gain for times between $t \approx 1.0$ ps and $t \approx 1.4$ ps. However, in a following phase, the gain quickly recovers due to the ultra-fast quantum-dot excited state-ground state relaxation [28, 68, 102, 103] and thereby provides additional gain, which leads to the global maximum at $t \approx 2.7$ ps. This maximum coincides with the last part of the trailing pulse edge, where a small bump can be observed. The previously mentioned group velocity dispersion, however, results in a pulse period of $T \approx 75.16$ ps, i.e., one that is longer than the cold-cavity round-trip time, which stabilizes the fundamentally mode-locked pulse train against perturbations in the trailing-edge.

Focusing on the frequency dynamics of the net-gain spectrum, the saturation of the absorption section also leads to a decreasing blueshift of the net-gain spectrum for times shortly after the pulse maximum ($t \lesssim 2.5$ ps). With the fast recovery of both the gain and the absorption, the initial blueshift is quickly reestablished. The instantaneous frequency shift $\Delta\nu$ of the portrayed pulse, which is plotted in Fig. 4.6a, reflects the net-gain frequency dynamics by undergoing a redshift from $\Delta\nu \approx 6.0\,\text{GHz}$ to $\Delta\nu \approx -4.0\,\text{GHz}$. This process, however, is only partially caused by the net-gain frequency dynamics, since the amplitude-phase coupling, which is proportional to the quantum-dot excited state occupation, additionally reduces in strength, which also leads to a relative redshift of the emission.

The optical spectrum of the pulse train is shown in Fig. 4.6c. The spectrum has a FWHM of $\approx$1.1 THz and exhibits a slightly longer tail towards blueshifted frequencies. The frequencies that experience a positive net gain map very well to the shape of the optical spectrum. This property can be expected for pulses that are close to transform limited. However, such behavior is not guaranteed, since the net gain is only a small-signal measure, while the mode-locked pulses themselves induce strong nonlinear light-matter interactions.

Third-Order Harmonic Mode-Locking

Figure 4.7 shows the results obtained for third-order harmonic mode-locking (HML3) at $P = 1.08\,\text{A}$ and $U = 6.0\,\text{V}$ with the same representation as it has been used previously for fundamental mode-locking (see Fig. 4.6). The results with respect to the pulse shape, the frequency dynamics, and the net-gain dynamics appear qualitatively very similar to fundamental mode-locking. However, the chosen operation point yields pulses with $\mathcal{P}^{\text{max}} \approx 8\,\text{W}$ peak power and a $\Delta_{\text{pw}} \approx 480\,\text{fs}$ pulse width. Compared to fundamental mode-locking the shorter pulse width results from a steeper trailing-edge of the pulse. Associated with that, the first local maximum of the net gain is more pronounced than the second, where the second local maximum has been attributed to the refilling of the quantum-dot ground state occupation from the quantum-dot excited state. The three pulses circulating within the laser cavity keep the quantum-dot excited state more depleted on average and thereby attenuate the refilling mechanism. Similar to fundamental mode-locking, the optical spectrum in Fig. 4.7 reproduces the shape of the net-gain spectrum very well. Peculiarly, the optical spectrum appears narrower than the spectrum produced by fundamental mode-locking, although the HML3 pulses are shorter than the FML pulses. However, the overall width of FML optical spectrum (see Fig. 4.6c) is produced by the second local maximum in the net-gain spectrum, which only contributes to the trailing-edge of the pulse. This suggests that the additional excited modes in fundamental mode-locking operation are not properly synchronized and thus do not contribute to a pulse shortening.

Pseudo Space-Time Representation

The complete temporal pulse and net-gain dynamics of the fundamental mode-locking (FML), pulse-switching unstable fundamental mode-locking (FML_{PS}), and third-order harmonic mode-locking (HML3) states are presented in pseudo space-time representations in Fig. 4.8. The top row ((a1)–(a3)) shows optical power pseudo space-time plots of the respective mode-locking states indicated on top, which are identical to the pseudo space-time plots shown in Fig. 4.5. The bottom row ((b1)–(b3)) depicts the corresponding net-gain pseudo space-time plots, where the net gain has been evaluated at the mean optical frequency of the respective mode-locking state.

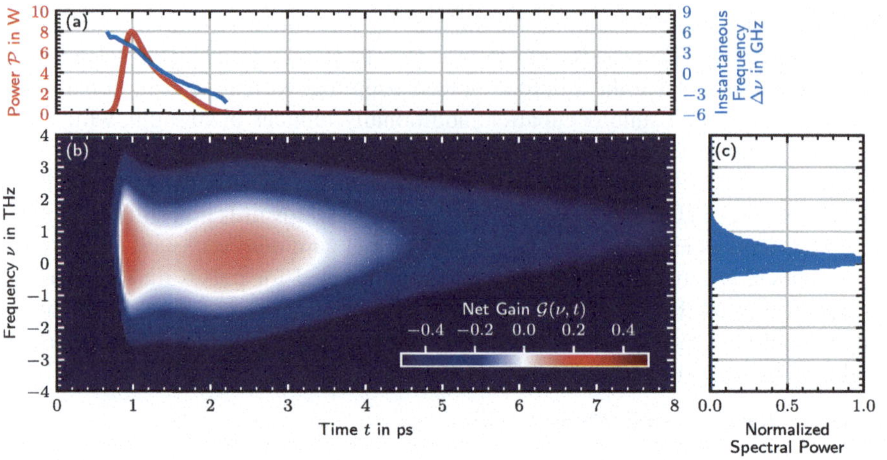

Fig. 4.7 Third-order harmonic mode-locking pulse shape and net gain at $P = 1.08\,\text{A}$ and $U = 6.0\,\text{V}$. **a** Time series slice portraying the pulse shape (red line) and instantaneous frequency shift along the pulse (blue line, only for optical powers $\gtrsim 100\,\text{mA}$). **b** Frequency resolved net gain calculated from the carrier distribution. Frequencies ν are relative to the mean quantum-dot ground state emission frequency of the gain sections. **c** Normalized optical spectrum calculated from the out-coupled field

Unlike Figs. 4.6 and 4.7, the net-gain dynamics are displayed both for a complete round-trip and over many consecutive round-trips.

As previously described, the emission of a pulse must coincide with positive net gain (red colors in (b1)–(b3)). Stability according to New [101] corresponds to negative net gain anywhere else. This framework can be used to understand the different emission states. In the case of fundamental mode-locking ((a1) and (b1)), positive net gain is only observed along the emission of pulse. The net gain outside of the pulse emission windows is nonetheless structured and already hints to the appearance of an instability and larger pump currents. Approximately 20 ps before the emission of a pulse ($\approx$55 ps after the emission of pulse) a negative-valued local maximum is observed, where the net gain rises to $\mathcal{G} \approx -0.5$. Moreover, multiple local minima can be found. The structure of the occurring time separations is determined by the geometry of the laser, e.g., the $\approx$20 ps correspond to the propagation time of a backward-moving pulse from the out-coupling facet to the absorber section.

Examining the pulse-switching unstable fundamental mode-locking state ((a2) and (b2)), which occurs at higher pump currents, the local maximum of the net gain $\approx$20 ps before a pulse emission ($\approx$55 ps after a pulse emission) has grown and periodically gives rise to new pulse trains. Moreover, the transition process itself is characterized by strongly positive net gain at the birth of the new pulse train. This produces a high power pulse that strongly saturates the absorber section and thereby causes a pronounced local maximum in the net gain at the leading edge of the new pulse train. This effect, however, is only transient and the laser emits pulses

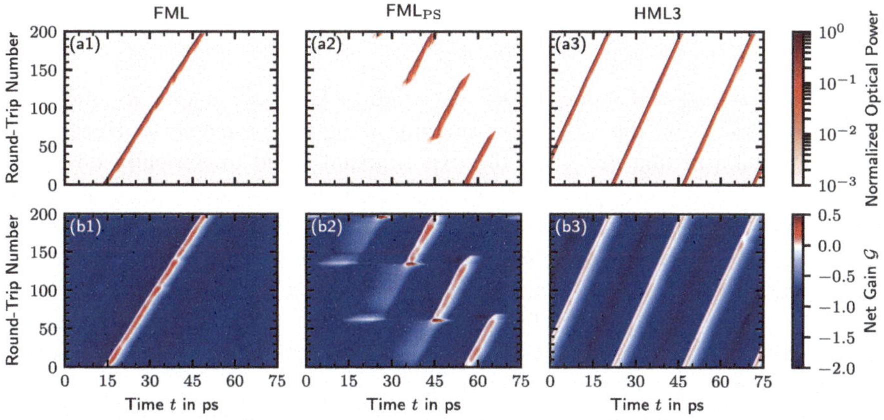

Fig. 4.8 Pseudo space-time representation of the pulse and net-gain dynamics. The top row shows pseudo space-time plots of the normalized optical power. The bottom row shows pseudo space-time plots of the net gain evaluated at the mean optical frequency. The columns correspond to fundamental mode-locking (FML, $P = 1.08$ A, $U = 6.0$ V), pulse-switching unstable fundamental mode-locking (FML$_{PS}$, $P = 1.08$ A, $U = 6.0$ V), and third-order harmonic mode-locking (HML3, $P = 1.08$ A, $U = 6.0$ V) as denoted on the top

at the fundamental repetition rate until the local net-gain window has grown again to support the birth of another pulse train. Based on the involved times, it can be concluded that the observed pulse-switching instability results from the geometric cavity configuration.

Finally, third-order harmonic mode-locking ((a3) and (b3)) produces a net-gain pseudo space-time plot that exhibits the least features. In between the emission of pulses, the net gain drops to a local minimum, which is roughly in the middle between two adjacent pulses. This simple structure is explained by the geometric configuration of the laser, where the absorber section is located at roughly one-third of the laser cavity. Third order harmonic mode-locking results in colliding pulses within the absorber section and thus an effective and symmetrical saturation mechanism.

Summary

In summary, the quantum-dot gain material results in pulse shape and net-gain dynamics that are more nuanced than those produced by quantum-well based lasers [59, 62, 64, 104]. The combination of the reverse bias induced redshift within the absorber section and ultra-fast quantum-dot ES-GS relaxation process generate a net-gain spectrum that evolves non-trivially along the pulse emission. Furthermore, the net gain outside of the pulse emission windows exhibits a structure that is governed by the geometric laser configuration and the counter-propagating pulses. Unlike the popular DDE-ring model [11, 23, 37, 59–61, 64], this feature in particular, which also leads to the presented pulse-switching instability, can only be properly described in the framework of the traveling-wave model.

4.3.3 Spatio-Temporal Dynamics

The full spatio-temporal modeling of the mode-locked laser allows to study the pulse propagation and thus the pulse shaping along the propagation. Hence, the specific mechanism that generates the experimentally and numerically observed ultra short and high-power pulses can be unraveled and understood in terms of the local interaction of the optical pulse with the respective gain and absorber sections. The simulations, moreover, provide access to the internal spatially inhomogeneous charge-carrier dynamics with their associated local gain and absorption saturation effects. This subsection develops techniques to visualize, evaluate, and interpret the high-dimensional spatio-temporal dynamics of the electric field and charge-carrier dynamics produced by the numerical model.

Fundamental Mode-Locking

A first glance at the possibilities provided by the full spatio-temporal modeling is presented in Fig. 4.9, where (a) shows the pulse propagation within the laser cavity via the color-coded optical power. The laser is operated in the fundamental mode-locking state at $P = 0.81$ A and $U = 6$ V (see Figs. 4.5 and 4.6). The three device sections are denoted on top of the figure. The propagation of the pulse and its reflections at the left and right laser facets produces a zig-zagging structure of the optical power, where the absolute value of the respective slopes corresponds to the group velocity. The dynamics are periodic in the time t with the round-trip time T_C, where one round-trip covers the complete back and forth propagation through the laser cavity.

The local interaction of the pulses along their propagation within the active medium affects the occupation numbers of the respective quantum dots. This effect is measured by the local resonant amplitude net gain

$$g(z,t,\omega) = \frac{g^{GS}(z)}{2}\left[2\rho^{\mathrm{GS}}(z,t) - 1\right] \frac{T_2^{\mathrm{eff}^2}}{(\Delta\omega_{\mathrm{GS}}(z) - \omega)^2 + T_2^{\mathrm{eff}^2}} - \frac{\alpha_{\mathrm{int}}(z)}{2}, \quad (4.38)$$

which adds the non-resonant losses α_{int} to the gain provided by the quantum-dot ground state. The integration of the local net gain along the pulse propagation for one round-trip yields the net gain of the mode-locked laser without the losses at the cavity facets. The local net gain, which corresponds to the optical pulse in Fig. 4.9a, is evaluated at the mean lasing frequency and plotted in Fig. 4.9b. Red colors indicate the amplification and blue colors the absorption of light. The propagating optical pulse leaves a clearly visible trace in the local net gain due to the saturation of the resonant quantum-dot populations, which manifest as a modified local net gain.

The saturation effect is especially prominent in the straight gain section, where the optical pulses completely saturate the resonant gain, such that the non-resonant waveguide losses lead to an overall absorptive (negative) net gain. The absorptive state of the straight gain section is, however, only short lived, since the fast quantum-dot ground-excited state carrier relaxation [28, 73, 102] quickly refills the ground state population and thereby restores the optical gain. The presented visualization,

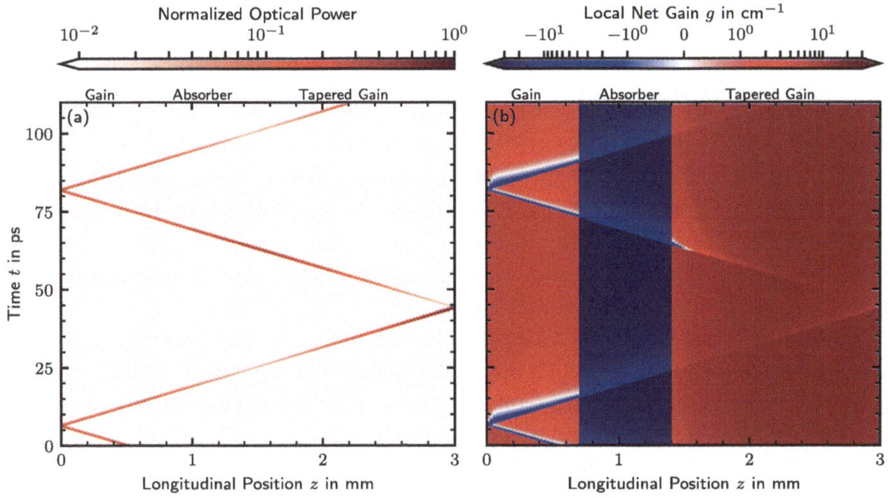

Fig. 4.9 Spatially resolved dynamics of the fundamentally mode-locked emission state. **a** Optical power normalized to the global maximum, which occurs at the out-coupling facet to the right. **b** Local net gain evaluated at the mean lasing frequency describing the amplification and absorption by the active medium and the non-resonant waveguide losses. Laser parameters: $P = 0.81$ A and $U = 6$ V

moreover, shows that the pulse, which is reflected at the left facet, experiences an effectively reduced local net gain very close to the facet, despite the strong gain saturation it has caused on its backward-moving journey due to the fast gain recovery. This highlights the very beneficial contribution of the fast quantum-dot excited state carrier reservoir, which ensures a large overall gain along the propagation. On that account, the gain recovery of the straight gain section takes considerably more time after the forward-moving pulse has passed, since the quantum-dot excited state reservoir is not replenished yet.

The saturation in the absorber section quantum-dot population is equally strong, but does not show up as clearly in the local net-gain representation, since the non-resonant waveguide losses keep the absorber section far from transparent. Nonetheless, the fast quantum-dot ground-excited state relaxation quickly disposes of any inversion that is generated by the absorption of the optical pulse and thereby provides almost the same absorption for the left and right-moving pulses.

The long tapered gain section, on the other hand, provides plenty of unsaturated gain via the extending transverse active region width. Thus, gain saturation effects are much less pronounced. Only the left-moving pulse achieves the saturation of the local net gain right before it enters the absorber section, where the tapered gain section is the thinnest. More importantly, however, the tapered geometry continuously provides more quantum dots within the transverse direction along the right-moving pulse propagation. Hence, the saturation is almost suppressed and optimal amplification of the complete pulse profile can be ensured. This very aspect represents the main motivation to construct a tapered gain geometry [8, 18, 19, 71].

Pulse-Shaping Mechanism

While the representation of the spatio-temporal dynamics in the stationary frame of the laser cavity Fig. 4.9 is appropriate to illustrate the propagation of the pulse and its local interaction with the active medium, it is not suitable to study the pulse shaping mechanisms. Since the pulse width and variations of the pulse shape are on sub-picosecond timescales, they are obscured by timescales of the pulse period, which is required to show the pulse propagation.

Hence, the co-moving frame, i.e., the characteristic curve of the traveling-wave equation (4.6), is adapted in which the relative time $t'(z)$ is constant along the propagation of light within the cavity. Considering the forward and backward directions, the definition for $t'(z)$ along one round trip reads

$$t'(z) = \begin{cases} t^-(z) = t + \frac{z}{v_g} & \text{for } E^-(z,t) \\ t^+(z) = t - \frac{z}{v_g} - \frac{l}{2v_g} & \text{for } E^+(z,t) \end{cases}, \tag{4.39}$$

where $l/2v_g$ corresponds to the propagation time from one side of the cavity to the other. In this new relative time, the propagation of a small perturbation along one round-trip occurs at the same time. The nonlinear interactions of the optical pulses with the gain medium, which also convey the pulse shaping, cause deviations of the (local) group velocity, which become visible in the this co-moving frame.

The nuances of the complete spatio-temporal pulse evolution along one round-trip can be shown by unfolding the laser cavity: Fig. 4.10a presents the propagation of $E^-(z,t)$ from the right out-coupling facet to the high-reflectivity coated facet on the left side of the plot and the propagation of $E^+(z,t)$ back to the out-coupling facet on the right side of the plot. The longitudinal position within the cavity is indicated by the horizontal axis, where negative positions z indicate the backward-moving $(-)$ field and positive positions z the forward-moving $(+)$ field. The respective gain and absorber sections are denoted on top of the plot and appear twice-once for the backward and once for the forward direction. The pulse power is color coded in red and normalized to the maximum power, which is found at the out-coupling facet. As a guide for the eye, the pulse maximum and the leading and trailing half-maximum are indicated by white lines. The time-domain pulse width can be defined by the FWHM of the pulse, which corresponds to the vertical distance between the top and bottom white line.

The corresponding local net gain $g(z,t)$ along the pulse propagation is plotted in Fig. 4.10b, where red colors indicate amplification and blue colors absorption. The optical pulse can be located by the maximum and half-maximum lines. Note that the integration along a horizontal line would yield the net gain at the central frequency without the facet losses. Additionally, the pulse width Δ_{pw} (FWHM, blue) and the normalized peak power (red) are extracted and shown separately in Fig. 4.10c. Note that the previous sections measured the pulse width via the FWHM of the intensity

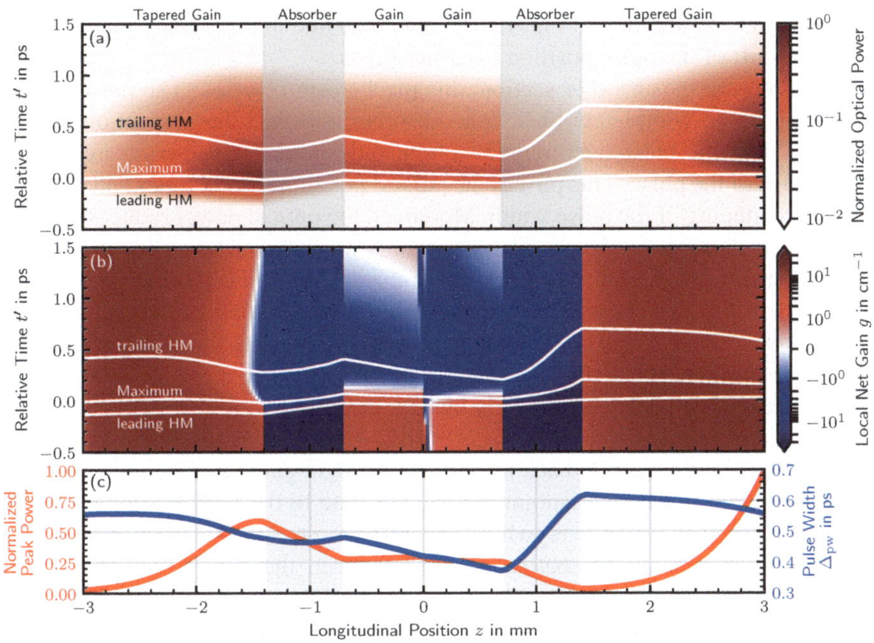

Fig. 4.10 Unfolded spatially resolved dynamics of the fundamentally mode-locked emission state. The pulse propagation has been unfolded, such that the left side of all subplots shows the backward-moving field (denoted by negative positions z) and the right side shows the forward-moving field. The top row indicates the respective device section. **a** Color-coded pulse shape along one complete round-trip in the co-moving frame. **b** Corresponding color-coded local net-gain evolution. Pulse maximum, leading, and trailing half-maximum (HM) are indicated with white lines in (**a**) and (**b**). **c** Evolution of the mean peak power (red) and the mean pulse width (blue) along the pulse propagation. Laser parameters: $P = 0.81$ A and $U = 6$ V. Adapted from [56]

autocorrelation function. Both measures, however, produce very similar results in the investigated case and perfectly agree regarding the pulse width at the out-coupling facet.

Starting at the left side of the plot, i.e., the out-coupling facet of the laser at $z = -3$ mm, this discussion first follows the backward-moving pulse through the tapered gain section, where the initially very weak pulse is amplified. In the beginning, the pulse power increases exponentially, but as the width of the tapered gain medium decreases in this direction, the number of quantum dots per given section reduces, the gain saturates more easily, and the net amplification reduces. The saturation, furthermore, produces an asymmetry in the amplification of the leading and trailing-edge of the pulse, which can be seen between $z \approx -2$ mm and $z \approx -1.4$ mm (the trailing half-maximum shifts closer to the pulse maximum). Especially right before the pulse enters the absorber section, it carries enough energy such that the pulse front alone fully bleaches the gain and the trailing-edge of the pulse is reduced in power by the non-resonant waveguide losses. This firstly leads to an effective reduction of the pulse width (white lines in Fig. 4.10a, b and blue line in Fig. 4.10c) and secondly to a slight shift of the pulse position to earlier times.

Upon entering the absorber section, this mechanism reverses: The pulse carries enough energy to completely bleach the absorber and thus the absorption at, and especially past, the pulse maximum is significantly weaker when compared to the leading edge of the pulse. This pushes the pulses' position to later times and causes a slight rebroadening of the pulse width. Moreover, the optical power reduces linearly to about half its value in the absorber section. Nevertheless, the pulse, which then enters the short gain section at the left side of the device, still carries enough energy to easily bleach it. The same mechanism as in the narrow part of the tapered gain section leads to a reduction of the pulse width and again shifts the pulse position to later times. The pulse peak power remains almost constant in the short gain section, since gain and waveguide losses nearly balance each other.

Upon being reflected at the back facet of the laser, which is located in the middle of Fig. 4.10, the pulse travels back through the straight gain section, where the fast carrier relaxation from the quantum-dot excited state to the ground state [28, 103, 105] has restored the gain everywhere except right next to the facet. Notice that due to the high reflectivity coating on the left facet, the power only marginally drops. There, waveguide losses still dominate, as indicated by the small vertical blue region in the middle of Fig. 4.10b. Along its return through the straight gain section, the forward-moving pulse is further shortened by the interplay of saturable gain and waveguide losses, while the peak power stays roughly constant. In total, the pulse width reduces within the complete passage through the straight gain section from $\Delta_{pw} \approx 470$ fs by $\approx$100 fs to $\Delta_{pw} \approx 370$ fs as can be seen Fig. 4.10c (from $z = -0.7$ mm to $z = 0.7$ mm).

Entering the absorber section, the forward-moving pulse carries less energy compared to the previous backward-moving pulse and therefore the pulse front as well as the pulse maximum are reduced before the absorber saturates. This results in a strong increase of the pulse width from $\Delta_{pw} \approx 370$ fs to $\Delta_{pw} \approx 620$ fs, which is accompanied by a significant shift of the pulse position to later times. Finally, the forward-moving pulse enters the tapered gain section again, where the increase of quantum dots along the propagation prevents the saturation of the gain. Thus, an optimal amplification of the pulse is ensured (increasing peak power) before reaching the out-coupling facet. Hence, the optical power, which is at a global minimum upon entering the tapered gain section, exponentially increases to emit the $\approx$10 W peak power pulses. Furthermore, the pulse width reduces from $\Delta_{pw} \approx 620$ fs to $\Delta_{pw} \approx 560$ fs along the tapered gain section.

In summary, the passively mode-locked pulses broaden in the absorber section and shorten in the gain sections, which is contrary to the common understanding of the pulse-shaping mechanism in semiconductor mode-locked lasers [8, 71, 106–109]. Specific to the three-section quantum-dot laser, the short straight gain section does not contribute to the out-coupled power, but rather functions as a pulse-shortening section. The out-coupled optical power is almost exclusively generated in the long tapered gain section towards the out-coupling facet. The pulse shaping can be described by tracking the pulse peak power and pulse width along the pulse

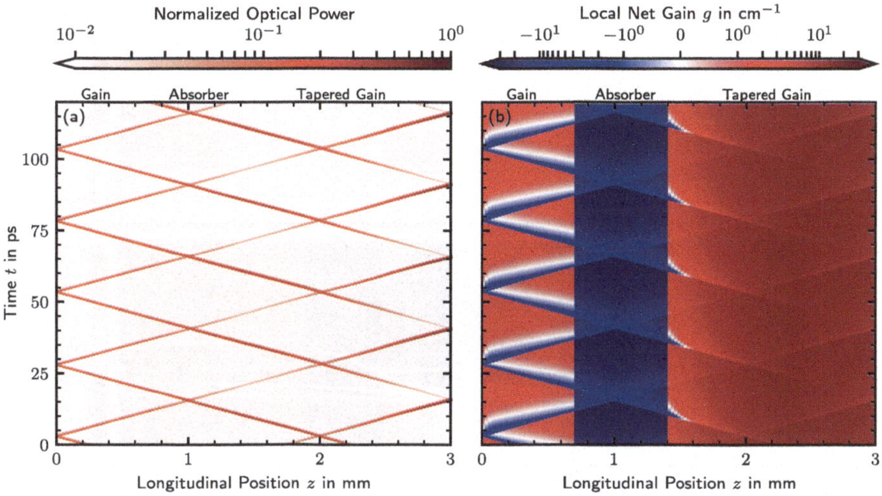

Fig. 4.11 Spatially resolved dynamics of the third-order harmonic mode-locking emission state. **a** Optical power normalized to the global maximum, which occurs at the out-coupling facet to the right. **b** Local net gain describing the amplification/absorption by the active medium and the non-resonant waveguide losses. Laser parameters: $P = 1.08$ A, $U = 6$ V, and $D = 10^{-6}$ cm ps$^{-1}[\eta]^{-1}$

propagation as shown in Fig. 4.10c. Both can be understood in terms of the underlying light-matter interaction by plotting the corresponding local net gain as shown in Fig. 4.10b.

Third-Order Harmonic Mode-Locking

While the fundamental mode-locking state is potentially of the highest interest due to its excellent pulse performance and temporal stability (see Sect. 4.4), the third-order harmonic mode-locking state is nonetheless relevant to the understanding of the emission dynamics. The naive application of the guidelines for the construction of monolithically integrated passively mode-locked lasers [12, 71, 110] points towards a colliding pulse effect in the absorber section that favors third-order harmonic mode-locking with short pulses [15, 111–113].

On that account, the pulse-shaping analysis that was just introduced is now applied to third-order harmonic mode-locking with the laser operated at $P = 1.08$ A and $U = 6$ V (see Figs. 4.5 and 4.7). The respective spatio-temporal dynamics of the optical power and the local net gain in the stationary device frame are presented in Fig. 4.11.

The three equidistant pulses circulating within the laser cavity create a unique pattern, where pulses intersect each other at the longitudinal cavity positions $z = 1$ mm and $z = 2$ mm. The pulse emission, and thus the created spatio-temporal pattern itself, is periodic with approximately one-third of the fundamental round-trip time. The interaction of the colliding pulses is mediated via the active medium. More specifically, the forward and backward-moving pulses both couple to the same quantum-dot

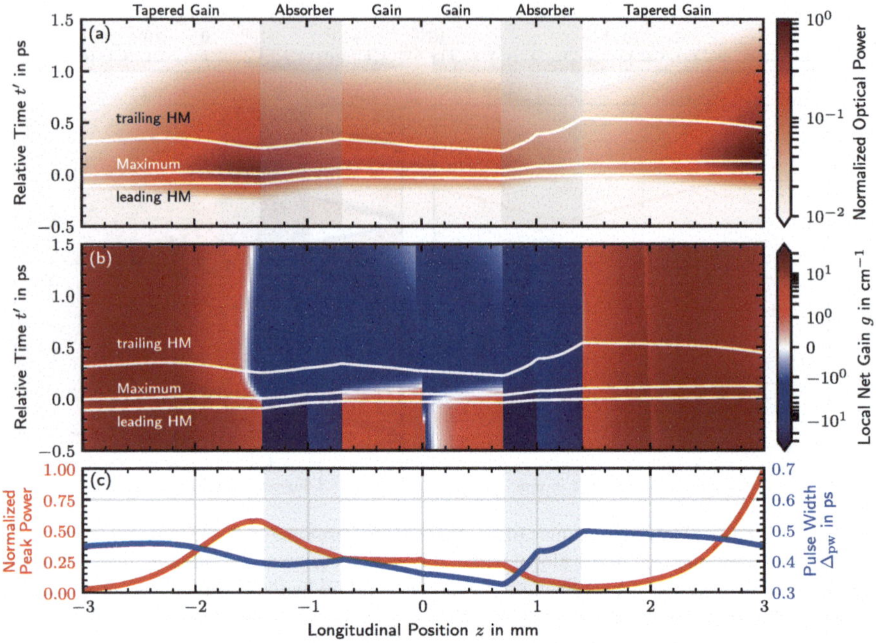

Fig. 4.12 Unfolded spatially resolved dynamics of the third-order harmonic mode-locked state. The pulse propagation has been unfolded such that the left side of all subplots shows the backward-moving field (denoted by negative positions z) and the right side shows the forward-moving field. The top row indicates the respective device section. **a** Color-coded pulse shape along one complete round-trip in the co-moving frame. **b** Corresponding color-coded local net-gain evolution. Pulse maximum, leading, and trailing half-maximum (HM) are indicated with white lines in (**a**) and (**b**). **c** Evolution of the mean peak power (red) and the mean pulse width (blue) along the pulse propagation. Laser parameters: $P = 1.08$ A and $U = 6$ V

ground state occupation numbers and thus their quantitative light-matter interaction depends on the saturation induced by the respective other pulses.

Compared to the fundamental mode-locking state, gain and absorber saturation effects are more pronounced in the straight gain section and the absorber section. The three pulses keep the gain section more saturated on average, which translates into a less efficient replenishment of the quantum-dot ground state population via relaxation from the excited state. This manifests in larger regions with negative net gain in Fig. 4.11b. The absorber section, on the other hand, still experiences a fast recovery, but is subject to colliding pulses, which reduce the absorption in the vicinity of the pulse intersection.

To study the pulse shaping of the third harmonic mode-locked emission state, the pulse propagation and local net-gain dynamics are presented in the co-moving frame and unfolded cavity, along with the mean pulse peak power and mean pulse width, in Fig. 4.12 similar to Fig. 4.10. As a guide for the eye, the pulse maximum and the leading and trailing half-maxima are indicated by white lines. Despite three pulses circulating within the laser cavity, the differences compared to fundamental mode-locking are rather subtle. While the pulse peak power profile along the pulse propagation is almost exactly the same, the pulse width generally takes on smaller values and shows one unique feature: The increase of the pulse width in the absorber section in the forward direction exhibits a kink at $z \approx 1$ mm, which is caused by the intersecting pulse that moves in the backward direction. Peculiarly, the backward-moving pulse at $z \approx -1$ mm does not experience such a kink. This difference is attributed to a much larger pulse power of the backward-moving pulse, which in turn causes a much larger saturation of the gain medium.

In conclusion, the pulse-shaping mechanism in third-order harmonic mode-locking operation does not significantly differ from the fundamental mode-locking operation. The similarity is explained by the fast quantum-dot ground and excited state coupling, which quickly restores the largest part of the gain and absorption of the respective sections. The fast quantum-dot ground and excited state relaxation, moreover, diminishes the colliding pulse effect in the absorber section. Hence, fundamental mode-locking is not at a big disadvantage in that respect, but benefits from an overall stronger gain.

4.4 Comparison to Experimental Measurements

This section presents the experimental characterization of the three-section tapered quantum-dot laser. The investigated operation parameter space is restricted to the absorber reverse bias $U = 6.0$ V, which represents the reference value used in Sects. 4.5 and 4.6. The measured results in terms of the observed emission states, the performance characteristics, and the pulse train stability are compared to the simulated results. The indeterminate simulation parameters, i.e., those that are not fixed by the device geometry or specific measurements, have been tuned such that the simulations best reproduce the experiments. Consistency among both the experiment and the simulation is crucial to gain confidence in the proposed numerical model. Hence, a good agreement between indicates an appropriate laser model with reasonable assumptions and approximations.

4.4.1 Experimental Setup and Characterization

A schematic of the experimental setup is shown in Fig. 4.13. The three-section tapered quantum-dot laser is operated via the pump (gain) current P and the absorber reverse bias U. The resulting laser emission is collimated and sent through an optical isolator to prevent unwanted back reflections, which would affect the dynamics of the laser [37, 52, 66, 67]. Parts of the optical beam are diverted via two beam splitters: The characterization of the optical pulse width Δ_{pw} is performed by a nonlinear intensity autocorrelator under the assumption of sech^2 shaped pulses. The average emitted optical power $\langle \mathcal{P} \rangle_t$ is recorded by a power meter. After fiber coupling the remaining

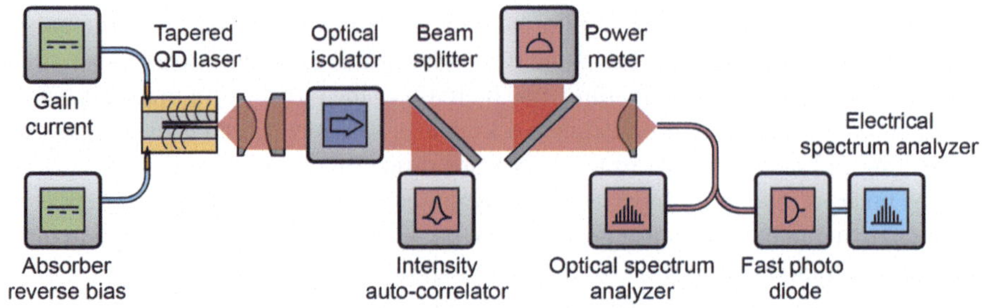

Fig. 4.13 Experimental setup. The three-section tapered quantum-dot laser is driven by the gain current P (homogeneously applied to both gain sections) and the absorber reverse bias U. The laser emission is characterized by a power meter, a nonlinear intensity autocorrelator, and an optical and electrical spectrum analyzer. Schematic by Breuer [56]

laser emission, the radio-frequency analysis of the pulse repetition frequency ν_0 and the corresponding linewidth $\Delta\nu$ is performed by a direct detection configuration using a fast photo-detector connected to an electrical spectrum analyzer. The resulting power spectral densities span the frequencies ν from 0 to 20 GHz. Combining the different measurements, the mean pulse peak power $\langle \mathcal{P}^{\max} \rangle$ is estimated according to

$$\langle \mathcal{P}^{\max} \rangle = \frac{\langle \mathcal{P} \rangle_t f_{\text{psf}}}{\nu_0 \Delta_{\text{pw}}}, \tag{4.40}$$

where f_{psf} is a pulse-shape factor that takes the assumed sech^2 pulse shape into account [22, 114].

The pulse train stability is analyzed based on the obtained power spectral density. The long-term timing jitter σ_{lt}, i.e., the temporal stability, is estimated from the fundamental harmonic with the Kéfélian method [115] (see Sect. 3.3). The amplitude stability, i.e., the amplitude jitter σ_{A}, is quantified by the relative standard deviation of pulse peak power fluctuations

$$\sigma_{\text{A}} = \frac{\sqrt{\langle \mathcal{P}_k^{\max^2} \rangle - \langle \mathcal{P}_k^{\max} \rangle^2}}{\langle \mathcal{P}_k^{\max} \rangle}. \tag{4.41}$$

Experimentally the amplitude jitter is accessible via the Wiener-Khinchin theorem: Integrating the low-frequency part of the power spectral density up to half the repetition rate ν_0 and normalizing the result by the power in the fundamental harmonic yields

$$\sigma_{\text{A}} = \sqrt{\frac{\int_{\nu_{\text{low}}}^{\nu_0/2} \mathcal{S}_{|E|^2}\, \text{d}\nu}{\int_{\nu_0 \pm \nu_0/2} \mathcal{S}_{|E|^2}\, \text{d}\nu}}. \tag{4.42}$$

Similar to the pulse-period fluctuations δT, the amplitude fluctuations are inherently bounded by the laser dynamics and thus small frequencies ν become increasingly less important as they approach $\nu = 0\,\text{Hz}$. Thus, the lower integration boundary can be chosen as $\nu_{\text{low}} = 50\,\text{MHz}$ [23, 116].

Mode-locking stability in the experiment is defined as an amplitude jitter $\sigma_{\text{A}} \lesssim 3\%$ and a long-term timing jitter $\sigma_{\text{lt}} \lesssim 250\,\text{fs}$, which corresponds to 0.33% of the pulse repetition period T_{C}.

4.4.2 Juxtaposition of Experimental and Simulated Results

In the experiment and the simulation accordingly, the pump current P is scanned from 750 to 1100 mA for the constant absorber reverse bias $U = 6.0\,\text{V}$. The obtained power spectral densities and intensity autocorrelation traces are presented in Fig. 4.14, where the left column shows the experimental and the right column the simulated results. The power spectra are normalized to the zero-frequency component ($\nu = 0$) and are shown in the top row. The autocorrelation traces are normalized to their zero-lag value and are shown in the bottom row. Based on those two indicators, the emission state is determined and denoted on top of the left and right column.

Scanning the pump current in the experiment, Q-switched mode-locking (QSML) is observed from $P \approx 750\,\text{mA}$ to $P \approx 780\,\text{mA}$, fundamental mode-locking (FML) from $P \approx 780\,\text{mA}$ to $P \approx 890\,\text{mA}$, pulse-switching unstable fundamental mode-locking (FML_{PS}) from $P \approx 890\,\text{mA}$ to $P \approx 990\,\text{mA}$, and third-order harmonic mode-locking (HML3) from $P \approx 990\,\text{mA}$ to $P \approx 1100\,\text{mA}$. The same sequence of emission states is observed in the simulated emission dynamics, however, with the transition from FML to FML_{PS} shifted to $P \approx 910\,\text{mA}$.

Q-switched mode-locking is identified by prominent low-frequency components in the power spectrum, which correspond to the frequency of the Q-switching modulation as well as autocorrelation trace that does not drop to zero in between pulses. Peculiarly, the QSML pulses are spaced by one-third of a round trip (see Fig. 4.5), leading to only a small peak at the fundamental frequency at $\nu_0 \approx 13.24\,\text{GHz}$ but a strong peak in the autocorrelation trace at $\tau \approx 25\,\text{ps} \approx T_{\text{C}}/3$. The occurrence of this inter-pulse spacing is explained by a colliding pulse mechanism [15, 111–113]. Two of three evenly spaced pulses meet in the absorber section, which is roughly located at the one-third position, and thus saturate it more efficiently.

The transition to fundamental mode-locking (FML) produces a power spectral density with a narrow and well-defined peak at $\nu \approx 13.24\,\text{GHz}$, corresponding to the cavity round-trip time. The low-frequency range of the spectrum does not exhibit characteristic frequencies and is comparatively clear. The corresponding intensity autocorrelation trace only exhibits a narrow peak at $\Delta t = 0.0\,\text{ps}$ and is zero everywhere else from $\Delta t = -38\,\text{ps}$ to $\Delta t = 38\,\text{ps}$, which indicates that only pulses with the fundamental repetition rate ν_0 are emitted. The pulse-switching instability of the fundamental mode-locking state (FML_{PS}) produces no visible change in the autocorrelation traces but becomes apparent in the power spectral densities: The instability

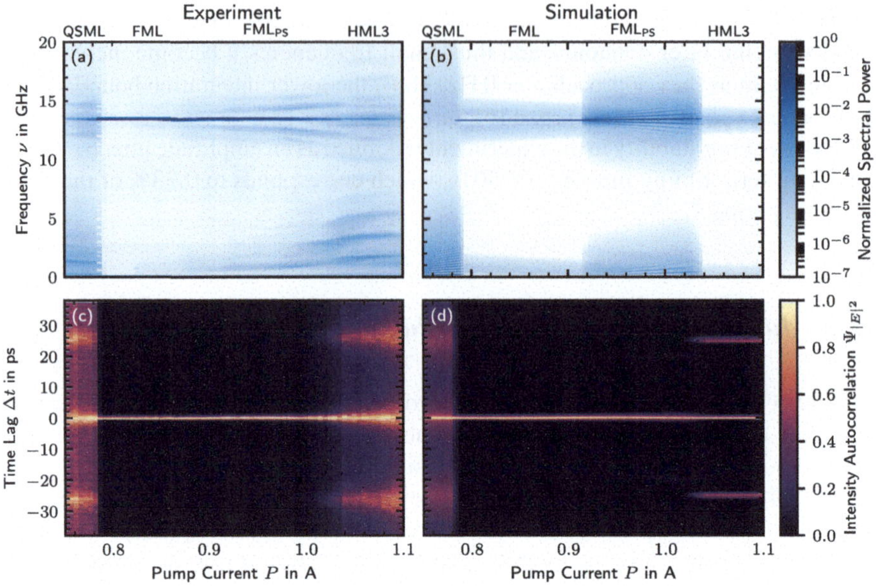

Fig. 4.14 Emission dynamics of the QD tapered laser as a function of the pump current P. Experimentally measured results are shown on the left and simulated results on the right. **a**, **b** Color-coded normalized power spectral densities of the emitted intensity. **c**, **d** Color-coded intensity autocorrelation function. The observed emission states Q-switched mode-locking (QSML), fundamental mode-locking (FML), fundamental mode-locking subject to a pulse-switching instability FML_{PS}, and third-order harmonic mode-locking (HML3) are denoted on top of (**a**) and (**b**). The measurements have been performed by the AG Breuer [56]. Laser parameters: $U = 6\,V$ and $D = 10^{-6}\,cm\,ps^{-1}[\eta]^{-1}$

routinely creates a new pulse train while destroying the old pulse train (see Fig. 4.5), which manifests in distinct low-frequency components around ≈1 GHz and the corresponding sidebands of fundamental repetition rate peak at $\nu \approx 13.24\,GHz$. Within the FML_{PS} regime, an increasing pump current P results in a higher mean repetition rate of the pulse-switching instability and thus greater separations between the main peak and the sidebands.

Lastly, third-order harmonic mode-locking is identified by a relatively weak frequency component close to the fundamental repetition rate ν_0 in the power spectral densities and an autocorrelation trace that exhibits characteristic peaks at approximately one-third of the round-trip time $\Delta t \approx \pm 25\,ps$. The first dominant peak in the power spectrum is produced at $\nu \approx 40.0\,GHz$, which is beyond the frequency band of the detection setup, but can be seen in the simulation (see Fig. 4.5). Within the HML3 region, the simulated pulse trains exhibit both cleaner power spectra (less low-frequency components) and autocorrelation traces (faster drop to zero between pulses). Both indicate a more stable operation compared to the experimentally observed pulse train.

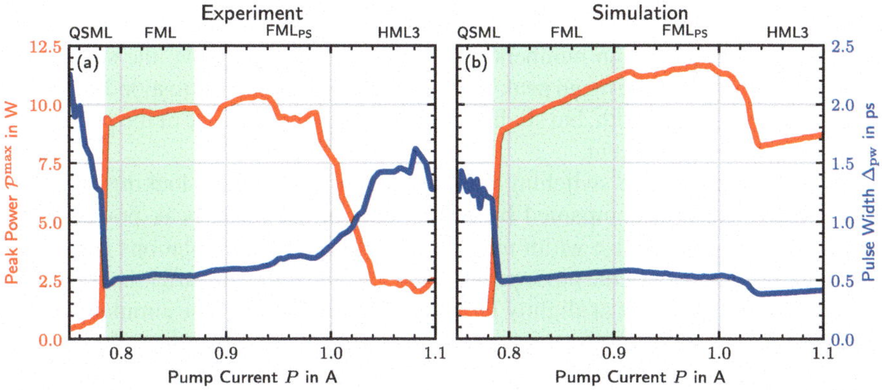

Fig. 4.15 Performance of the tapered quantum-dot laser. Measured and simulated pulse peak power (red) and pulse width (blue) are shown on the left (**a**) and on the right (**b**), respectively. The mode-locking states are denoted on top: Q-switched mode-locking (QSML), fundamental mode-locking (FML), pulse-switching unstable fundamental mode-locking (FML_{PS}), and third-order harmonic mode-locking (HML3) are observed for an increasing pump current. Stable FML is highlighted by a green-shaded background. The measurements have been performed by the AG Breuer [56]. Laser parameters: $U = 6\,\text{V}$ and $D = 10^{-6}\,\text{cm}\,\text{ps}^{-1}[\eta]^{-1}$

Nonetheless, the agreement among the experimentally observed and the simulated dynamics, as characterized by the power spectral density and intensity autocorrelation, is extraordinary and therefore suggests that the modeling approach reproduces all relevant features.

Pulse Performance

In the next step, the pulse performance of the mode locked laser is analyzed in terms of the pulse peak power and the pulse width. The results are presented in Fig. 4.15, where (a) shows the experimental measurements and (b) the simulations. In both cases, the peak power is plotted in red and the pulse width in blue. The respective mode-locking states, as conceived in Fig. 4.14, are denoted on top of (a) and (b). The region of stable fundamental mode-locking (FML) is highlighted by a green-shaded background.

Within the Q-switched mode-locking region (QSML) weak and broad pulses with peak powers below $\mathcal{P}^{\max} \approx 2\,\text{W}$ and pulse widths above $\Delta_{\text{pw}} \approx 1.2\,\text{ps}$ are observed for both the experimental and the simulated results. The pulse width, as determined via the FWHM of intensity auto-correlation, however, likely underestimates the time-domain pulse width, since the autocorrelation does not drop to zero between pulses (see Fig. 4.14). Moreover, the intensity autocorrelation function averages over all pulse shapes, which do severely vary with time (see Fig. 4.5).

These issues do not occur in the fundamental mode-locking regime, where pulses as short as $\Delta_{\text{pw}} \approx 500\,\text{fs}$ with a peak power of about $\mathcal{P}^{\max} \approx 10\,\text{W}$ are observed. Within that region, the pulse width increases with the pump current from $\Delta_{\text{pw}} \approx 500\,\text{fs}$ to $\Delta_{\text{pw}} \approx 600\,\text{fs}$. This trend can be attributed to an enlarged trailing-edge of the pulse (see Figs. 4.6 and 4.10) due to a faster recovery of optically active quantum-dot ground state occupation numbers [68]. Within the FML region, the simulation excellently reproduces the mean peak power (averaged over the corresponding pump currents) and the pulse width, but slightly overestimates the increase of the pulse peak power with the pump current.

The transition to pulse-switching unstable fundamental mode-locking (FML_{PS}) in the experiment is accompanied by minor jerks in the pulse peak power and a further increase of the pulse width up to $\Delta_{pw} \approx 800$ fs. The simulations reproduce the qualitative behavior of the pulse peak power, e.g., the initial dip after the transition to FML_{PS} emission, albeit at slightly larger pulse peak powers. The simulated pulse widths within the FML_{PS} region, however, decrease with an increasing pump power down to a value of $\Delta_{pw} \approx 500$ fs.

In the third-order harmonic mode-locking (HML3) region, peak power values drop to $\mathcal{P}^{max} \approx 2.5$ W and the pulse width increases to $\Delta_{pw} \approx 1.5$ ps in the experiment. The simulated pulses in the HML3 region carry the same energy, however, differ from the experimental characterization as they are much shorter ($\Delta_{pw} \approx 500$ fs) and consequently higher in peak power ($\mathcal{P}^{max} \approx 8.0$ W). The discrepancy in the pulse peak power and pulse width is explained by the difference in pulse train stability, which is evident in the measured and simulated spectra plotted in Fig. 4.14a, b. The measured power spectral densities exhibit pronounced low-frequency contributions, contrary to the simulated spectra, which are much cleaner at those small frequencies.

Despite the differences in the HML3 regime, the overall excellent agreement of the pulse performance figures produced by the experiment and the simulation, suggest that the model parameters have been adequately chosen to quantitatively describe the three-section tapered quantum-dot laser.

Pulse Train Stability

Lastly, the pulse train stability characterization in terms of the amplitude and timing jitter is presented in Fig. 4.16, where, again, the left column shows the experimental results and the right column the simulated results. The analysis is restricted to the fundamental mode-locking (FML) and the pulse-switching unstable fundamental mode-locking FML_{PS} regime. Regions of stable FML are highlighted by a green-shaded background.

Subfigures (a) and (b) in the top row present the amplitude jitter, where the black lines correspond to the power spectral density evaluation method according to Eq. (4.42). The simulated results shown in (b) also include the relative standard deviations of the pulse maxima $\{\mathcal{P}_k^{max}\}$ (orange line) and the pulse energies $\{\epsilon_k^{max}\}$ (purple line).

At the lower stability boundary of the fundamental mode-locking regime, the experiment produces a strong initial drop of the amplitude jitter to excellent values of approximately $\sigma_A \approx 5 \times 10^{-4}$. In the middle of the FML regime at the pump current $P \approx 820$ mA, the amplitude jitter rises again to values of $\sigma_A \approx 0.01$. The transition to pulse-switching unstable mode-locking is characterized by a somewhat erratic behavior of the amplitude jitter at further increased values above $\sigma_A \approx 0.05$.

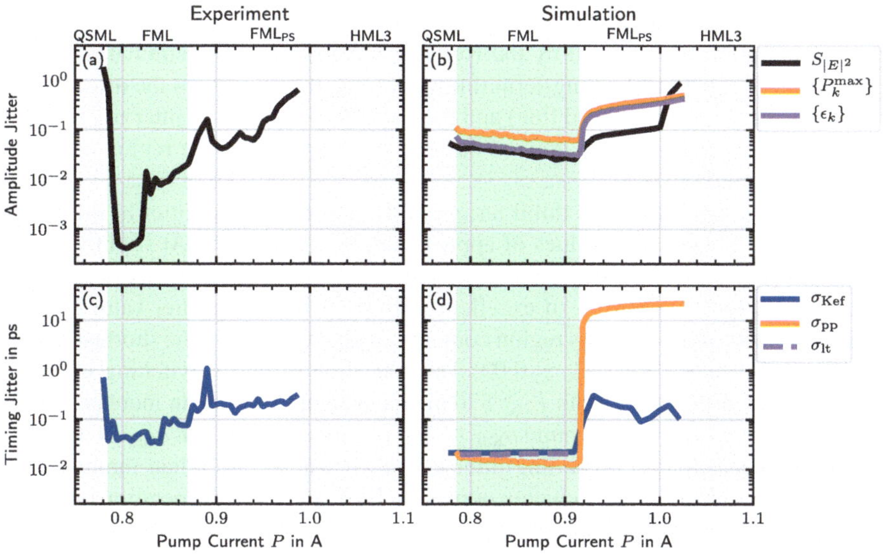

Fig. 4.16 Pulse-train stability of the tapered quantum-dot laser in the FML and FML_{PS} operation regime. Measured results are shown on the left and simulated results on the right. Subfigures **a**, **b** present the amplitude jitter obtained from the power spectrum (black), the pulse maxima $\{\mathcal{P}_k^{\max}\}$ (orange), and the pulse energies $\{\epsilon_k\}$ (purple). Subfigures **c**, **d** present the long-term timing jitter obtained via the Kéfélian method σ_{Kef} (blue) and the direct time-domain method σ_{lt} (purple) and the pulse-to-pulse timing jitter (orange). The observed emission states are denoted on top of (**a**) and (**b**): Q-switched mode-locking (QSML), fundamental mode-locking (FML), pulse-switching unstable fundamental mode-locking FML_{PS}, and third-order harmonic mode-locking (HML3). Stable FML is highlighted by a green-shaded background. The measurements have been performed by the AG Breuer [56]. Simulated data points have been obtained from 100 realizations, each 5 μs long. Laser parameters: $U = 6\,\text{V}$ and $D = 10^{-6}\,\text{cm ps}^{-1}[\eta]^{-1}$

In the simulations the amplitude jitter, as determined by the spectral method, starts of at $\sigma_{\text{A}} \approx 0.05$ at the lower stability boundary and monotonically decreases to $\sigma_{\text{A}} \approx 0.03$ at the upper stability boundary of the FML state. The transition to pulse-switching unstable fundamental mode-locking leads to an initial increase of the amplitude jitter to $\sigma_{\text{A}} \approx 0.1$ and a secondary increase to $\sigma_{\text{A}} \approx 0.1$ at larger pump currents. Peculiarly, the relative standard deviation of the pulse maxima $\{\mathcal{P}_k^{\max}\}$ (orange line) produces amplitude jitter values, which are about a factor of two larger than the amplitude jitter values obtained from the spectral method. The relative standard deviation of the pulse energies $\{\epsilon_k\}$ (purple line), however, agrees much better with the spectral estimate. Hence, the spectral estimation of the amplitude jitter, which is utilized in the experimental analysis, must be rather interpreted as a measure of the pulse energies, i.e. the pulse areas, than the pulse peak powers. In the transition to FML_{PS} operation both the standard deviation of the pulse maxima and the pulse energies exhibits a larger jump than the spectral measure. The substantial increase of the two direct estimates is attributed to transition processes, where a new pulse train grows, while the old pulse train decays simultaneously (see Fig. 4.5).

Subfigures Fig. 4.16c, d show the timing jitter, where the blue lines plot the long-term timing jitter as estimated by the Kéfélian method σ_{Kef}. The simulated results (d) additionally include the long-term timing jitter σ_{lt} obtained from the direct time-domain method (purple dashed line) and the pulse-to-pulse timing jitter σ_{pp} (orange line). The direct time-domain method has only been applied to the region of stable fundamental mode-locking.

The experimental results exhibit an initial drop of the timing jitter at the lower FML stability boundary to values of approximately $\sigma_{\mathrm{Kef}} \approx 30$ fs. At slightly larger pump currents, the timing jitter increases again, but nonetheless a 90 mA range from $P \approx 780$ mA to $P \approx 870$ mA of excellent stability with timing jitter values below $\sigma_{\mathrm{Kef}} = 100$ fs is observed. This region coincides with the region of the shortest pulses and amplitude jitter values $\sigma_{\mathrm{A}} \lesssim 0.03$. The transition to pulse-switching unstable mode-locking at pump currents $P \gtrsim 870$ mA is accompanied by an increase of the timing jitter to mean values around $\sigma_{\mathrm{Kef}} \approx 200$ fs with one outlier at $\sigma_{\mathrm{Kef}} \approx 1000$ fs.

The simulated long-term timing jitter evaluated with the Kéfélian method σ_{Kef} reproduces the experimental behavior across the FML and $\mathrm{FML}_{\mathrm{PS}}$ operation regimes. Within the fundamental mode-locking regime the long-term timing jitter remains approximately constant at $\sigma_{\mathrm{Kef}} \approx 20$ fs. The identical behavior within FML regimes is also obtained via the direct time-domain method as indicated by σ_{lt} (purple dashed line). The pulse-to-pulse timing jitter, on the other hand, decreases from $\sigma_{\mathrm{pp}} \approx 18$ fs to $\sigma_{\mathrm{pp}} \approx 10$ fs with the pump current P. The difference between the long-term and pulse-to-pulse timing jitter is caused by pulse-period fluctuations (see Sect. 3.5) and is further discussed in Sect. 4.6.3.

The pulse-switching instability for pump currents $P \gtrsim 915$ mA ($\mathrm{FML}_{\mathrm{PS}}$) causes the simulated long-term timing jitter obtained via the Kéfélian method to increase and fluctuate between $\sigma_{\mathrm{Kef}} = 100$ fs and $\sigma_{\mathrm{Kef}} = 300$ fs. The pulse-to-pulse timing jitter σ_{pp}, however, increases by over three orders of magnitude to values above $\sigma_{\mathrm{pp}} = 10$ ps. Similar to the simulated amplitude jitter, the pulse-to-pulse timing jitter, i.e., the standard deviation of the pulse periods $\{T_k\}$, is drastically inflated by the transition process, during which two pulse trains coexist (see Fig. 4.5).

Having presented the pulse-train stability characterization, it is important to stress that both the amplitude and timing jitter results obtained within the pulse-switching unstable fundamental mode-locking regime must be interpreted with care. As shown in Fig. 4.5, the pulse-switching instability regularly creates a new pulse train, which leads to the death of the previous pulse train. Both pulse trains share no phase coherence and are therefore not suitable for applications that utilize field interference effects. Hence, especially the long-term timing jitter, which describes time scales much longer than the lifetime of the individual pulse trains, is not a suitable measure for the timing stability. Nevertheless, the power spectral densities of the incoherent superposition of the individual pulse trains can be processed with the spectral amplitude and timing jitter estimation methods. By analyzing the resulting stability figures for qualitative changes, regions of stable mode-locking operation can be identified. The procedure has been used to identify the fundamental mode-locking regime as presented in Fig. 4.16a, b. Complementary simulations of the laser dynamics then allow for the exact classification of the instability and thus for the proper interpretation of the amplitude and timing jitter values outside the region of stable emission.

In conclusion, the proposed model is able to reproduce the experimentally observed mode-locking emission states, the pulse performance characteristics, and the pulse train stability figures to a more than satisfactory level. The comprehensive agreement between experiment and simulation provides the confidence that the internal pulse generation and shaping processes are well described. The extrapolation of the proposed laser model's validity to parameters outside of the just presented pump current scan is the foundation for the numerical investigation and optimization of the laser in the subsequent sections.

Comparison to other Published Devices

In order to appreciate the pulse performance figures of the three-section tapered quantum-dot laser, a comprehensive overview of monolithically integrated passively mode-locked semiconductor lasers is presented in Table 4.2. Arranged by the publication date from top to bottom, the repetition rate, largest pulse peak power, shortest pulse width, narrowest repetition-rate linewidth, and lowest timing jitter are reported for representative publications. This work's results are shown in the last row. Note that the quoted values are not necessarily achieved at one, but rather at multiple operation points and describe the pulse properties as emitted by the solitary laser. Hence, neither any pulse amplification and compression nor any repetition-rate linewidth reduction scheme has been applied. Unless otherwise noted, all device are based on GaAs Stranski-Krastanov semiconductor quantum dots and utilize straight waveguides.

The comparison to the selected results reveals that the three-section laser's pulse peak power is only surpassed by the device presented by Ref. [125] ($\mathcal{P}^{\max} \approx 18.3$ W vs. $\mathcal{P}^{\max} \approx 10$ W), although at a much slower repetition rate and with much broader pulses. Moreover, only the device presented in Ref. [8] achieves a shorter pulse width than the three-section laser ($\Delta_{\mathrm{pw}} \approx 360$ fs vs. $\Delta_{\mathrm{pw}} \approx 500$ fs), albeit at smaller pulse peak powers. Both devices [8, 125] rely on tapered waveguide structures to achieve these performance figures.

The listed repetition-rate linewidths range from 400 Hz to 370 kHz and generally improve with the time of publication. The three-section laser's 16.0 kHz repetition-rate linewidth is therefore neither especially good nor especially bad. Using the Kéfélian method, the long-term timing jitter is computed from the repetition rate and its respective linewidth. The results are presented in the jitter column with values that range from $\sigma_{\mathrm{lt}} \approx 6.0$ fs to $\sigma_{\mathrm{lt}} \approx 381.2$ fs. Hence, the three-section laser's timing jitter of $\sigma_{\mathrm{lt}} \approx 33.3$ fs belongs to the best performances and is only surpassed by a few devices. Among them, only the lasers presented in Refs. [8, 22] achieve similar pulse peak powers and pulse widths. Similar to this work's laser, those two devices also utilize ten layers of quantum dots and a long tapered gain section.

In conclusion, the three-section tapered quantum-dot laser exhibits good to exceptional pulse performance figures. The unique character of the three-section laser is, however, comprised by the simultaneous occurrence of those pulse train properties.

Table 4.2 Performance comparison of monolithically integrated passively mode-locked semiconductor lasers

Repetition rate (GHz)	Peak power (W)	Pulse width (ps)	Linewidth (kHz)	Jitter (fs)	References	Notes
7.4	2.0	3.3	–		[18]	Tapered/QW
7.4	–	17.0	370.0	381.2	[117]	
5.0	1.7	3.2	–		[118]	
24.0	0.5	0.78	–		[20]	Tapered
5.0	0.5	3.9	–		[119]	
14.2	2.25	0.36	10.0	23.4	[8]	Tapered
10.0	0.05	2.0	0.5	8.9	[120]	
5.0	0.1	10.0	26.8	184.8	[121]	
5.0	–	12.1	50.0	252.3	[52]	
5.25	0.1	10.9	46.2	225.4	[53]	
14.65	2.0	3.3	–		[19]	Tapered
5.1	0.15	4.0	19.0	151	[38]	
16.0	4.2	1.37	17.5	26.1	[22]	Tapered
9.0	–	1.3	80.0	132.0	[122]	On silicon
14.6	3.8	0.94	2.4	11.0	[23]	Tapered
20.0	–	4.0	1.8	6.0	[123]	On silicon
9.4	0.028	1.7	0.4	8.8	[124]	On silicon
2.89	18.3	11.1	–		[125]	Tapered/QW
13.24	9.9	0.5	16.0	33.4	[56]	This work

The linewidth refers to the fundamental repetition-rate peak in the power spectrum. The jitter refers to the long-term timing jitter computed with the Kéfélian method. The quoted pulse performance values are directly collected from the emitted pulses with neither any additional compression and amplification nor any external repetition-rate linewidth reduction scheme

4.5 Emission Dynamics in the Operation Parameter Space

A pulse train emitted from a passively mode-locked laser is generated by the interaction of the electric field with the gain and absorber sections of the device. The pulse generation requires suitable saturation energies and recovery rates in the respective device sections. On the one hand, those properties depend on the characteristics of the active medium and the cavity design, which are both determined by the device design and fixed with the manufacturing of the device. On the other hand, the saturation energies and recovery rates can be manipulated via the external driving conditions, namely the pump current P and the reverse bias U, which are applied to the gain and absorber sections. Passive mode-locking is typically achieved by properly tuning both of these parameters. Hence, the parameter subspace (P, U), which is spanned by the pump current and the reverse bias, is the most natural choice in order to investigate the emission dynamics.

This section further explores the emission dynamics and pulse performance of the three-section tapered quantum-dot laser. The one-dimensional pump current scan, which has previously been presented the in Sect. 4.4, is expanded to larger pump current ranges and to absorber reverse biases other than the experimentally studied $U = 6\,\text{V}$. The dependence on the two operation parameters is presented in Sect. 4.5.1, which is followed by an in-depth analysis of the effects induced by the reverse bias in Sect. 4.5.2. Lastly, the stability of the observed emission states against spontaneous emission noise is presented in Sect. 4.5.3.

4.5.1 Pump Current and Reverse Bias Dependence

The pump current P supplies the energy to the active region, which is required to achieve an inversion in the gain sections, i.e., to achieve amplification of light. The pump current, furthermore, effectively enhances the gain recovery at the lasing threshold by increasing the unsaturated gain. The reverse bias U, which is applied to the absorber section, enhances the escape rates of charge-carriers from the quantum-dot excited state to the surrounding quantum-well reservoir and thereby controls the recovery of the saturable absorber. It, furthermore, produces a redshift of the optically active transitions, which leads to a detuning with respect to the gain spectrum and thus an overall asymmetric net-gain spectrum (see Sect. 4.3.2).

To start off, Fig. 4.17a presents a map of the detected dynamics as a function of the pump current P and the reverse bias U. To obtain each data point in Fig. 4.17a, the laser is prepared in the off state, i.e., without external driving, before a respective set (P, U) of pump current and reverse bias is applied. After a sufficiently long transient time, the laser emission settles to an attracting solution and a finite time series is recorded and used for further analysis. The black dotted line indicates the reference reverse bias $U = 6\,\text{V}$, which has been used for the comparison with the experimental results.

Among the resulting emission dynamics, the following mode-locking states are identified: Q-switched mode-locking (QSML, pink), fundamental mode-locking (FML, blue), fundamental mode-locking with a pulse-switching instability (FML_{PS}, light blue), two non-identical pulses (MP2, green), third-order harmonic mode-locking (HML3, red), and third-order harmonic mode-locking with a strong trailing-edge plateau (HML3_{TEP}, orange). In cases of no mode-locking, white denotes the off-state, i.e., no lasing, teal denotes continuous-wave (cw) lasing, and gray (labeled with 'else') indicates that the classification has failed since the laser emits non-periodic complex oscillatory patterns. A detailed illustration of all mode-locking states is presented in Fig. 4.5 and discussed in Sect. 4.3.

The results are discussed in terms of pump current scans in dependence of the reverse bias U. Generally, the available mode-locking states become more diverse with an increasing absorber bias U. For reverse biases up to $U \approx 2\,\text{V}$, only the off and cw states are observed with their pump current boundary at $P \approx 1\,\text{A}$. As the laser is started in the off state, this boundary represents the lasing threshold. For reverse biases between $U \approx 2\,\text{V}$ and $U \approx 3\,\text{V}$, third-order harmonic mode-locking emission with strong trailing-edge plateaus appears between the off and the cw state, while the lasing threshold starts shifting towards smaller pump currents. In the

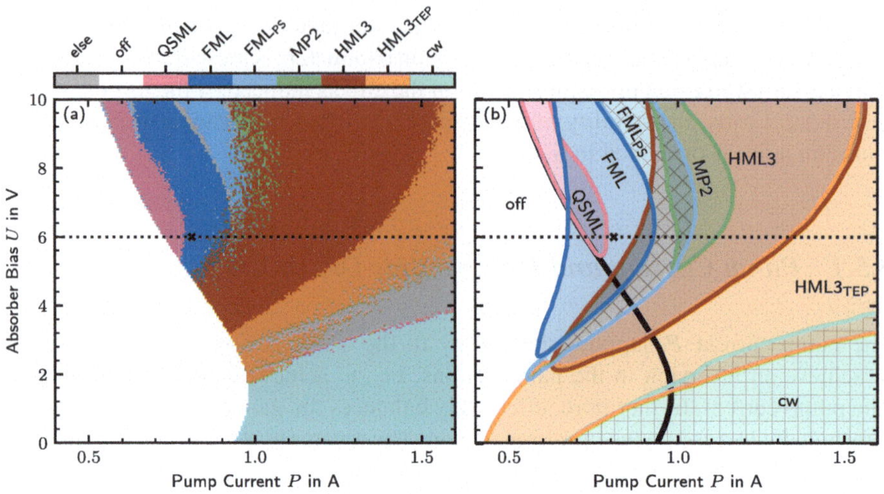

Fig. 4.17 Emission states observed in the operation parameter space (P, U): off, Q-switched mode-locking (QSML), fundamental mode-locking (FML), fundamental mode-locking with a pulse-switching instability (FML_{PS}), two non-identical pulses (MP2), third-order harmonic mode-locking (HML3), third-order harmonic mode-locking with a strong trailing-edge plateau ($HML3_{TEP}$), and cw. **a** Laser (initial conditions) prepared in the off state. **b** Stable regions obtained from combined simulations with the laser prepared in the off, FML, HML3, and MP2 state, respectively. The black dotted lines and black crosses indicate parameters used for further investigation. Hatches highlight the FML_{PS} and cw states. Adapted from [58], with the permission of AIP Publishing

reverse bias interval between $U \approx 3\,\text{V}$ and $U \approx 4.5\,\text{V}$, third-order harmonic mode-locking emission is observed in between the off and the $HML3_{TEP}$ states. For further increased reverse biases between $U \approx 4.5\,\text{V}$ and $U \approx 5.2\,\text{V}$, fundamental mode-locking appears between the off and the HML3 state. The transition to HML3 does not exhibit a clean boundary, but is noisy with FML and HML3 emission intertwined, which indicates a multistability between those two states. For reverse biases above $U \approx 5.2\,\text{V}$, Q-switched mode-locking starts in-between the lasing threshold and the fundamental mode-locking state. Moreover, at the multistable transition from FML to HML3, states with two non-identical pulses appear as well. For a further increasing reverse bias, the lasing threshold reduces to $P \approx 0.5\,\text{A}$ at $U = 10\,\text{V}$, which is almost half the value compared to $U = 0\,\text{V}$. Furthermore, with an increasing reverse bias U, the region of observed fundamental mode-locking shifts to smaller pump currents, while the region of HML3 states extends to larger pump currents before transitioning to $HML3_{TEP}$ states. The greatest dynamical complexity, i.e., multitude of observed emission states, is found for reverse biases $U \gtrsim 6.0\,\text{V}$.

In the case of multiple stable emission states, i.e., multiple attractors in phase space, as indicated by the noisy parts of Fig. 4.17a, the final state obtained by numerical integration depends on the initial conditions and the stochastic realization of the simulated spontaneous emission noise. Preparing the laser in the off state, as shown in Fig. 4.17a, therefore first of all probes the stability of the off state and thus yields an accurate depiction of the lasing threshold. The stability boundaries of the other emission states, however, are only accurately described if no multistability is present at a given parameter combination $\{P, U\}$.

This problem can be mitigated by combining the results from multiple simulations with purpose-built specific initial conditions. For example, preparing the laser in the FML state with $U = 6$ V and $P = 0.81$ A yields a reasonable approximation of the stability boundaries of the FML state.[4] Even though this case is designed to probe the stability of the FML state, it also yields the region of FML_{PS} emission, since its attractor is likely close to the initially prepared FML state in phase-space. The same argument applies to the HML3 and HML3_{TEP} states.

Utilizing this 'poor man's continuation', simulations using the off, FML, HML3, and MP2 state as initial conditions are combined to extract the stability boundaries of the off, QSML, FML, FML_{PS}, HML3, HML3_{TEP}, and cw states. The respective regions of stability computed from all simulations are shown in Fig. 4.17b and reveal a complex structure of multistabilities. At low reverse biases $U \lesssim 2$ V, both the HML3_{TEP} and the cw state are multistable with the off state. A bistability between the cw and the HML3_{TEP} states themselves also exists at larger pump currents and reverse biases. The lower extent of the FML, FML_{PS}, and HML3 stable regions appears in a region around $P \approx 0.6$ A and $U \approx 2$ V, where all are multistable with the off state.

The FML_{PS} state always appears beyond the upper pump-current stability boundary of the FML state, except for absorber biases around $U \approx 2$ V where no FML is found yet. This suggests that both are connected by the stability limiting bifurcation of the FML state. Hence, no bistability between the two is observed. Similarly, the HML3 state always appears at the lower stability boundary of the HML3_{TEP} state without bistability, which also suggests a connecting bifurcation between those two states. The FML and FML_{PS} states, however, can be multistable with the HML3 state, which indicates their independence from each other. In relatively simple delay-differential equation models, the fundamental and different harmonic mode-locking emission states are born in independent Hopf bifurcations of the cw-lasing state [59, 61, 62, 126]. Hence, it can be reasoned that the FML and HML3 emission states, which are observed for the three-section laser, are created similarly.

The QSML state only appears at reverse biases above $U \approx 5.3$ V just beyond the lasing threshold, where it is bistable with the FML state. This changes, however, for absorber biases above $U \approx 8.7$ V, where the QSML state transitions into the FML state without bistability. This, and the fact that the QSML emission bursts are characterized by three pulses circulating in the laser cavity, indicates that the QSML and FML states are not directly connected via any bifurcation.

At pump currents around $P \approx 1.0$ A the MP2 state appears for absorber biases above $U \approx 5$ V. For an increasing absorber bias, the pump current extend of the stable regions first increases up to $U \approx 7.5$ V and then decreases again. For all reverse biases

[4] This technique fails if the basin of attraction of the respective emission either moves away or shrinks too much. This likely happens close to the stability determining bifurcation points. Hence, the stability boundaries, i.e., bifurcation lines, are slightly underestimated.

the MP2 state is bistable with the HML3 state and also with FML_{PS} state for lower pump currents. There is no evidence which suggests that the MP2 emission state is directly related to any other observed emission state.

Focusing on the fundamental mode-locking emission state, the range of pump currents for which it is observed appears at $U \approx 2.6$ V, increases with the absorber bias up to $U \approx 6.4$ V, and then decreases again. The bistability with the off solution is only found below $U \approx 6.8$ V with the HML3 state below $U \approx 7.6$ V and the QSML state below $U \approx 8.7$ V.

The black dotted line along the absorber bias $U = 6.0$ V, which represents the experimentally investigated absorber bias, crosses a region of maximal multistability. The black cross denotes a parameter combination $U = 6.0$ V and $P = 0.81$ A, where only the FML state is stable. This operation point has been chosen to exemplify the timing jitter in Chap. 3, the net-gain dynamics in Fig. 4.6, and the spatio-temporal pulse shaping mechanism in Fig. 4.10. A very different scenario could be accessed at an absorber bias $U \approx 10$ V, where all multistabilities between the considered states disappear.

In conclusion, a diverse map of laser emission states has been presented. The characteristics of the absorber section, controlled by the absorber reverse bias, govern the mode-locking states, which can be obtained by tuning the pump current. Fundamental mode-locking only appears at reverse biases $U \gtrsim 2.6$ V and for pump currents between $P \approx 0.55$ A and $P \approx 0.95$ A. Multistabilities are generally more pronounced at small reverse biases and completely disappear at $U \approx 10.0$ V.

Comparison to Literature

Among the semiconductor quantum-dot based monolithically integrated edge-emitting mode-locked lasers, this work's three-section laser is unique due to its positioning of the absorber section and simultaneous use of tapered gain section. Nonetheless, some of the observed characteristics shall be compared to the existing literature:

While some devices require only marginal reverse biases in order to produce mode-locked pulse trains [8, 51, 127], a number of devices achieve mode-locking only for reverse biases $U \gtrsim 3$ V [19, 22, 23, 31, 128, 129], similar to the investigated three-section laser. This behavior has also been produced by various different modeling approaches [11, 14, 23, 63, 130]. However, all cited references report either cw emission or unstable mode-locking for smaller reverse biases, while the three-section laser also exhibits a unique region of third-order harmonic mode-locking for intermediate reverse biases. This feature is attributed to the positioning of the absorber section.

Subthreshold mode-locking, as observed for smaller reverse biases in this work, is well known in external cavity mode-locked lasers [131–135]. However, it has also been simulated for monolithically integrated quantum-dot lasers [61] and experimentally observed [8, 136].

The last considered emission characteristic addresses the shape of the FML region. For the three-section laser, the pump current range of stable FML emission first increases, then reaches a maximum and decreases again as presented in Fig. 4.17. Moreover, the lower stability boundary is concave a with local pump current maxi-

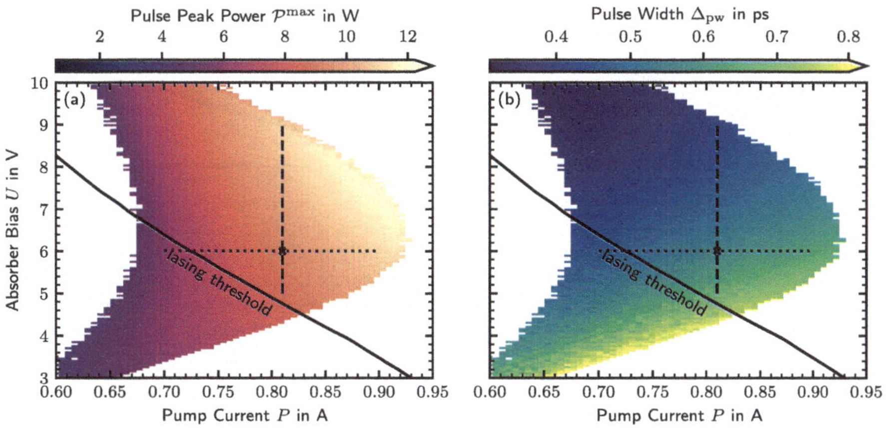

Fig. 4.18 Fundamental mode-locking performance characteristics in the pump current and reverse bias parameter space (P, U): **a** pulse peak power and **b** pulse width of the FML state. Each data point has been generated with the laser prepared in the FML state. The solid black lines indicate the lasing threshold. The black dotted lines denote the parameter range scanned in Fig. 4.15 and the black dashed lines the parameter range scanned in Fig. 4.20

mum at intermediate reverse biases. This behavior has been reported for a multitude of experiments [8, 12, 22, 31, 61, 129, 137]. Most reported simulated results [11, 61, 68, 130], however, do not exhibit this behavior. With only two exceptions [23, 63] to that, this discrepancy points to a relevant physical mechanism that is mostly likely missing in most modelling approaches. This issue will be further discussed in Sect. 4.5.2.

Pulse Performance Characteristics

To further asses the mode-locking characteristics, the pulse peak power and pulse width of the FML emission states are shown in Fig. 4.18a, b, respectively. For each data point, the laser has been prepared in the FML state and thus includes parameter combinations (P, U), which are multistable with the off, QSML, and HML3 emission states. The pulse peak power is directly computed as the mean of all recorded output-power pulse maxima. The pulse width is computed from the intensity autocorrelation assuming a sech^2 pulse shape deconvolution. These figures of merit could also be evaluated for the other states, but are far less representative of the mode-locking performance due to the lacking stability or regularity of the pulse trains. The lasing threshold, as extracted in Fig. 4.17, is indicated by black solid lines. The black dotted and dashed lines denote parameter ranges, which are used in Figs. 4.19 and 4.20, respectively.

Both the pulse peak power and the pulse width exhibit a relatively simple behavior: The pulse peak power increases with the pump current P and the reverse bias U and the pulse width increases with the pump current P and decreases with the reverse bias U. Both operation parameters, however, have their distinct impact as the pump current P imprints a much stronger dependence on the pulse peak power than the

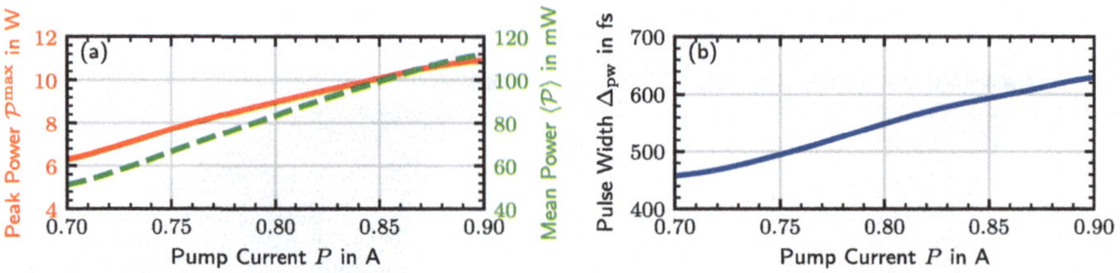

Fig. 4.19 Mode-locking performance in the fundamentally mode-locked state along the pump current P at $U = 6.0$ V: **a** pulse peak power (red solid line) and mean power (green dashed line) and **b** pulse width. The plotted parameter range is indicated by the black dotted lines in Fig. 4.18

reverse bias U. The reverse bias, on the other hand, indirectly controls the achievable peak powers by determining the accessible pump currents.

The maximum peak power in the FML region is found with $\mathcal{P}^{max} \approx 13.0$ W at $P \approx 0.92$ A and $U \approx 6.4$ V, i.e., the largest achievable pump current. The minimum pulse peak power $\mathcal{P}^{max} \approx 3.0$ W is found at $P \approx 0.6$ A and $U \approx 3.0$ U, i.e., at the smallest achievable pump current. The shortest pulses in the FML region exhibit a width of $\Delta_{pw} \approx 330$ fs and are found at the largest reverse biases U and the smallest accessible pump currents P. The minimum is observed at $P \approx 0.64$ A and $U \approx 10.0$ V. The broadest pulses, on the other hand, are observed at the smallest reverse biases and the largest available pump currents. The respective pulse width maximum is now found at $P \approx 0.75$ A and $U \approx 4.0$ U with a value of $\Delta_{pw} \approx 800$ fs. Hence, the best achievable pulse peak power and pulse width are achieved at different operating points. It is further important to note that the evolutions of both the pulse peak power and the pulse width are neither affected by the lasing threshold, i.e., the multistability with the off state, nor any other multistability with another lasing state.

In the next step, the changes in the pulse performance characteristics are highlighted by one-dimensional scans along the pump current P and absorber reverse bias U. Figure 4.19 presents a scan from $P = 0.7$ A to $P = 0.9$ A at $U = 6$ V and Fig. 4.20 a scan from $U = 5$ V to $U = 9$ V at $P = 0.81$ A. The two investigated parameter ranges are indicated by black dotted and dashed lines in Fig. 4.18, respectively, and intersect at the point ($P = 0.81$ A, $U = 6$ V). In both figures, (a) plots the pulse peak power (red solid line) and the mean power (green dashed line) and (b) plots the pulse width.

Along the pump current scan presented in Fig. 4.19, the pulse peak power increases slightly sublinearly from $\mathcal{P}^{max} \approx 6.4$ W to $\mathcal{P}^{max} \approx 11.0$ W, while the mean power linearly increases from $\langle \mathcal{P} \rangle \approx 50$ mW to $\langle \mathcal{P} \rangle \approx 123$ mW. The sublinear increase of the peak power is caused by a broadening of the pulse width from $\Delta_{pw} \approx 460$ fs to $\Delta_{pw} \approx 630$ fs, which has been attributed to a faster intra-dot relaxation process (see Sect. 4.4.2).

Along the absorber reverse bias U scan, which is presented in Fig. 4.20, the pulse peak power increases linearly from $\mathcal{P}^{max} \approx 8.7$ W to $\mathcal{P}^{max} \approx 11.2$ W, while simultaneously the mean power decreases sublinearly from $\langle \mathcal{P} \rangle \approx 95$ mW to $\langle \mathcal{P} \rangle \approx 77$ mW.

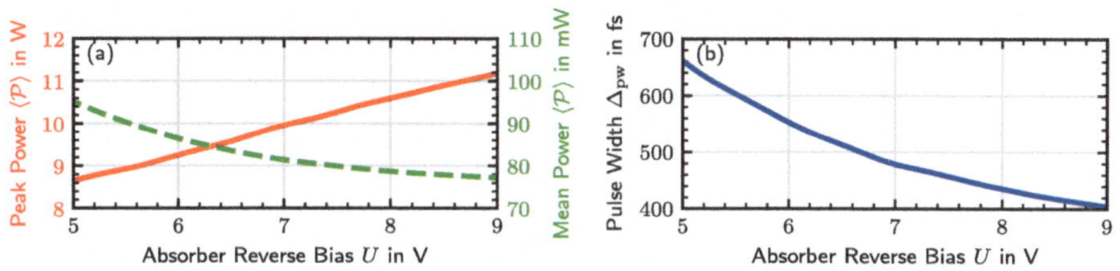

Fig. 4.20 Mode-locking performance in the fundamentally mode-locked state along the reverse bias U at $P = 0.81$ A: **a** pulse peak power (red solid line) and mean power (green dashed line) and **b** pulse width. The plotted parameter range is indicated by the black dashed lines in Fig. 4.18

Their opposing behavior is linked to a constant pump current and a redistribution of the optical power into shorter pulses. Hence, the pulse width reduces from $\Delta_{\mathrm{pw}} \approx 660$ fs to $\Delta_{\mathrm{pw}} \approx 400$ fs along the increasing reverse bias U.

Comparing the two different operation parameters P and U, the pulse peak power reacts quantitatively stronger to the pump current P, while the pulse width reacts quantitatively stronger to the reverse bias U. It might appear that the overall performance increases with a larger absorber reverse bias U. However, as shown in Fig. 4.17, large reverse biases also shift the upper pump current stability boundary to smaller pump currents and thus limit the maximum pulse peak power. For that reason, the experimentally utilized absorber reverse bias $U \approx 6.0$ provides the largest peak powers as well as access to relatively short pulses.

Pump Current Dependence of the Pulse-Shaping Mechanism

As shown in Fig. 4.10 in Sect. 4.3.3, the pulse-shaping along the pulse propagation can be both illustrated and quantified in terms of the spatially resolved pulse width. This insight is now used to study the pump current dependence on the pulse-shaping mechanism. The effect of an increasing pump current on the pulse-shaping can be expected to be twofold: Firstly, the stronger pumping increases the effective recovery rate of the gain sections. Secondly, the increased gain produces stronger pulses, which in turn cause a stronger saturation of the gain and absorber sections.

The investigated pump current range covers the region of stable fundamental mode-locking emission at the absorber reverse bias $U = 6$ V (see Figs. 4.17 and 4.18). The results are presented in Fig. 4.21, where (a) plots the pulse width Δ_{pw} as a function of the longitudinal position z within the unfolded laser cavity and the pump current P. The previously shown pulse width evolution Fig. 4.10c corresponds to a horizontal cut at $P = 0.81$ V. The pulse width in each row has been normalized to the out-coupled pulse width. Figure 4.21b shows the corresponding local pulse width changes $\partial_z \Delta_{\mathrm{pw}}$, where the derivative has been evaluated on the discretized grid.

For an increasing pump current, the pulse-shaping qualitatively remains the same at most of the positions along the pulse propagation. The only exception is found for the backward-traveling pulse in the absorber section, where the pulses first shorten and then broaden again. At low pump currents, the pulse shortening dominates, while at larger pump currents $P \gtrsim 0.8$ A the pulse broadening dominates the net

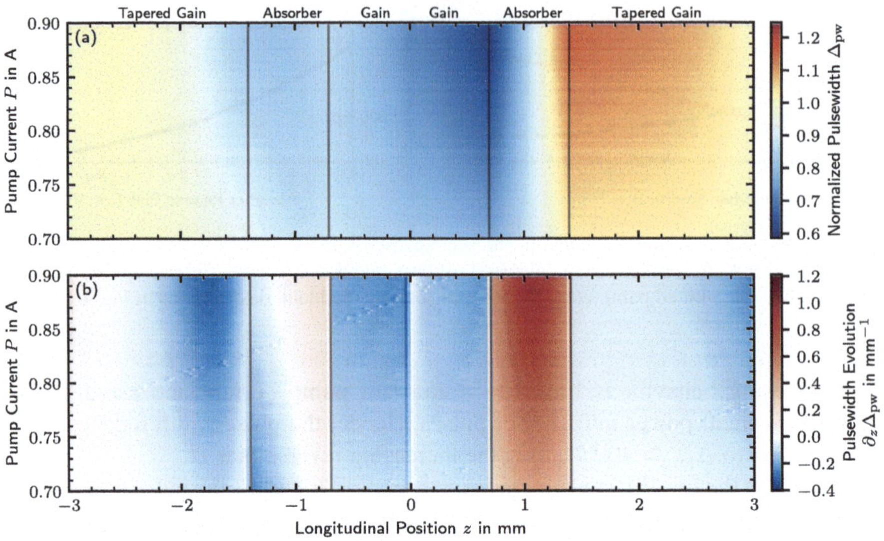

Fig. 4.21 Pulse shaping in the fundamental mode-locking state along the pump current P. The left side of both subplots shows the backward-moving field (denoted by negative positions z) and the right side shows the forward-moving field. **a** Presents the color-coded pulse width, which has been normalized to the respective out-coupled pulse width. **b** Plots the changes $\partial_z \Delta_{\text{pw}}$ of the normalized pulse width with respect to the longitudinal coordinate z. Laser parameters: $U = 6\,\text{V}$ and $D = 10^{-6}\,\text{cm}\,\text{ps}^{-1}[\eta]^{-1}$

contribution to the pulse-shaping. The transition between shortening and broadening further moves closer to the tapered gain section for an increasing pump current as it can be seen in Fig. 4.21b.

Qualitatively, however, the increasing pump current enhances all individual contributions: The normalized pulse width only varies between $\Delta_{\text{pw}} \approx 0.8$ and $\Delta_{\text{pw}} \approx 1.04$ at the lowest pump current $P = 0.7\,\text{A}$ and between $\Delta_{\text{pw}} \approx 0.6$ and $\Delta_{\text{pw}} \approx 1.2$ at the largest pump current $P = 0.9\,\text{A}$. Peculiarly, the global minimum is always found for the forward direction at the interface between the short gain and the absorber section and the global maximum is always found at the following interface between the absorber and the tapered gain section.

The regions that respond most sensitively to the increasing pump current are the narrow part of the tapered gain section (between $z \approx -2.0\,\text{mm}$ and $z \approx -1.4\,\text{mm}$) along the backward direction and the absorber section along the forward direction. Regarding the former, the local minimum of the pulse width change decreases from $\partial_z \Delta_{\text{pw}}(z) \approx -0.1\,\text{cm}^{-1}$ at the smallest pump current to $\partial_z \Delta_{\text{pw}}(z) \approx -0.4\,\text{cm}^{-1}$ at the largest pump current, which produces a much more pronounced pulse shortening. Regarding the latter, the local maximum of the pulse width change increases from $\partial_z \Delta_{\text{pw}}(z) \approx 0.45\,\text{cm}^{-1}$ to $\partial_z \Delta_{\text{pw}}(z) \approx 1.2\,\text{cm}^{-1}$, which leads to a much stronger pulse broadening. Both effects can be attributed to a stronger saturation of the respective gain and absorber section due to the larger pulse power with an increased pump current. Furthermore, comparing the magnitude of the pulse width changes, the pulse broadening clearly dominates, which leads to an out-coupled pulse width that increases with the pump current as experimentally observed (see Fig. 4.15).

4.5.2 *Understanding the Absorber Reverse Bias*

As presented in the previous subsection, the absorber reverse bias U has a strong impact on the observable emission dynamics and is therefore investigated in more detail in this section. On that account, the effect of tuning the reverse bias of the absorber section, as introduced in the previous model derivation Sect. 4.2, is twofold: It firstly decreases the carrier sweep-out time via enhanced thermionic escape rates and secondly causes a transition frequency redshift due to the quantum confined Stark effect [9, 94–97]. The former effect increases the absorber recovery rate and can increase the total absorption if the recovery time is on the order of the pulse width or faster. The latter effect increases the effective maximal gain by shifting the absorption peak away from the gain peak but creates an asymmetry net-gain spectrum thereby.

This subsection separates both effects by keeping either the absorber redshift $\Delta\omega_{\text{GS}}^{\text{abs}}$ or the absorber QD excited state recovery rate $\gamma_{\text{ES}}^{\text{abs}}$ constant while tuning the other. The respective fixed parameter corresponds to an absorber reverse bias of $U = 6.0\,\text{V}$, i.e., $\Delta\omega_{\text{GS}}^{\text{abs}} = 2.2\,\text{ps}^{-1}$ and $\gamma_{\text{ES}}^{\text{abs}} = 1.0\,\text{ps}^{-1}$.

Absorber Recovery Rate

Figure 4.22 presents the dynamics obtained from pump current and absorber recovery rate scans in the same manner as the previous pump current and absorber bias analysis. Subfigure (a) shows the emission states obtained with the laser prepared in the off state for each parameter combination. Hence, the lasing threshold (transition from off to lasing) is accurately represented. Regions of multistability are located where the transitions between states are comparatively noisy. The lasing threshold itself, i.e., the transition from the off to a lasing state, only marginally shifts to larger pump currents as the recovery rate is increased: At the slowest considered recovery rate $\gamma_{\text{ES}}^{\text{abs}} = 0.05\,\text{ps}^{-1}$, the lasing threshold is found at $P \approx 0.66\,\text{A}$, which then increases to $P \approx 0.73\,\text{A}$ at the fastest considered recovery rate $\gamma_{\text{ES}}^{\text{abs}} = 7.5\,\text{ps}^{-1}$.

Further examining the pump current scans (horizontal slices) for increasing absorber recovery rates $\gamma_{\text{ES}}^{\text{abs}}$, the mode-locking states appear in the same order as they do for an increasing absorber reverse bias. At absorber recovery rates below $\gamma_{\text{ES}}^{\text{abs}} \approx 0.06\,\text{ps}^{-1}$, the absorber stays permanently bleached for all pump currents beyond the lasing threshold, which results in continuous-wave (cw, teal) emission. For larger recovery rates, the HML3_{TEP} state (orange) appears in between the off and cw state and increases in width with a further increasing absorber recovery rate. At $\gamma_{\text{ES}}^{\text{abs}} \approx 0.2\,\text{ps}^{-1}$, the absorber recovers fast enough to absorb the trailing-edge of the HML3_{TEP} state, which allows for the HML3 (red) state to slot in between the off and HML3_{TEP} state. For increasing absorber recovery rates, the HML3 emission

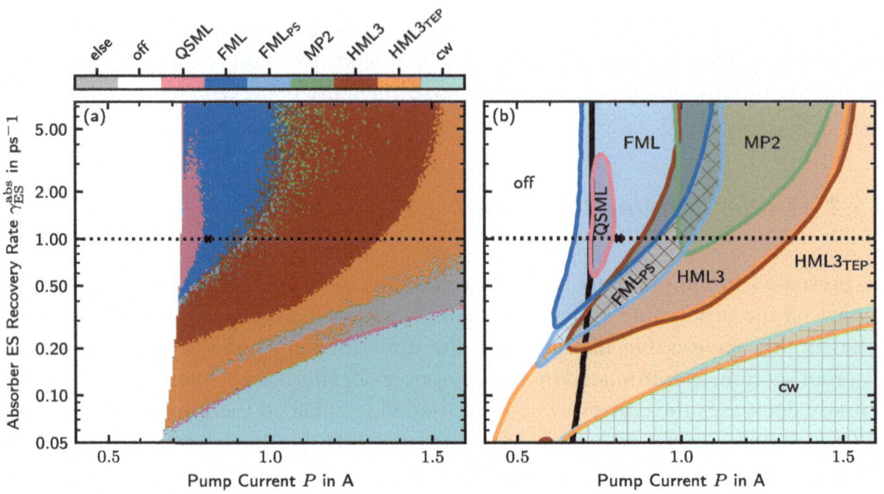

Fig. 4.22 Emission states observed in the parameter space (P, $\gamma_{\text{ES}}^{\text{abs}}$): off, Q-switched mode-locking (QSML), fundamental mode-locking (FML), fundamental mode-locking with a pulse-switching instability (FML_{PS}), two non-identical pulses (MP2), third-order harmonic mode-locking (HML3), third-order harmonic mode-locking with a strong trailing-edge plateau (HML3_{TEP}), and cw. **a** Laser (initial conditions) prepared in the off state. **b** Stable regions obtained from combined simulations with the laser prepared in the off, FML, HML3, and MP2 state. The black dotted lines correspond to $U = 6.0\,\text{V}$

state is observed for increasingly larger pump currents and is always followed by the HML3_{TEP} state beyond its upper pump current stability boundary. Hence, the complete suppression of the trailing-edge plateau always breaks down at sufficiently strong pump currents.

The fundamental mode-locking emission state (FML, blue) is observed at absorber recovery rates above $\gamma_{\text{ES}}^{\text{abs}} \approx 0.4\,\text{ps}^{-1}$, where at lower pump currents it becomes favorable for the laser to only support one circulating pulse within the laser cavity to minimize losses in the now more quickly recovering absorber section. Additionally, the closely related FML_{PS} state (light blue) appears beyond the upper pump current stability boundary of the FML state. With an increasing absorber recovery rate, the upper pump current boundary shifts from $P \approx 0.7\,\text{A}$ to $P \approx 1.0\,\text{A}$ at $\gamma_{\text{ES}}^{\text{abs}} \approx 2.0\,\text{ps}^{-1}$.

In between, however, Q-switched mode-locking (QSML, pink) appears at $\gamma_{\text{ES}}^{\text{abs}} \approx 0.5\,\text{ps}^{-1}$, reaches its maximal pump current extend at $\gamma_{\text{ES}}^{\text{abs}} \approx 1.2\,\text{ps}^{-1}$ and disappears again at absorber recovery rates above $\gamma_{\text{ES}}^{\text{abs}} \approx 4.0\,\text{ps}^{-1}$. Lastly, the two non-identical pulses state (MP2, green) first appears at $\gamma_{\text{ES}}^{\text{abs}} \approx 1.0\,\text{ps}^{-1}$ in the noisy transition region between FML and HML3 and becomes more prominent as the absorber recovery rate is increased.

A thorough depiction of the stability boundaries of the various mode-locking states is presented in Fig. 4.22b. The stability information is extracted by combining scans for which the laser has been prepared in the off, FML, HML3, and MP2 states, respectively, for each data point. Similar to the (P, U) parameter space, a complex structure of overlapping stability boundaries is also unraveled in the $(P, \gamma_{\mathrm{ES}}^{\mathrm{abs}})$ parameter space.

Depending on the absorber recovery rate, the off state is multistable with the HML3$_{\mathrm{TEP}}$, HML3, FML$_{\mathrm{PS}}$, or FML state. The respective pump current range for which the off state is multistable with a lasing state, however, reduces from $\approx$230 mA at $\gamma_{\mathrm{ES}}^{\mathrm{abs}} = 0.05\,\mathrm{ps}^{-1}$ to $\approx$30 mA at $\gamma_{\mathrm{ES}}^{\mathrm{abs}} = 7.5\,\mathrm{ps}^{-1}$. The subthreshold stability of the FML state also implies that the QSML state is always bistable with the FML state. This further supports that those states share no direct connection, i.e., one is not born in a bifurcation of the respective other state. The FML$_{\mathrm{PS}}$ state, on the other hand, always appears beyond the upper FML stability boundary or by itself at lower absorber recovery rates, where no stable FML emission exists. The pump current range, which produces FML$_{\mathrm{PS}}$ emission, however, reduces with an increasing absorber recovery rate from $\approx$130 mA at $\gamma_{\mathrm{ES}}^{\mathrm{abs}} = 0.5\,\mathrm{ps}^{-1}$ to $\approx$20 mA at $\gamma_{\mathrm{ES}}^{\mathrm{abs}} = 7.5\,\mathrm{ps}^{-1}$. This reduction is explained by a more efficient absorption of noise-induced perturbations, which are supported by excess gain, in between pulses. Furthermore, an overlap between regions of stable FML and HML3 is first observed at $\gamma_{\mathrm{ES}}^{\mathrm{abs}} \approx 0.5\,\mathrm{ps}^{-1}$ and from there on steadily increases with the absorber recovery rate. Lastly, the MP2 state is only observed for absorber recovery rates above $\gamma_{\mathrm{ES}}^{\mathrm{abs}} \approx 0.8\,\mathrm{ps}^{-1}$. The respective lower pump current stability boundary remains constant at $P \approx 1.0$ A, while the upper stability boundary increases with the absorber recovery rate up to $P \approx 1.47$ A at $\gamma_{\mathrm{ES}}^{\mathrm{abs}} = 7.5\,\mathrm{ps}^{-1}$. Note that the MP2 state is at no parameter combination the only stable solution: At lower pump currents it is multistable with the FML, FML$_{\mathrm{PS}}$, and HML3 state and at larger pump currents only with the HML3 state.

In summary, a sufficiently fast absorber recovery rate ($\gamma_{\mathrm{ES}}^{\mathrm{abs}} \gtrsim 0.3\,\mathrm{ps}^{-1}$) is necessary to observe fundamental mode-locking at all. Moreover, a fast absorber recovery rate increases the pump current range, at which fundamental mode-locking is observed. Increasing the absorber recovery rate, furthermore, leads to reduced subthreshold multistabilities and increases the dynamical diversity above threshold.

Absorber Redshift

The results obtained from a pump current and absorber redshift scans are presented in Fig. 4.23, where (a) shows the emission state, which has been obtained with the laser prepared in the off state for all data points. The absorber redshift is scanned from $\Delta\omega_{\mathrm{GS}}^{\mathrm{abs}} = 0.0\,\mathrm{ps}^{-1}$ to $\Delta\omega_{\mathrm{GS}}^{\mathrm{abs}} = 4.0\,\mathrm{ps}^{-1}$, which corresponds to a reverse bias scan from $U = 0$ V to $U \approx 11$ V. The absorber recovery rate is kept constant at $\gamma_{\mathrm{ES}}^{\mathrm{abs}} = 1.0\,\mathrm{ps}^{-1}$, which corresponds to an absorber reverse bias $U = 6.0$ V. This reverse bias, furthermore, corresponds to an absorber redshift of $\Delta\omega_{\mathrm{GS}}^{\mathrm{abs}} \approx 2.2\,\mathrm{ps}^{-1}$, which is indicated by horizontal black dotted lines.

For an increasing absorber redshift, the lasing threshold decreases from $P_{\mathrm{th}} \approx 1.2$ A at $\Delta\omega_{\mathrm{GS}}^{\mathrm{abs}} = 0.0\,\mathrm{ps}^{-1}$ down to $P_{\mathrm{th}} \approx 0.5$ A at $\Delta\omega_{\mathrm{GS}}^{\mathrm{abs}} = 10.0\,\mathrm{ps}^{-1}$. This effected is rooted in the increased effective gain, since the spectral absorption maximum is shifted away from the spectral amplification maximum. Furthermore, along the increasing absorber redshift, the observable mode-locking states, especially just

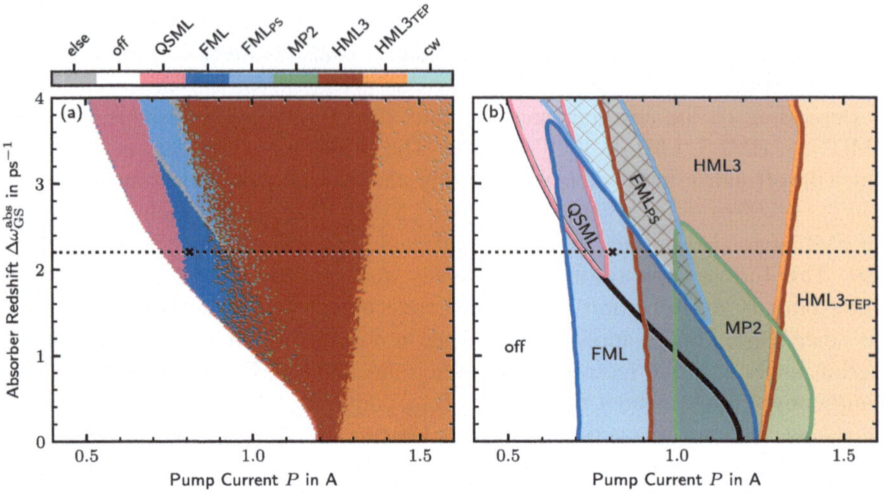

Fig. 4.23 Emission states observed in the $(P, \Delta\omega_{GS}^{abs}$ parameter space: off, Q-switched mode-locking (QSML), fundamental mode-locking (FML), fundamental mode-locking with a pulse-switching instability (FML_{PS}), two non-identical pulses (MP2), third-order harmonic mode-locking (HML3), and third-order harmonic mode-locking with a strong trailing-edge plateau ($HML3_{TEP}$). **a** Laser (initial conditions) prepared in the off state. **b** Stable regions obtained from combined simulations with the laser prepared in the off, FML, HML3, and MP2 state. The horizontal black dotted lines correspond to $U = 6.0$ V

beyond the lasing threshold, are subject to multiple changes. Below $\Delta\omega_{GS}^{abs} \approx 1.0\,\mathrm{ps}^{-1}$ the HML3 stabilizes behind the lasing threshold and transitions to the $HML3_{TEP}$ state at larger pump currents. Between $\Delta\omega_{GS}^{abs} \approx 1.0\,\mathrm{ps}^{-1}$ and $\Delta\omega_{GS}^{abs} \approx 1.8\,\mathrm{ps}^{-1}$ the laser operates in the FML state just beyond the threshold. At higher pump currents the FML state transitions into the FML_{PS} state and becomes bistable with the HML3 as indicated by the noisy transition region. Above $\Delta\omega_{GS}^{abs} \approx 1.0\,\mathrm{ps}^{-1}$, the QSML state appears first after the lasing threshold is crossed. The pump current range, in which QSML is observed, then steadily increases with the absorber redshift. On the other side, the increasing absorber redshift leads to a shift of the FML upper pump current stability boundary, i.e., the transition to FML_{PS} emission, to lower pump currents. As a result, the FML emission pump current range (lower to upper stability boundary) steadily decreases with the absorber redshift until the FML state becomes completely quenched between the QSML and FML_{PS} states at $\Delta\omega_{GS}^{abs} \approx 3.3\,\mathrm{ps}^{-1}$. The FML_{PS} emission state, on the other hand, becomes more prominent for absorber redshifts above $\Delta\omega_{GS}^{abs} \approx 2.5\,\mathrm{ps}^{-1}$, since the apparent bistability with the HML3 state reduces, which leaves a pump current range of $P \approx 100$ mA in which only FML_{PS} is observed.

The stability boundaries of the various emission states are portrayed in Fig. 4.23b, where the stability information is extracted from simulations with the laser prepared in the off, FML, HML, and MP2 state. Similar to the (P, U) and the (P, γ_{ES}^{abs}) parameter spaces, the absorber redshift $\Delta\omega_{GS}^{abs}$ has a profound impact on the observable dynamics. At small absorber redshifts, a large region of subthreshold multistability is found. At zero absorber redshift, the FML state stabilizes at $P \approx 0.7\,\mathrm{A}$, the HML3 state at $P \approx 0.9\,\mathrm{A}$ and the MP2 state at $P \approx 1.0\,\mathrm{A}$, while the lasing threshold is only found at $P \approx 1.2\,\mathrm{A}$. This yields a pump current range of $\approx 0.5\,\mathrm{A}$ of subthreshold multistability and a multistability between the off, FML, HML3, and MP2 state just below the lasing threshold.

For an increasing absorber redshift, the lower pump current stability boundaries of the FML, HML3, and MP2 states only marginally shift to smaller pump currents, while the lasing threshold (black solid line) shifts much more strongly and crosses all three lower stability boundaries, such that no more subthreshold multistability is observed for absorber redshifts above $\Delta\omega_{\mathrm{GS}}^{\mathrm{abs}} \approx 2.6\,\mathrm{ps}^{-1}$.

An increasing absorber redshift, furthermore, causes the upper stability boundaries of the FML and MP2 states to shift to lower pump currents up the point, where they meet and connect with the lower stability boundary. Thus, no MP2 emission is observed for absorber redshifts above $\Delta\omega_{\mathrm{GS}}^{\mathrm{abs}} \approx 2.5\,\mathrm{ps}^{-1}$ and no FML emission for absorber redshifts above $\Delta\omega_{\mathrm{GS}}^{\mathrm{abs}} \approx 3.7\,\mathrm{ps}^{-1}$. Unlike the FML and MP2 states, the HML3 emission state increases its stable pump current range with an increasing absorber redshift, while simultaneously the overlap with other emission states, i.e., multistability, is reduced.

The $\mathrm{FML}_{\mathrm{PS}}$ and QSML emission states, on the other hand, only appear at sufficient absorber redshifts. The QSML state is observed beyond the lasing threshold for absorber redshifts above $\Delta\omega_{\mathrm{GS}}^{\mathrm{abs}} \approx 1.9\,\mathrm{ps}^{-1}$ and the $\mathrm{FML}_{\mathrm{PS}}$ is found beyond the upper FML stability boundary for absorber redshifts above $\Delta\omega_{\mathrm{GS}}^{\mathrm{abs}} \approx 1.3\,\mathrm{ps}^{-1}$. Hence, both can be interpreted as instabilities, which are induced by the absorber redshift, i.e., the detuning between the absorber and gain section. The $\mathrm{FML}_{\mathrm{PS}}$ state in particular always appears to be connected to the FML state, which lends the interpretation that it is born in a bifurcation of the FML state. The QSML state, however, always appears at the lasing threshold and does not seem to be directly connected to any other mode-locking state.

In summary, small absorber redshifts produce large regions of subthreshold multistabilities as well as large regions of FML emission. Increasing the absorber redshift moves the lasing threshold to smaller pump currents and thereby eliminates the subthreshold lasing states, but also destabilizes the FML and MP2 emission states. FML emission is only monostable at intermediate absorber redshifts and is therefore predominately found in those regions if the laser has been prepared in the off state.

Based on those results, it can be concluded that the reverse bias induced absorber section redshift is a crucial modeling parameter for the three-section quantum-dot laser at hand. Note that the absorber redshift can not be described in the conventional ring geometry delay-differential equation framework, where all spectral features are summarized by one lumped-element Lorentzian filtering element [37, 59–61, 63, 64]. Hence, the importance of this effect justifies the chosen modeling hierarchy. Taking a look at the literature, a number of published traveling-wave based laser models allow for the inclusions of the absorber redshift [14, 65, 67–69], but are not used to systematically study its effects. The only exception is Ref. [9], which reports optimal mode-locking performance for intermediate redshift values. Those results are in accordance with this work, where intermediate reverse bias values produce the best mode-locking characteristics (see Fig. 4.17).

Conclusion

Finally, the insights gained from the separate studies of the absorber recovery rate $\gamma_{\mathrm{ES}}^{\mathrm{abs}}$ and the absorber redshift $\Delta\omega_{\mathrm{GS}}^{\mathrm{abs}}$ can be combined in order to understand the mode-locking dynamics obtained in the pump current and reverse bias (P, U) parameter space. For both small absorber redshifts and recovery rates, pronounced subthreshold multistabilities exist and, henceforth, are also observed at small absorber reverse biases. Regarding the FML state, however, the absorber recovery time and redshift have opposite effects. While an increasing recovery rate stabilizes the FML state, an increasing absorber redshift destabilizes the FML state. Therefore, a window of stable FML emission is observed along an increasing absorber reverse bias, which exhibits a maximum pump current stability range at intermediate reverse biases ($U \approx 6.0$ V). The interplay between the recovery time and the redshift thus produces the characteristic shape of the FML emission region, which has been discussed in Sect. 4.5.1.

A similar argument also applies to the MP2 state. The HML3 state, on the other hand, does not follow this behavior, most likely due to the geometric configuration of the laser. The positioning of the absorber section allows for colliding pulses in HML3 operation, which gives this state an unique advantage over the other mode-locking states.

4.5.3 Spontaneous Emission Noise Induced Dynamics

The stability of the various emission states in the presence of spontaneous emission noise is investigated in this last subsection. The spontaneous emission contribution to the lasing mode can be explicitly included in the model if the lifetimes of the optically active states and the β-factor are known [69, 73]. Measuring or calculating the β-factor for this tapered guided laser cavity unfortunately turns out to be a non-trivial task [99]. Hence, the approach of this work is to adjust the spontaneous emission noise parameter D, such that the observed dynamics line up with the experimental data.

The effect of the spontaneous emission noise is twofold: Firstly, sufficient noise can prevent marginally stable states from being observed by preventing the system from settling on the respective attractor.[5] Secondly, spontaneous emission noise leads to fluctuations of the pulse positions and pulse amplitudes.

The former effect is known as pulse timing jitter and will be studied and discussed in detail in the following Sect. 4.6 of this chapter. The final chosen spontaneous

[5] A dynamical system can not settle on an attractor if a stochastic process dominates the relaxation on the attractor and drives the system's state beyond the basin of attraction.

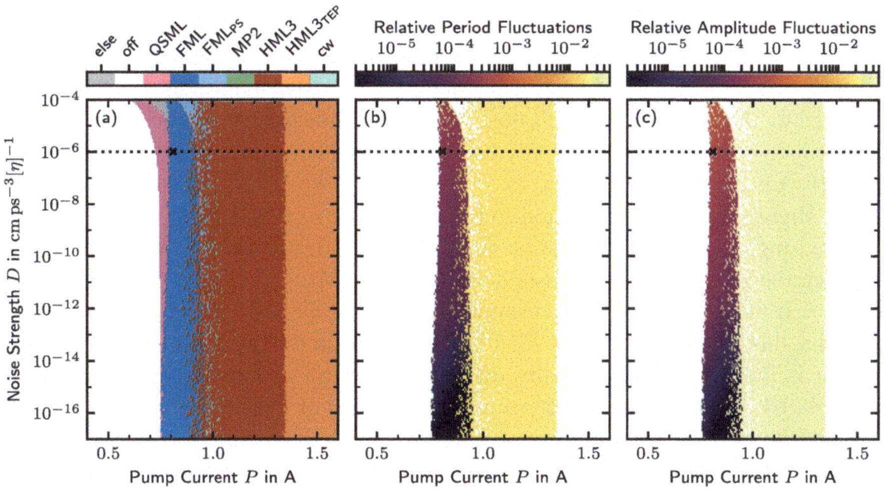

Fig. 4.24 Mode-locking characteristics in the pump current and spontaneous emission noise strength parameter space (P, D): **a** emission states obtained after turn-on: off, Q-switched mode-locking (QSML), fundamental mode-locking (FML), fundamental mode-locking with a pulse-switching instability (FML_{PS}), two non-identical pulses (MP2), third-order harmonic mode-locking (HML3), and third-order harmonic mode-locking with a strong trailing-edge plateau ($HML3_{TEP}$). Subfigures **b**, **c** show relative (to the mean) pulse period and pulse amplitude fluctuations, respectively, of the FML and HML3 states. The horizontal black dotted lines correspond to the default noise strength $D = 1.0^{-6}\,\mathrm{cm\,ps^{-1}}[\eta]^{-1}$ and the black cross indicates parameters used for further investigation

emission noise strength then aims to reproduce the observed emission states as well as the timing jitter values of the experiment.

To study the impact of the spontaneous emission noise on the observable emission states, scans of the pump current and spontaneous emission noise strength (P, D) parameter space are performed. For each data point, the laser has been prepared in the off state. The reverse bias has been set to $U = 6\,\mathrm{V}$ (indicated by the black dotted lines in Fig. 4.17). The recorded emission states are presented in Fig. 4.24a via the same color code as in the previous subsections. The spontaneous emission noise strength is scanned over 13 orders of magnitude from $10^{-17}\,\mathrm{cm\,ps^{-1}}[\eta]^{-1}$ to $10^{-4}\,\mathrm{cm\,ps^{-1}}[\eta]^{-1}$.

At small noise strengths the laser immediately emits in the FML state beyond the lasing threshold and no QSML is observed. Despite very small noise strengths, a noisy region in between homogeneous FML and HMl3 regions, where the emission states vary between the FML, FML_{PS}, MP2, and HML3, is found. This indicates a subtle sensitivity of the transient systems' trajectory to the spontaneous emission noise if the laser has been prepared in the off state. Thus, small perturbations shortly after the turn-on can lead to different final states if multistabilities exist.

For spontaneous emission noise strengths $D \gtrsim 10^{-14}\,\mathrm{cm\,ps^{-1}}[\eta]^{-1}$ QSML appears just beyond the lasing threshold, i.e., in between the off and the FML state.

For an increasing noise strength the QSML pump current range increases as both the lasing threshold shifts the lower pump currents and the transition to FML shifts to larger pump currents. The decreased lasing threshold is due to the additional photons produced by the spontaneous emission. This effect, however, is almost negligible at intermediate noise strengths and only becomes pronounced for noise strengths above $D \approx 10^{-7}\,\text{cm ps}^{-1}[\eta]^{-1}$.

The slight shift of the transition to the FML state to higher pump currents with an increasing noise strength must be attributed to the transient dynamics after turn-on, since the FML state is multistable with the QSML state at the chosen reverse bias $U = 6\,\text{V}$. Together with the disappearance of the QSML state at very low noise strengths, this motivates the classification of QSML in this device as a noise driven state. It is therefore fundamentally different from the simulated QSML states in Refs. [59, 64, 104], which are deterministic solutions of a DDE system.

While the next section presents an in-depth analysis of the timing jitter, Fig. 4.24b, c provide a brief overview of the noise-induced pulse-train fluctuation in the FML and HML3 state. Figure 4.24b plots the relative (to the mean) pulse period standard deviation and Fig. 4.24c the relative (to the mean) pulse amplitude (pulse peak power) standard deviation. Both measures exhibit a qualitatively similar behavior across the two parameters.

In the FML region the fluctuations first increase with the noise strength and then reach a plateau for noise strengths above $D \approx 10^{-10}\,\text{cm ps}^{-1}[\eta]^{-1}$. Within the initial increase, the fluctuations scale linearly with the square root of the noise strength. Reviewing Eq. (4.24), the spontaneous emission noise enters the system with the square root of the noise strength parameter D. On that account, the fluctuations within the initial increase represent the linear response of the system to small noise-induced perturbations. Consequently, the stalling increase and plateau at larger noise strengths is then created in a nonlinear regime. Both regimes, furthermore, differ in their pump current dependence: At small noise strengths, the fluctuations decrease with an increasing pump current, while at larger pump currents the fluctuations appear almost constant along an increasing pump current.

The HML3 region, on the other hand, exhibits a behavior totally different from the FML region. Both the period and amplitude fluctuations are constant and relatively large across the noise strength and pump current parameters. This indicates an underlying deterministic dynamical effect, which prevents clean mode-locking and dominates the effects of the stochastic spontaneous emission noise. Due to the large fluctuations in the HML3 region, the timing jitter analysis in the following section focuses exclusively on the FML region.

Finally, the parameter choice of $D = 10^{-6}\,\text{cm ps}^{-1}[\eta]^{-1}$ as the default spontaneous emission noise strength is motivated by achieving timing jitter values comparable to the experiment and simultaneously producing similar regions of mode-locking dynamics, i.e., reproducing the experimentally measured sequence of QSML, FML, FML_{PS}, and HML3 emission states.

4.6 Timing Jitter Analysis

Having introduced, developed, and benchmarked a diverse timing-jitter characterization toolbox in Chap. 3, this section presents its application to the three-section quantum-dot laser in the fundamental mode-locking regime with varying laser parameters.

4.6.1 *Spontaneous Emission Noise Dependence of the Timing Jitter*

The first laser parameter to be investigated is the spontaneous emission noise strength D. This quantity is not an operation parameter that can be easily tuned in the experiment, but rather results from lifetimes of the optical transitions (Einstein coefficient) and the β-factor, i.e., the probability that a spontaneously emitted photon ends up in one of the lasing modes. However, the calculation of the β-factor is a non-trivial task for complex tapered laser geometries [21, 99]. Hence, rather than trying to calculate the β-factor, the approach chosen in this work is to adjust the spontaneous emission noise strength D such that the simulated timing jitter reproduces the experimental results, which are presented in Sect. 4.4. Note, however, that the choice of the spontaneous emission noise strength D is also constrained by the stability of the various mode-locking states as presented in Sect. 4.5.3.

Even though the performance analysis of the various long-term timing jitter estimation methods presented in Sect. 3.6 favors the Kéfélian and pulse-period fluctuation autocorrelation method, all long-term timing jitter estimation methods are utilized in this section for the sake of completeness and to to provide additional characteristics.

On that account, the results obtained from the direct time-domain (blue solid line, Sect. 3.2), the von der Linde/Otto (orange dashed line, Sect. 3.3.1), the Kéfélian (green dashed line, Sect. 3.3.2), and the hybrid von der Linde/Otto (red dashed line, Sect. 3.4) methods are presented in Fig. 4.25. The long-term timing jitter obtained from the direct time-domain method produces the reference values against which the other methods must be tested. The laser is operated at $P = 0.81$ A, $U = 6$ V (indicated by black crosses in Figs. 4.17, 4.22, 4.23 and 4.24). Each data point has been obtained from 100 time series realizations, each 5 μs long ($\approx$65150 pulses).

The discussion firstly focuses on the results, which are obtained from the direct time-domain method: For an increasing spontaneous emission noise strength D, the long-term timing jitter experiences two regimes: For noise strengths $D \lesssim 10^{-12}\,\mathrm{cm\,ps^{-1}}[\eta]^{-1}$, the long-term timing jitter increases with the square root of the noise strength, corresponding to a slope of $1/2$ in the log-log plot. Since the spontaneous emission noise enters Eq. (4.24) with a square root, this regime is char-

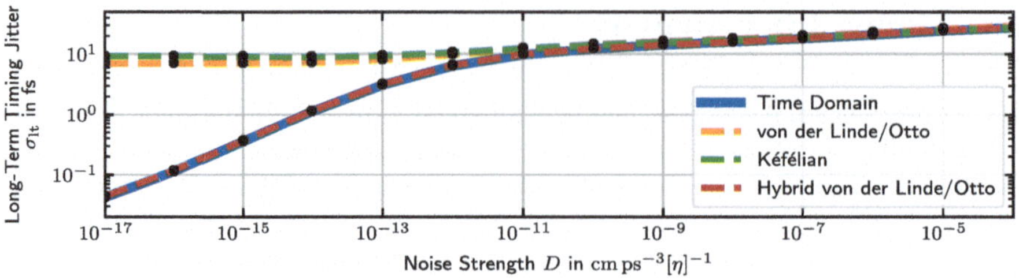

Fig. 4.25 Spontaneous emission noise strength dependence of the long-term timing jitter. Comparison between the different estimation methods: Direct time-domain method from Δt_n (blue line), von der Linde/Otto method from the 20ths harmonic of the power spectrum $S_{|E|^2}$ (orange dashed line), Kéfélian method obtained from the 20ths harmonic (green dashed line), and hybrid von der Linde/Otto method from the spectrum $S_{\Delta t_n}$ (red dashed line). Individual data points (denoted by black dots) obtained from 100 realizations, each 5 µs long. rms timing jitter integration range from 5 MHz to 6 GHz. Laser parameters: $P = 0.81$ A and $U = 6$ V

acterized by a linear response of the long-term timing jitter to the noise-induced perturbations.[6]

For spontaneous emission noise strengths $D \gtrsim 10^{-12}$ cm ps$^{-1}[\eta]^{-1}$, on the other hand, the long-term timing jitter levels off and increases much slower with a slope of $\approx 1/10$ in the log-log plot of Fig. 4.26b. Thus, the long-term timing jitter only increases from $\sigma_{\mathrm{lt}} \approx 10$ fs by a factor of three to $\sigma_{\mathrm{lt}} \approx 30$ fs, while the spontaneous emission noise strength increases by seven orders of magnitude from $D = 10^{-11}$ cm ps$^{-1}[\eta]^{-1}$ to $D \gtrsim 10^{-4}$ cm ps$^{-1}[\eta]^{-1}$. Based on this observation, the timing jitter represents a nonlinear response of the laser to spontaneous emission noise-induced perturbations for noise strengths $D \gtrsim 10^{-12}$ cm ps$^{-1}[\eta]^{-1}$.

Comparing the results obtained from the von der Linde/Otto, Kéfélian, and hybrid von der Linde/Otto method to the reference values produced by the direct time-domain method reveals that, for the given evaluation parameters, only the hybrid von der Linde/Otto method reproduces the proper long-term timing jitter values over the complete spontaneous emission noise strength interval. The von der Linde/Otto and Kéfélian methods, which rely on the power spectral density $S_{|E|^2}$, vastly overestimate the long-term timing jitter at small noise strengths D and only agree with the direct time-domain method for noise strengths $D \gtrsim 10^{-10}$ cm ps$^{-1}[\eta]^{-1}$. The failure of the two spectral methods at low noise strengths is caused by spectral leakage, as discussed in Sect. 3.3. At low noise strengths, and thus small timing jitter values, the spectral resolution becomes insufficient and thereby amplifies the spectral leakage issues. For the sake of comparing the various methods, the erroneous timing jitter estimates produced by the two spectral methods are specifically shown to highlight

[6] A semi-analytical method [138, 139] for the calculation of the long-term timing jitter could, in principle, be applied to the laser in the linear regime. The spatial discretization of the laser model at hand, however, creates a large number of dynamical equations and would therefore severely complicate the implementation and the computations.

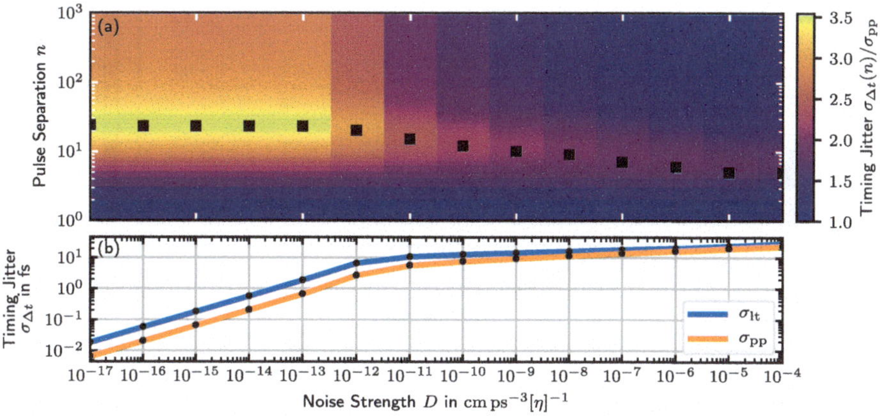

Fig. 4.26 Spontaneous emission noise dependence of the timing jitter: **a** color-coded pulse separation dependent timing jitter $\sigma_{\Delta t}(n)$ normalized to the respective pulse-to-pulse timing jitter σ_{pp}. Maxima are indicated by black squares. **b** Long-term timing jitter σ_{lt} (blue line, corresponding to the top row in (**a**)) and **b** pulse-to-pulse timing jitter σ_{pp} (orange line, corresponding to the bottom row in (**a**)). Individual data points (denoted by black dots) are obtained from 100 realizations, each 5 μs long. Laser parameters: $P = 0.81$ A and $U = 6$ V

the importance of the reference value produced by the direct time-domain method and the versatility of the hybrid von der Linde/Otto method.

A closer look at the data obtained from the direct time-domain method is taken in Fig. 4.26. Subfigure (a) shows the color-coded pulse separation dependent timing jitter $\sigma_{\Delta t}(n)$, which has been normalized to the pulse-to-pulse timing jitter σ_{pp}. (b) Shows the long-term timing jitter σ_{lt} (blue line) corresponding to the top row in (a) and the pulse-to-pulse timing jitter σ_{pp} (orange line) corresponding to the bottom row in (a), both without the normalization.

Besides the slope of the timing jitter increase, the linear and nonlinear regimes can also be well distinguished by changes of the pulse separation n dependent timing jitter $\sigma_{\Delta t}(n)$, presented in Fig. 4.26a. In the linear regime $D \lesssim 10^{-12}$ cm ps$^{-1}[\eta]^{-1}$ the timing jitter $\sigma_{\Delta t}(n)/\sigma_{pp}$ does not depend on the spontaneous emission noise strength D. However, once the nonlinear regime is reached, the maxima of $\sigma_{\Delta t}(n)$, indicated by black squares, shift to smaller n and decrease in magnitude. Both features suggest a change in the pulse-period correlations induced by the nonlinear response.

Similarly, the change in regimes can also be identified via the ratio σ_{lt}/σ_{pp} of the long-term and pulse-to-pulse timing jitter (top row in Fig. 4.26a), which results from the pulse-period correlations as detailed in Sect. 3.5. Within the linear regime, the ratio remains constant at $\sigma_{lt}/\sigma_{pp} \approx 2.8$. Once the nonlinear regime is entered, however, the ratio decreases to $\sigma_{lt}/\sigma_{pp} \approx 1.2$ at $D = 10^{-4}$ cm ps$^{-1}[\eta]^{-1}$.

Lastly, the results from the pulse-period fluctuation autocorrelation method applied to the spontaneous emission noise strength scan are presented in Fig. 4.27. The respective autocorrelation functions $\Psi_{\delta T}(n)$ are color coded and plotted for the first $100\,n$ in (a). As anticipated, based in the pulse separation dependent timing

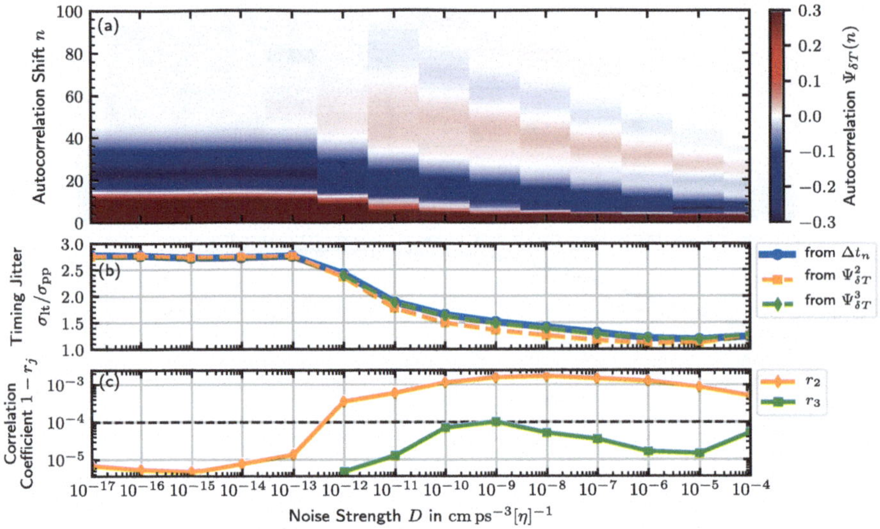

Fig. 4.27 **a** Pulse-period fluctuation autocorrelation function $\Psi_{\delta T}$ as a function of the spontaneous emission noise strength D. **b** Long-term timing jitter σ_{lt} normalized by the pulse-to-pulse timing jitter σ_{pp} calculated from $\{\Delta t_n\}$ (blue solid line) and fitted model function for $j = 2, 3$ (orange and blue dashed lines, respectively). **c** Deviation of the Pearson correlation coefficient r_j from unity. Small values indicate a high quality of the model function $\Psi_{\delta T}^j$ and the fit parameters. The black dashed line denotes the threshold, beyond which $\Psi_{\delta T}^3$ was used. Laser parameters: $P = 0.81\,\text{A}$, $U = 6\,\text{V}$

jitter $\sigma_{\Delta t}(n)$ (see Fig. 4.26a), the autocorrelation function qualitatively and quantitatively changes with the transition from the linear to the nonlinear regime. At small spontaneous emission noise strengths $D \lesssim 10^{-12}\,\text{cm}\,\text{ps}^{-1}[\eta]^{-1}$, the autocorrelation exhibits one drop to negative values with a minimum of $\Psi_{\delta T}(n) \approx -0.3$ at $n \approx 25$, after which the autocorrelation function relaxes to $\Psi_{\delta T}(n) \approx 0$ for $n \gtrsim 40$. Within the linear regime, the autocorrelation function does not depend on the noise strength D. Hence, changes in the long-term timing jitter σ_{lt} are exclusively caused by changes of the pulse-to-pulse timing jitter σ_{lt}.

Once the nonlinear regime is entered additional features in the pulse-period fluctuation autocorrelation function appear: After the drop below zero, the autocorrelation does not relax to zero, but exhibits another rise to positive values and for intermediate noise strengths $10^{-10}\,\text{cm}\,\text{ps}^{-1}[\eta]^{-1} \lesssim D \lesssim 10^{-6}\,\text{cm}\,\text{ps}^{-1}[\eta]^{-1}$ another small dip below zero. Moreover, with an increasing spontaneous emission noise strength D, the autocorrelation function features, e.g., the zero crossings and the local extrema, move to smaller autocorrelation shifts n. The final relaxation to $\Psi_{\delta T} \approx 0$, however, is shifted to larger n due to the additional features. Thus, the long-term timing jitter σ_{lt} changes within the nonlinear regime both due to changes of the pulse-to-pulse timing jitter σ_{pp} and due to changes of the pulse-period correlations.

To calculate the long-term timing jitter from the respective pulse-period fluctuation autocorrelation functions, the recipe proposed in Sect. 3.5 is applied. Figure 4.27b plots the ratios $\sigma_{\mathrm{lt}}/\sigma_{\mathrm{pp}}$ obtained from the direct time-domain method (blue solid line) and the pulse-period fluctuation autocorrelation method with the fitted model function $\Psi^{j}_{\delta T}$ with $j = 2$ (orange dashed line) and $j = 3$ (green dashed line). The ratio $\sigma_{\mathrm{lt}}/\sigma_{\mathrm{pp}}$ with $j = 3$ has only been computed for spontaneous emission noise strengths, where the Pearson correlation coefficient r_2 exceeds the proposed threshold $r^{\mathrm{th}}_j = 1-10^{-4}$, while the ratio $\sigma_{\mathrm{lt}}/\sigma_{\mathrm{pp}}$ with $j = 2$ has been computed for all noise strengths for comparison. Additionally, the deviation of the Pearson correlation coefficient from unity $1 - r_j$ is depicted in Fig. 4.27c for $j = 2$ (orange line) and $j = 3$ (green line), where applicable. The threshold r^{th}_j is indicated by a black dashed line.

Within the linear regime, $\sigma_{\mathrm{lt}}/\sigma_{\mathrm{pp}}$ obtained from the direct time-domain method remains constant at $\approx$2.8, as previously described, and is perfectly reproduced by the pulse-period fluctuation autocorrelation method with $j = 2$. The corresponding correlation coefficient deviations are on the order of $1 - r_2 \approx 10^{-5}$. The transition to the nonlinear regime complicates the respective autocorrelation functions by features, which can not be properly reproduced by $\Psi^{2}_{\delta T}$. Hence, the correlation coefficient deviations increase to $1 - r_2 \approx 10^{-3}$ and the resulting ratios $\sigma_{\mathrm{lt}}/\sigma_{\mathrm{pp}}$ are underestimated. The transition to the nonlinear regime thus invokes the use of the higher order $j = 3$, which then sufficiently reproduces all features of $\Psi_{\delta T}$ such that $\sigma_{\mathrm{lt}}/\sigma_{\mathrm{pp}}$ is accurately reproduced for all $D \gtrsim 10^{-12}\,\mathrm{cm\,ps^{-1}}[\eta]^{-1}$. The corresponding correlation coefficients r_3 accordingly satisfies $r_3 \gtrsim r^{\mathrm{th}}_j$ for all noise strengths D. The decrease of the ratio $\sigma_{\mathrm{lt}}/\sigma_{\mathrm{pp}}$ in the nonlinear regime can mostly be attributed to faster initial drop of the autocorrelation function $\Psi_{\delta T}$.

Model Function Validity at Low Spontaneous Emission Noise Strengths

Due to the qualitatively different spontaneous emission noise strength regimes, the pulse-period autocorrelation method is also illustrated in detail for $D = 10^{-14}\,\mathrm{cm\,ps^{-1}}[\eta]^{-1}$ in Fig. 4.28. Subfigure (a) shows the computed autocorrelation function $\Psi_{\delta T}$ (solid black line) and fitted model functions for $j = 1$ and $j = 2$ (blue and orange dashed lines, respectively). The model functions $\Psi^{1,2}_{\delta T}$ were fitted by using the first 100 autocorrelation shifts n.

A drop of the autocorrelation function with a zero crossing at $n \approx 13$ and a local minimum at $n \approx 23$ is observed, which is followed by a simple relaxation to zero. The behavior is accurately reproduced by the model function for $j = 2$ and only approximately for $j = 1$. The latter case especially fails to reproduce the simple relaxation to zero and exhibits an additional overshoot to positive correlation values instead.

The resulting pulse separation dependent timing jitter is shown in Fig. 4.28b. Starting with a pulse-to-pulse jitter of $\sigma_{\mathrm{pp}} \approx 0.21$ fs, the directly computed timing jitter (solid gray line) first rises to $\approx$0.75 fs and then relaxes to its long-term value $\sigma_{\mathrm{lt}} \approx 0.57$ fs. While the qualitative behavior is captured well by both model functions, the maximum and the long-term value are only correctly described by $j = 2$. While the long-term timing jitter estimate provided by model function $\Psi^{1}_{\delta T}$ only marginally

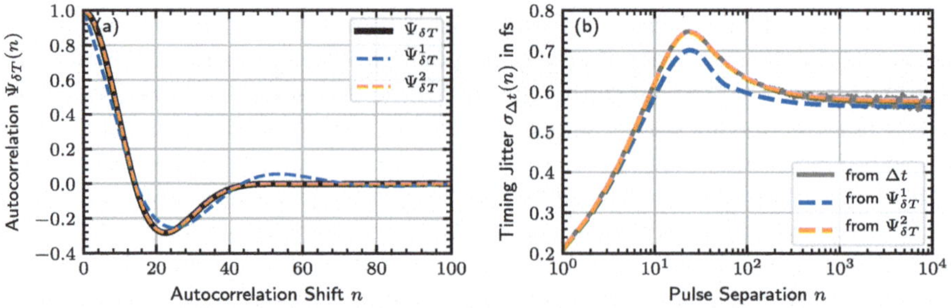

Fig. 4.28 **a** Autocorrelation function of the pulse-period fluctuations calculated from $\{\delta T_k\}$ (solid black line) and fitted model functions $\Psi^j_{\delta T}$ for $j = 1$ and $j = 2$ (blue and orange dashed lines, respectively). **b** Timing jitter as a function of the pulse separation calculated from $\{\Delta t_n\}$ and the fitted model functions. Laser parameters: $P = 0.81$ A, $U = 6$ V, and $D = 10^{-14}$ cm ps$^{-1}[\eta]^{-1}$

differs from the value obtained from the direct time-domain method, it is important to note that this results from multiple deviations $\Psi^1_{\delta T}$, which cancel each other.

In conclusion, the proposed model function and the pulse-period fluctuation autocorrelation method have proven to be valid over a large range of spontaneous emission noise strengths.

4.6.2 *Timing Jitter Along the Pulse Propagation*

In the next step, the local contributions to the long-term timing jitter of the emitted pulse train are studied. While the long-term timing jitter of the circulating pulse within the laser cavity must be the same at every point along the propagation, the pulse-to-pulse timing jitter σ_{pp} and the pulse-period fluctuation can vary. Hence, Fig. 4.29 comprehensively presents the evolution of timing fluctuations along the unfolded cavity, where (a) depicts the pulse-to-pulse timing jitter σ_{pp} (red line) and correlation contribution ratio σ_{lt}/σ_{pp} (blue line), (b) the color-coded pulse separation dependent timing jitter $\sigma_{\Delta t}(n)$ normalized to the pulse-to-pulse timing jitter σ_{pp}, and (c) the pulse-period fluctuation autocorrelation functions. The results have been obtained from 2000 time series, where each is 80 ns long ($\approx 2.13 \times 10^6$ pulses).

Starting the discussion with the pulse-to-pulse timing jitter (red line in Fig. 4.29a), an improvement is observed in the gain sections and a deterioration in the absorber section. The pulse-to-pulse jitter values at the very ends, i.e., $z = \pm 3$ mm, correspond to the out-coupled pulse-to-pulse timing jitter. Starting with the backward-traveling pulse at the out-coupling facet (left side of the plot), the timing jitter reduces from $\sigma_{pp} \approx 16.7$ fs to $\sigma_{pp} \approx 13.6$ fs in the gain section and then increases to $\sigma_{pp} \approx 17.0$ fs in the absorber section. Within the straight gain section, the timing jitter reduces again to $\sigma_{pp} \approx 13.7$ fs before it rises to its global maximum at $\sigma_{pp} \approx 21.0$ fs at the end of the absorber section and finally reduces again to its out-coupling value of

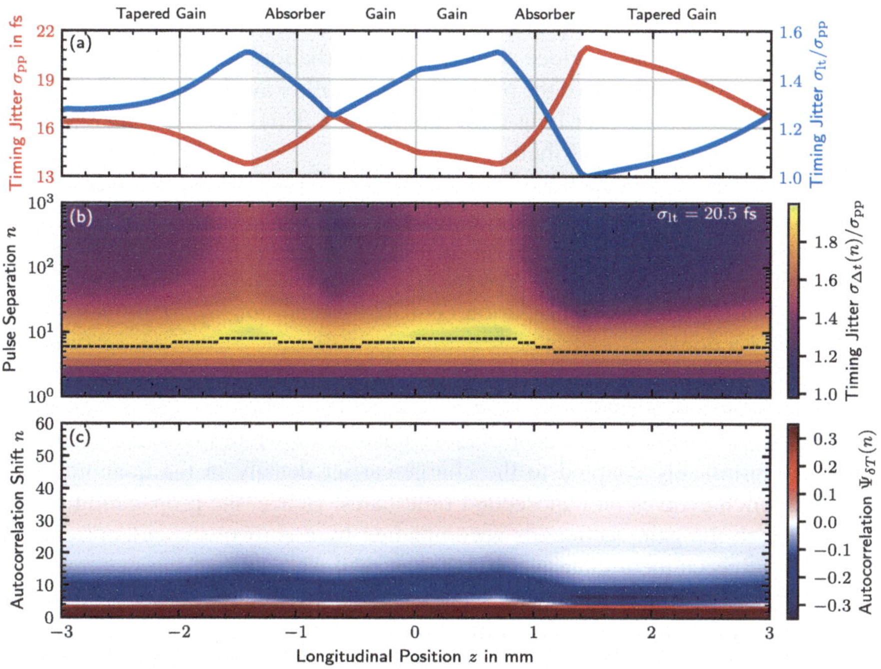

Fig. 4.29 Spatially resolved timing jitter evolution along one round trip in the unfolded cavity. The left side of all subplots shows the backward-moving field (denoted by negative positions z) and the right side shows the forward-moving field (denoted by positive positions z). Subfigure **a** depicts the pulse-to-pulse timing jitter σ_{pp} (red line) and correlation contribution ratio $\sigma_{\mathrm{lt}}/\sigma_{\mathrm{pp}}$ (blue line). Subfigure **b** plots the color-coded pulse separation dependent timing jitter $\sigma_{\Delta t}(n)$ normalized to the pulse-to-pulse timing jitter σ_{pp}. Subfigure **c** presents the pulse-period fluctuation autocorrelation functions. The results have been obtained from 2000 time series, each 80 ns long. Laser parameters: $P = 0.81$ A, $U = 6$ V, and $D = 10^{-6}$ cm ps$^{-1}[\eta]^{-1}$

$\sigma_{\mathrm{pp}} \approx 16.7$ fs. The correlation contribution ratio $\sigma_{\mathrm{lt}}/\sigma_{\mathrm{pp}}$ (blue line in in Fig. 4.29a) must compensate the changes of the pulse-to-pulse timing jitter to yield a position independent long-term timing jitter. It therefore inversely mirrors the behavior of the pulse-to-pulse timing jitter.

The evolution of the pulse-to-pulse timing jitter along the propagation can be understood by acknowledging that it is related to the shifts of the pulse position due to the interplay of the saturable gain and absorption and the non-saturable absorption, as seen in Fig. 4.10a, b via the respective recovery processes. If a pulse arrives slightly too early with respect to the previous pulse within a gain section, the available gain is slightly smaller since it has not recovered as much. Thus, the saturation effect is stronger and the pulse is further shifted towards earlier times. This reduces the standard deviation of the pulse distances, which defines the pulse-to-pulse timing jitter. Similarly, if a pulse arrives slightly too late, the gain has recovered more and the pulse is not shifted as much towards earlier times, which again reduces the pulse-to-pulse timing jitter. Hence, perturbations from the ideal round-trip time (clock time) in the gain sections produce succeeding perturbations that point in the same direction in time, and thus drive the pulse positions further away from the ideal pulse train and act towards a higher long-term timing jitter. This effect also manifests in the increasing correlation contribution ratio $\sigma_{\mathrm{lt}}/\sigma_{\mathrm{pp}}$.

In the absorber section, on the other hand, the pulse-position shifting mechanism works exactly in the other direction, since saturation effects push the pulses towards later and not towards earlier times. Hence, the pulse-to-pulse timing jitter increases in the absorber section, but the pulse positions are naturally pushed back towards the ideal pulse train. Hence, the correlation contribution ratio $\sigma_{\mathrm{lt}}/\sigma_{\mathrm{pp}}$ improves within the absorber section.

The gain and absorber recovery induces a memory effect that manifests in the pulse separation dependent timing jitter $\sigma_{\Delta t}(n)$, which is shown in Fig. 4.29b and in the pulse-period fluctuation autocorrelation function shown in Fig. 4.29c. The gain recovery is ultimately coupled to the charge-carrier density in the quantum-well reservoir, which evolves on nanosecond timescales. Hence, the propagation of the pulses through the gain sections is accompanied by a shift of the autocorrelation features to larger autocorrelation shifts Δn as well as a shift of the maximum in the timing jitter $\sigma_{\Delta t}(n)$ to larger pulse separations n. In the absorber section, on the other hand, the recovery is determined by the quantum-dot excited state escape rates (carrier sweep-out rates). Depending on the reverse bias, the escape rates vary from tens of picoseconds to hundreds of femtoseconds [94, 96] and are therefore always significantly faster than the gain sections quantum-well dynamics. Hence, the autocorrelation features move to smaller autocorrelation shifts Δn and the timing jitter maxima move to smaller pulse separations n within the absorber section.

In conclusion, the timing jitter evolution along the pulse propagation is tightly bound to the pulse-shaping mechanism. The gain and absorber saturation dynamics not only determine the pulse shaping, but also how pulses react to spontaneous emission induced perturbations of their positions. For the investigated pulse-shaping mechanism of the three-section tapered quantum-dot laser, the gain sections act towards a smaller pulse-to-pulse but a larger long-term timing jitter and the absorber section acts towards a larger pulse-to-pulse and a smaller long-term timing jitter. The resulting long-term timing jitter σ_{lt} then results from the interplay of all contributions.

4.6.3 Timing Jitter in the Operation Parameter Space

In this last subsection, the long-term timing jitter σ_{lt} is studied in the operation parameter space of the mode-locked laser, i.e., the pump current P and absorber reverse bias U. The strength of the pulse-period fluctuation autocorrelation function method is demonstrated by computing the long-term timing jitter σ_{lt} in the strong spontaneous emission noise regime at $D = 10^{-6}\,\mathrm{cm\,ps}^{-1}[\eta]^{-1}$ for the fundamental mode-locking regime depicted in Fig. 4.17b. The results are shown in Fig. 4.30a. Each data point has been computed from 2.5×10^5 pulses, which by extrapolating

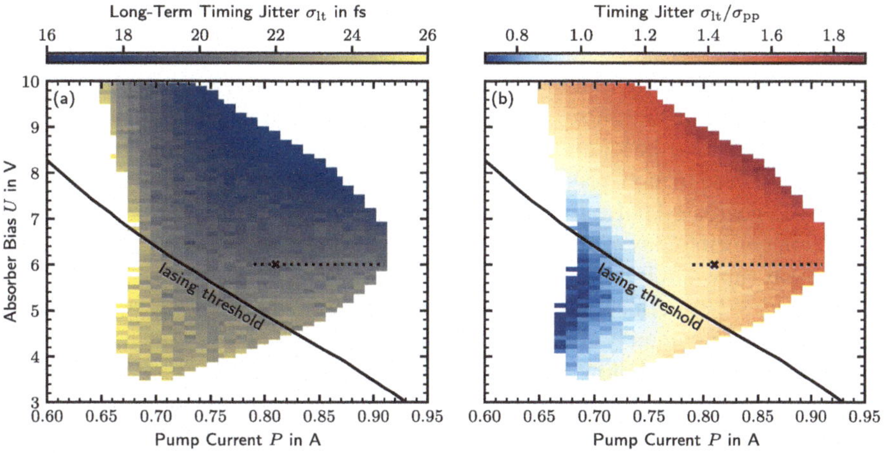

Fig. 4.30 **a** Long-term timing jitter as a function of the pump current P and the absorber bias U computed with the pulse-period fluctuation autocorrelation method. **b** Long-term timing jitter normalized to the pulse-to-pulse timing jitter visualizing finite time correlations. 2.5×10^5 pulses have been used for each data point. The solid black lines indicate the lasing threshold and the black dotted lines the detailed scan presented in Fig. 4.31. The timing jitter at the black crosses is discussed in Chap. 3. Laser parameters: $D = 10^{-6}\,\mathrm{cm\,ps^{-1}}[\eta]^{-1}$. Adapted from [58], with the permission of AIP Publishing

the results presented in Sect. 3.6 provides a relative error below 2%. Recognizing that the pulse-performance characteristics do not change by orders of magnitude within the FML regime (see Fig. 4.18) suggests that comparable relative timing jitter estimate errors can be expected at different operating points.

Note that the resulting long-term timing jitter map covers a slightly smaller region compared to the region of fundamental mode-locking as reported in Figs. 4.17b and 4.18. For strong spontaneous emission noise strengths and small pump currents, irregular losses of coherence with the pulse train invalidate the long-term timing jitter as a suitable measure for the timing stability.[7] Hence, those operation points have been excluded from the presented scans.

The strongest long-term timing jitter values amount to $\sigma_{lt} \approx 26$ fs and are observed at small pump currents and reverse biases close to the stability boundary. The smallest timing jitter values amount to $\sigma_{lt} \approx 16$ fs and are observed at large pump currents and reverse biases. With respect to the median, the jitter values vary by ≈30%, which

[7] Similar to the fundamental mode-locking state subject to a pulse-switching instability (FML_{PS}), the spontaneous emission noise irregularly sparks a new competing pulse train, which does not exhibit any field coherence with respect to the existing pulse train. The long-term timing jitter in that case still produces a figure of merit, with respect to the pulse intensity, which, however, does not reflect the properties of a coherent pulse train.

is only a statistically significant result in this case due to the small relative standard error of the pulse-period fluctuation autocorrelation method.[8]

The dynamics leading to the observed long-term jitter are further highlighted in Fig. 4.30b by plotting the ratio $\sigma_{\mathrm{lt}}/\sigma_{\mathrm{pp}}$, which represents the contributions of the pulse-period correlations to the long-term timing jitter. A stronger driving of the laser (P and U) is found to lead to more pronounced correlation contributions (yellow/red colors), but also to a decreased pulse-to-pulse jitter σ_{pp} (not shown). The latter outweighs the correlation contributions $\sigma_{\mathrm{lt}}/\sigma_{\mathrm{pp}}$ and overall leads to smaller long-term timing jitter values σ_{lt} as shown in Fig. 4.30a.

At $U \approx 6\,\mathrm{V}$ both effects cancel each other out and lead to a long-term jitter σ_{lt} that remains approximately constant along the pump current (see Fig. 4.16). At small P and U, ratios $\sigma_{\mathrm{lt}}/\sigma_{\mathrm{pp}}$ below 1.0 (blue colors) are observed, i.e., a long-term timing jitter that is smaller than the pulse-to-pulse jitter due to correlation effects. This region coincides with the region of subthreshold stability of the fundamental mode-locking state.

On a final note, the Kéfélian method could have been equally used to produce the long-term timing jitter estimates in the considered (P, U) parameter space with a similar relative error to computational cost ratio. The pulse-period fluctuation autocorrelation method, however, provides additional information in terms of the complete pulse separation dependent timing jitter $\sigma_{\Delta t}(n)$.

On that account, a closer look at the experimentally investigated parameter range is taken (see Fig. 4.16). Hence, a detailed timing jitter analysis incorporating the direct time-domain method is carried out for $U = 6\,\mathrm{V}$. The pump current is further restricted to $0.79\,\mathrm{A} \leqq P \leqq 0.91\,\mathrm{A}$, which represents the regime, where the fundamental mode-locking state stabilizes if the laser has been prepared in the off state (see Fig. 4.17). The scanned pump current range is indicated by the black dotted lines in Fig. 4.30.

The results are presented in Fig. 4.31, where (a) shows the long-term timing jitter σ_{lt} (blue line), the pulse-to-pulse timing jitter σ_{pp} (orange line), and their ratio $\sigma_{\mathrm{lt}}/\sigma_{\mathrm{pp}}$ (purple line). Figure 4.31b shows the color-coded pulse separation dependent timing jitter $\sigma_{\Delta t}(n)$ normalized to the pulse-to-pulse timing jitter σ_{pp}. Figure 4.31c plots the color-coded pulse-period fluctuation autocorrelation functions $\Psi_{\delta T}(n)$ as a function of the autocorrelation shift n.

As already observed in Fig. 4.30, the long-term timing jitter remains approximately constant at $\sigma_{\mathrm{lt}} \approx 21\,\mathrm{fs}$ along the pump current P. The pulse-to-pulse timing jitter, on the other hand, decreases from $\sigma_{\mathrm{pp}} \approx 17.5\,\mathrm{fs}$ to $\sigma_{\mathrm{pp}} \approx 13.0\,\mathrm{fs}$ along the pump current scan. The decrease of the pulse-to-pulse timing jitter can be attributed to the pulse power, which increases with the pump current. Thus, the amount of spontaneously emitted photons has relatively less weight in comparison to the coherent pulses and therefore presents a relatively small perturbation. The stronger driving, i.e., the larger pump current, on the other hand, modifies the pulse-period correla-

[8] The direct time-domain method exhibits a relative standard error of $\approx 15\%$ for a sample set with 2.5×10^5 per data point. Hence, only the values close to the stability boundaries would exhibit improvements/deteriorations that exceed the standard error.

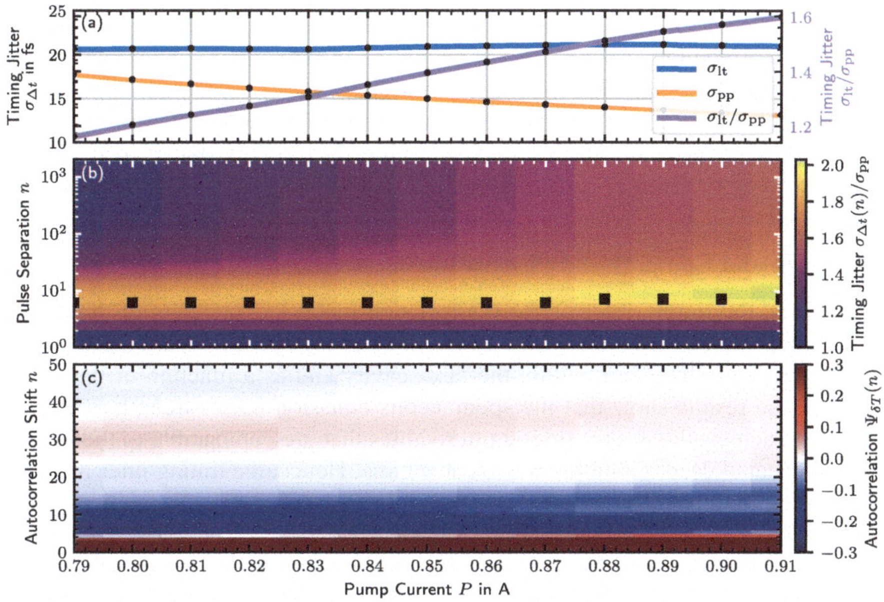

Fig. 4.31 Pump current dependence of the timing jitter: **a** long-term timing jitter σ_{lt} (blue line, corresponding to the top row in (**b**)), pulse-to-pulse timing jitter σ_{pp} (orange line, corresponding to the bottom row in (**a**)), and their ratio σ_{lt}/σ_{pp} (purple line). **b** Color-coded pulse separation dependent timing jitter $\sigma_{\Delta t}(n)$ normalized to the pulse-to-pulse timing jitter σ_{pp}. **c** Color-coded pulse-period fluctuation autocorrelation functions $\Psi_{\delta T}$. Individual data points (denoted by black dots) are obtained from 100 realizations, each 5 μs long. Laser parameters: $U = 6\,\text{V}$ and $D = 10^{-6}\,\text{cm}\,\text{ps}^{-1}[\eta]^{-1}$

tions such that the effect of the reduced pulse-to-pulse jitter σ_{pp} is canceled in the long-term jitter σ_{lt}.

The changes in pulse-period correlations manifest themselves in the pulse-separation dependent timing jitter $\sigma_{\Delta t}(n)$ shown in (b). The timing jitter $\sigma_{\Delta t}(n)$ shows the same qualitative behavior for all pump currents P, i.e., an initial increase to a local maximum around $n \approx 7$, followed by a simple relaxation to the respective long-term value σ_{lt}. At smaller pump currents, however, the local maximum is less pronounced and the following relaxation reduces the timing jitter by a significant amount. At larger pump currents, on the other hand, the maximum appears more pronounced and the following relaxation only marginally reduces the timing jitter, thus leading to pump current independent long-term timing jitter values.

The ratio σ_{lt}/σ_{pp} (purple line in (a)), furthermore, quantifies the contributions of the pulse correlations to the long-term jitter σ_{lt}. For an increasing pump current P, this ratio increases from $\sigma_{lt}/\sigma_{pp} \approx 1.17$ to $\sigma_{lt}/\sigma_{pp} \approx 1.57$. The pulse-period correlations are further illustrated in Fig. 4.30c via their autocorrelation functions and enable the understanding of the changes of the ratio σ_{lt}/σ_{pp}. With an increasing pump current P, the initial large dip to negative correlations $\Psi_\delta(n)$ extends further to larger n, which would suggest a decreasing ratio $\sigma_{\mathrm{lt}}/\sigma_{\mathrm{pp}}$. Similarly, the following overshoot to positive correlations around $n \approx 30$ decreases in magnitude, which also leads to a reduction of $\sigma_{\mathrm{lt}}/\sigma_{\mathrm{pp}}$. However, both effects are dominated by a slower initial drop-off of the autocorrelation function, which in total leads to the increased ratio $\sigma_{\mathrm{lt}}/\sigma_{\mathrm{pp}}$. Thus, stronger driving of the laser and higher energy pulses increase the positive correlations at timescales of a few pulses and thereby increase the correlation contribution to the long-term timing jitter for the considered laser parameters.

Conclusion and Comparison to Literature

In summary, the timing jitter of the fundamental mode-locking state of the three-section quantum-dot laser has been characterized in the operation parameter space, along the pulse propagation within the laser cavity and as a function of the noise strength. The results show that the spontaneous emission noise strength, which is required to achieve long-term timing jitter values that are comparable to the experiment, is located deep within the nonlinear regime. Hence, the timing jitter must be evaluated numerically and can not be computed via the established analytical [140, 141] or semi-analytical methods [138, 139, 142].

However, even though being computationally more expensive, applying the direct time-domain timing jitter estimation method, which has been presented in Chap. 3, produces additional information that can be utilized to interpret and understand the underlying mechanics. On that account, the timing jitter analysis along the pulse propagation yields that the gain sections decrease the pulse-to-pulse jitter but increase the long-term jitter via positive correlations. The absorber section increases the pulse-to-pulse jitter but decreases the long-term jitter via negative correlations.

These results carry over and allow to understand the obtained operation parameter dependence of the timing jitter. For the three-section laser at hand, the long-term timing jitter reduces for stronger pump currents and reverse biases. This behavior results from the interplay of the pulse-to-pulse timing jitter and the pulse-period fluctuation correlations. Their individual contributions non-trivially depend on the operation parameters and thus must be explicitly calculated.

The comparison to the published literature shows that qualitatively similar results, i.e., timing jitter values that improve with stronger driving conditions, have been obtained for traveling-wave equation based models [13, 66, 130]. Note, however, that the cited results have only been presented for very few operation parameter sets, likely due to the computational limitations. Experimental investigations have reported qualitatively similar results [23] as well as timing jitter values that deteriorate with the pump current [124]. Contrary to this work, simulations based on the ring-cavity delay-differential equation modeling approach [59] have repeatedly shown timing jitter values, which strongly increase close to the respective stability boundaries of the chosen emission state [37, 104, 138, 139, 142, 143].

4.7 Tuning the Taper Angle

Having thoroughly characterized the simulated emission dynamics of the three-section tapered quantum-dot laser and successfully reproduced the experimental results, the remainder of this chapter addresses the geometric device optimization towards better pulse performance. In that respect, the well-characterized and understood model, which has been validated via the experiment, can be utilized to numerically study device designs and save the expensive and time-consuming trial and error process of fabricating large sets of test samples.

This section chooses the angle Θ of the long tapered gain section as the device parameter to be investigated. The three-section laser model implements the tapered geometry by assuming that the electric field profile adiabatically follows the transverse expansion of the active medium and roughly maintains a Gaussian transverse mode profile (see Sect. 4.2). Changes in the geometric overlap between the field and the active medium are phenomenologically modeled by a modified relative transverse confinement factor Γ. A similar phenomenological approach is taken for the non-resonant waveguide losses α_{int}. To ensure that the adiabatic expansion of the electric field is a good approximation, the maximum considered taper angle is capped at $\Theta = 3°$. This value falls in-between the taper angle values, which have been reported for quantum-dot based devices, which exemplarily cover $\Theta = 2°$ [19, 22, 23], $\Theta = 3°$ [22], $\Theta = 3.6°$ [8, 20], and $\Theta = 4°$ [86]. From a modeling perspective, Refs. [21, 99] report that beam propagation modeling techniques show an excellent adiabatic expansion of the electric field modes up to taper angles $\Theta \approx 0.6°$ and slightly increased radiative losses at the section interfaces for larger taper angles. The modeling approach in this work, however, neglects those nuances by assuming that they only produce subtle changes in the mode-locking dynamics.

Apart from the size of the active region, the impact of tuning the taper angle Θ on the waveguide properties within the chosen model framework is summarized in Fig. 4.32. The blue line plots the accumulated non-resonator waveguide field losses

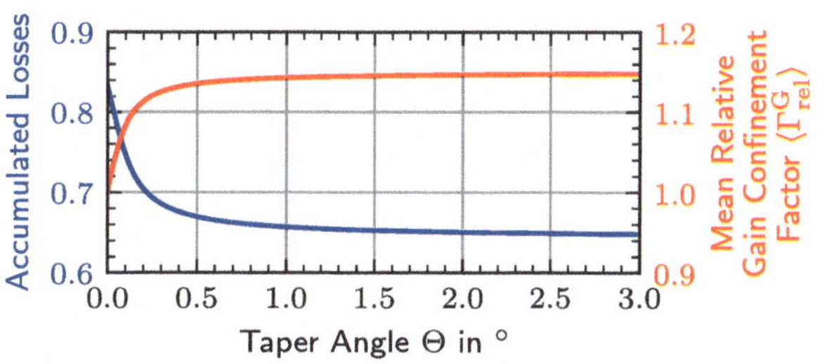

Fig. 4.32 Impact of the taper angle Θ on the waveguide properties. The blue line plots the accumulated field losses κ per round trip and the red line the mean transverse optical confinement factor in the gain sections $\langle \Gamma^{G}_{\text{rel}} \rangle$

$$\kappa(\Theta) = \exp\left(-\int_0^l \mathrm{d}z\, \alpha_{\mathrm{int}}(z, \Theta)\right) \tag{4.43}$$

along one complete round trip. The red line plots the mean transverse confinement factor within the gain sections $\langle\Gamma^{G}_{\mathrm{rel}}\rangle$. The accumulated losses quantify the proportion of a test field that is absorbed along one complete round trip. For an increasing taper angle, the losses reduce from $\kappa \approx 0.85$ to $\kappa \approx 0.65$, where most of the improvement in the losses is already achieved for taper angles around $\Theta \approx 0.5°$. The mean transverse confinement factor hand leads to an increase of the small-signal gain,[9] which small perturbations experience. For an increasing taper angle, this quantity improves from $\langle\Gamma^{G}_{\mathrm{rel}}\rangle = 1.0$ to $\langle\Gamma^{G}_{\mathrm{rel}}\rangle \approx 1.15$, where, again, most of the improvement is already achieved for taper angles $\Theta \gtrsim 0.5°$. Hence, changes of the taper angle beyond $\theta \gtrsim 0.5°$ mostly impact the lasing dynamics by further increasing the active region area and thus the gain saturation energy.

4.7.1 Emission Dynamics and Pulse Performance

To investigate the influence of the taper angle Θ on the emission dynamics, pump current scans for taper angles between $\Theta = 0°$ and $\Theta = 3°$ are performed. The absorber reverse bias is kept constant at $U = 6\,\mathrm{V}$. The observed emission states are presented in Fig. 4.33, which is structured similar to Fig. 4.17. Subfigure (a) presents the emission states, which have been obtained with the laser prepared in the off state for every data point. Subfigure (b) maps the regions of stability of the respective emission states, which have been extracted from multiple scans with the laser prepared in state specific initial conditions. The black dotted lines and the black crosses indicate the parameters ($\Theta = 2.0°$), which have been used in the previous sections.

The results are discussed in terms of pump current scans and can be interpreted as if they were obtained from an ensemble of lasers with increasing taper angles. On that account, tuning the taper angle fundamentally changes the sequence of emission states that is observed along a given pump current scan. At taper angles close to $\Theta = 0.0°$, no lasing at all is observed as the overall waveguide losses are too large to be overcome by the gain at the considered pump currents. In a very small region around $\Theta \approx 0.1°$, cw-lasing (cyan region, squared hatches in (b)) is found with a large but quickly reducing threshold between $P \approx 1.0\,\mathrm{A}$ and $P \approx 1.5\,\mathrm{A}$.

For an increasing taper angle between $\Theta \approx 0.1°$ and $\Theta \approx 0.15°$, the threshold quickly reduces to $P \approx 0.7\,\mathrm{A}$, as the internal waveguide losses of the long tapered gain section further decrease (see Fig. 4.32). In this region, the laser first emits third-order harmonic mode-locking with a trailing-edge plateau ($\mathrm{HML3_{TEP}}$, orange region) immediately above threshold, which is followed by a transition to cw-lasing at higher pump currents. This transition is accompanied by a considerable region of bistability

[9] The differential gain coefficient g^{GS} linearly depends on the relative confinement factor.

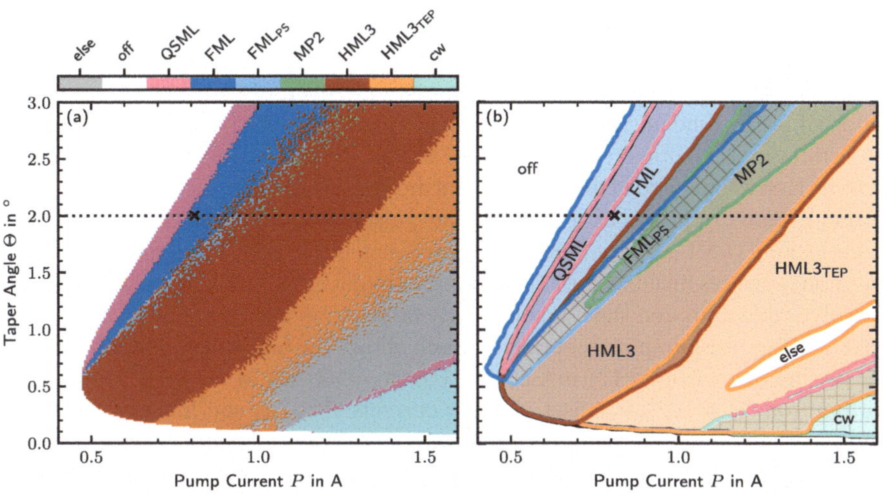

Fig. 4.33 Emission states observed in the pump current and taper angle (P, Θ) parameter space: off, Q-switched mode-locking (QSML), fundamental mode-locking (FML), fundamental mode-locking with a pulse-switching instability (FML_{PS}), two non-identical pulses (MP2), third-order harmonic mode-locking (HML3), third-order harmonic mode-locking with a strong trailing-edge plateau ($HML3_{TEP}$), and cw. **a** Laser (initial conditions) prepared in the off state. **b** Stable regions obtained from combined simulations with the laser prepared in the off, FML, HML3, and MP2 state, respectively. The black dotted lines and black crosses indicate the parameters used in the previous sections. Hatches highlight the FML_{PS} and cw states. Laser parameters: $U = 6.0\,\mathrm{V}$

with small pockets of Q-switched mode-locking (QSML, pink regions) at the lower cw emission boundary.

For taper angles between $\Theta \approx 0.15°$ and $\Theta \approx 0.5°$, the lasing threshold further reduces to its global minimum at $P_{\mathrm{th}} \approx 0.475\,\mathrm{A}$. Unlike at smaller taper angles, however, third-order harmonic mode-locking (HML3, red region) stabilizes beyond the lasing threshold. Further increasing the pump current then eventually causes a transition in which the trailing-edge plateau third-order harmonic mode-locking emission state is reestablished. This behavior then remains consistent for all larger taper angles Θ.

The significant drop of the lasing threshold in the taper angle regime from $\Theta \approx 0.0°$ to $\Theta \approx 0.5°$ is attributed to the reduced non-resonant waveguide losses and the improved transverse confinement factor. As shown in Fig. 4.32, however, both effects approximately saturate at taper angles around $\Theta \approx 0.5°$.

For taper angles above $\Theta \approx 0.5°$, the behavior of the laser drastically changes. Firstly, the lasing threshold starts shifting to larger pump currents ($P \approx 0.93\,\mathrm{A}$ for $\Theta = 3.0°$) for increasing taper angles, since the increased active region leads to a reduced pump-current density. The smaller pump current density results in decreased occupation numbers of the quantum dot and thus in less optical gain. This effect was already at play at the smaller taper angles, but was dominated by the decreasing waveguide losses and the increasing confinement factors. Secondly, after crossing the lasing threshold either Q-switched mode-locking or fundamental mode-locking (FML, blue region) is observed instead of third-order harmonic mode-locking. The region of stable Q-switched mode-locking is, moreover, always bistable with the region of stable fundamental mode-locking, which itself always extends below the lasing threshold and thus creates a region of subthreshold fundamental mode-locking.

If the laser is operated in the fundamental mode-locking state, increasing the pump current eventually produces a transition to pulse-switching unstable fundamental mode-locking (FML_{PS}, light blue and diagonal hatches in (b)). The region of FML_{PS} emission is always bistable with third-order harmonic mode-locking, which remains the only stable emission state at sufficiently large pump currents. For an increasing taper angle Θ, the pump current range, which allows for fundamental mode-locking linearly increases from $\approx$120 mA at $\Theta \approx 0.6°$ to $\approx$500 mA at $\Theta = 3.0°$. Moreover, starting from $\Theta \approx 0.7°$, a subset of pump currents P exists and linearly grows for increasing taper angles, where fundamental mode-locking is the only stable emission state. The pump current range, over which pulse-switching unstable fundamental mode-locking is observed, however, remains constant at $\approx$100 mA for all taper angles.

Lastly, for taper angles $\Theta \gtrsim 1.2°$, a multi-pulse emission state with two non-identical and non-equidistant pulses (MP2, green) appears at pump currents that are always larger than the lower HML3 stability boundary. This state is therefore always multistable with HML3 emission and, depending on the taper angle and pump current, also with the FML and FML_{PS} state. However, if not specifically seeded, it only sporadically stabilizes as can be seen in Fig. 4.33a.

In conclusion, the observed emission dynamics as a function of the taper angle can be separated into two regimes: Firstly, for taper angles $\Theta \lesssim 0.6°$, the laser prefers third-order harmonic mode-locking, which benefits from the colliding pulse mechanism [15, 71, 110, 111] in the absorber section. This regime, furthermore, experiences a strong lasing threshold dependence on the taper angle due to the waveguide losses reduction and the optical confinement factor increase. Secondly, for taper angles $\Theta \gtrsim 0.6°$, the observable emission states are greatly diversified and fundamental mode-locking can be found for pump currents close to the lasing threshold. This qualitative behavior is attributed to the increased active gain region, i.e., the increased gain saturation energy [8, 11], which favors one powerful single pulse instead of multiple pulses circulating within the laser cavity.

Pulse Performance Characteristics

In the next step, the pulse peak power and the pulse width of the fundamental mode-locking state are presented in Fig. 4.34a, b, respectively. The white and black solid lines in (a) and (b), respectively, denote the lasing threshold and the black dashed lines the parameter scan presented in Fig. 4.35. Both performance figures clearly benefit from an increasing taper angle: The achievable pulse peak power increases and the achievable pulse width decreases for large taper angles Θ. As reported in Sect. 4.5, the shortest pulses for a given pump current scan are always observed at the lower pump current stability boundary. Hence, the shortest pulses are found at the largest considered taper angle $\Theta \approx 3.0°$ with a width of $\Delta_{pw} \approx 380$ fs. On the other

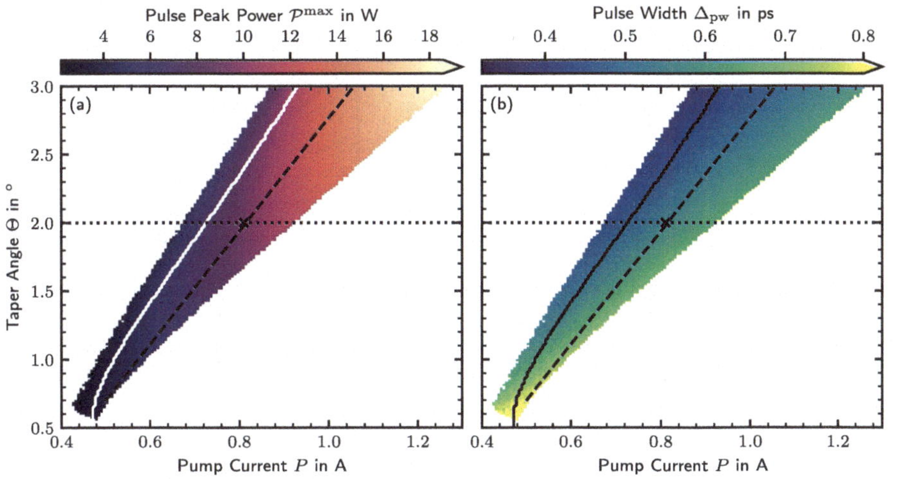

Fig. 4.34 Mode-locking performance characteristics in the pump current and taper angle (P, Θ) parameter space: Color-coded pulse peak power (**a**) and pulse width (**b**) of the fundamental mode-locking state. The white/black solid lines denote the lasing threshold. The black dotted lines and black crosses indicate the parameters used in the previous sections ($\Theta = 2.0°$). The black dashed lines indicate the parameter scan presented in Fig. 4.35. Laser parameters: $U = 6.0$ V

side, the longest pulses with a width of $\Delta_{pw} \approx 810$ fs are observed at the smallest taper angle $\Theta \approx 0.55°$, where FML emission can still be observed.

Unlike the pulse width, the best pulse peak power performance for a given pump current scan is always found at the upper stability boundary. For fundamental mode-locking, the observed pulse peak powers vary between $\mathcal{P}^{max} \approx 3.1$ W and $\mathcal{P}^{max} \approx 18.2$ W, where the former is obtained for the smallest taper angle $\Theta \approx 0.55°$ and the latter for the largest taper angle $\Theta \approx 3.0°$. The underlying mechanics that lead to the large differences in peak powers are twofold: Firstly, an increased taper angle yields a greater gain saturation energy. Thus, the laser can maintain a strong effective amplification of more powerful pulses. Secondly, an increased taper angle pushes the region of stable fundamental mode-locking to larger pump currents, which supply more energy to the amplification of powerful pulses.

The improvement of the pulse performance characteristics with an increasing taper angle in the fundamental mode-locking region is further studied in detail: Fig. 4.35 presents a one-dimensional scan across the pump current and taper angle parameter space, which has been indicated by the diagonal black dashed lines in Fig. 4.34. The chosen pump currents are placed in the region where fundamental mode-locking is the only stable emission state. Moreover, the relative distances to the lasing threshold and the upper Q-switched mode-locking stability boundary are kept constant. Increasing the pump current P along the taper angle Θ ensures a constant pump current density that is applied to the gain section. Hence, the unsaturated quantum-dot inversion of the gain section remains constant along the chosen scan. Figure 4.34a presents the pulse peak power (red line) and the mean optical power (green dashed line) and (b)

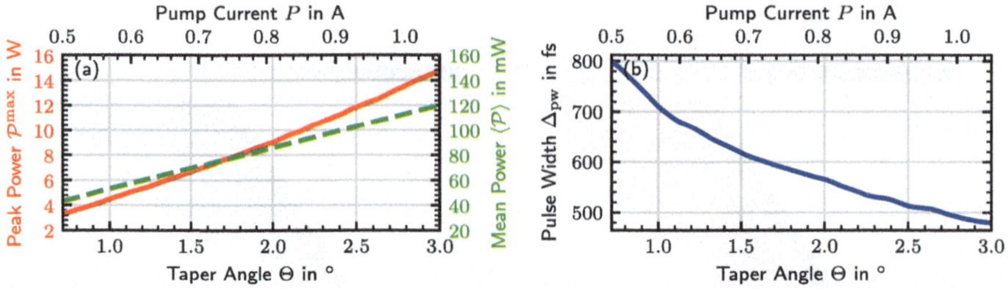

Fig. 4.35 Mode-locking performance in the fundamentally mode-locked state along Θ and P: **a** pulse peak power (red solid line) and mean power (green dashed line) and **b** pulse width. The relative distance to the lasing threshold is kept constant. The pump current is denoted on the top and the respective taper angle on the bottom. The presented parameter range is indicated by the dashed lines in Fig. 4.34. Laser parameter: $U = 6.0\,\mathrm{V}$

the pulse width. The pump current is denoted on the top and the respective taper angle on the bottom.

For an increasing taper angle and the accordingly increasing pump current, the emitted mean output power increases approximately linearly from $\langle \mathcal{P} \rangle \approx 44\,\mathrm{mW}$ to $\langle \mathcal{P} \rangle \approx 120\,\mathrm{mW}$. The linear increase, however, results from the interplay of multiple nonlinearities that involve the saturation of the gain and absorber sections, as well as the pulse shaping itself, instead of a simple input-output relation between the pump current and the mean output power. This becomes especially evident as the mean output power triples as the pump current doubles. This suggests an improvement of the internal efficiency as the taper angle is increased.

On the account of the involved saturation effects, the pulse peak power increases super-linearly from $\mathcal{P}^{\max} \approx 3.3\,\mathrm{W}$ to $\mathcal{P}^{\max} \approx 14.9\,\mathrm{W}$, which is inversely related to a sub-linear decrease of the pulse width from $\Delta_{\mathrm{pw}} \approx 800\,\mathrm{fs}$ to $\Delta_{\mathrm{pw}} \approx 480\,\mathrm{fs}$. Both can be attributed to more efficient gain and absorber saturation due to the increased pulse energy at larger taper angles.

In conclusion, larger taper angles allow for an improvement of both the pulse peak power and the pulse width. Unless low operation pump currents are desired, the presented results suggest to increase the taper angle as much as the transverse mode properties allow it. The analysis thereof is, however, beyond this works model. Suitable beam propagation simulation techniques have been proposed and demonstrated in Refs. [21, 99]. The results are consistent with the published literature [8, 18, 20, 21], which also report improved pulse peak powers and reduced pulse widths for increasing taper angles. However, contrary to the cited references, this work explains the pulse width reduction by saturation effects in the gain sections instead of the absorber section. This issue, in particular, shall be elaborated in the following subsection.

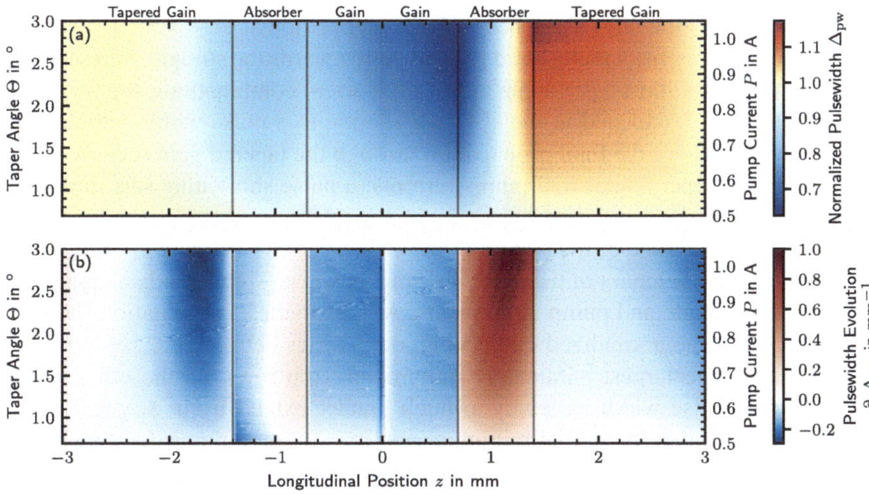

Fig. 4.36 Pulse shaping in the fundamental mode-locking state along the taper angle Θ. The left side of both subplots shows the backward-moving field (denoted by negative positions z) and the right side shows the forward-moving field (denoted by positive positions z). **a** Presents the color-coded pulse width, which has been normalized to the respective out-coupled pulse width. **b** Plots the changes $\partial_z \Delta_{\text{pw}}$ of the normalized pulse width with respect to the longitudinal coordinate z. The relative distance to the lower stability boundary of the fundamental mode-locking state is kept constant. The plotted parameter range is indicated by the black dashed lines in Fig. 4.34. Laser parameters: $U = 6\,\text{V}$ and $D = 10^{-6}\,\text{cm}\,\text{ps}^{-1}[\eta]^{-1}$

4.7.2 *Pulse-Shaping Along the Taper Angle and Pump Current*

To further study the impact of the taper angle on the pulse-shaping mechanism in the fundamental mode-locking state, the spatial evolution of the pulse along the pulse propagation is presented in Fig. 4.36. The scanned parameters are identical to those shown in Fig. 4.35 and have been indicated by the black dashed lines in Fig. 4.34. The results are visualized similar to Fig. 4.21: Subfigure (a) plots the pulse width Δ_{pw} as a function of the longitudinal position z within the unfolded laser cavity and subfigure (b) shows the corresponding local pulse width changes Δ_{pw}. Each row encodes a taper angle and pump current set, which is indicated on the left and right, respectively. The pulse width of each row has been normalized to the out-coupled pulse width.

For an increasing taper angle and pump current, the pulse-shaping qualitatively remains the same at most of the positions along the pulse propagation. Similar to the sole pump current scan presented in Fig. 4.21, only two exceptions are found: Firstly, the backward-traveling pulse first experiences a shortening and then a broadening in the absorber section. The transition between shortening and broadening, which can be best seen in Fig. 4.36b by the change from blue to red colors, first moves closer to the tapered gain section until it reaches its leftmost position at $\Theta \approx 1.5°$, $P \approx 0.7$ A and then moves away from the tapered gain section towards the straight gain section. Secondly, for the smallest taper angle and pump current combinations, the forward-moving pulses are broadened in the absorber section to a pulse width, which then remains constant along the final propagation through the tapered gain section. Only at slightly larger taper angles and pump currents, a pulse shortening sets in within the tapered gain section and thereby results in a local maximum of the pulse width (dark red) at the interface of the two sections.

The quantitative behavior of the pulse-shaping, however, significantly changes for increasing taper angle and pump current sets. At the smallest combination (bottom row in Fig. 4.36), the normalized pulse width only varies between $\Delta_{\text{pw}} \approx 0.95$ and $\Delta_{\text{pw}} \approx 1.02$. At the largest parameter combination (top row), on the other hand, the normalized pulse width varies by a much greater extend from $\Delta_{\text{pw}} \approx 0.64$ to $\Delta_{\text{pw}} \approx 1.17$. Note that the global minimum and maximum are nonetheless always found at the left and right interface of the absorber section for the forward-moving pulse ($z = 0.7$ mm and $z = 1.4$ mm, respectively).

The increased variations of the pulse width are attributed to more pronounced gain saturation effects by the more powerful pulses. Figure 4.36b shows that increased taper angle and pump current combinations produce much more pronounced pulse shaping responses in almost all positions along the propagation. Especially the narrow part of the tapered gain section along the backward-moving pulse and the absorber section along the forward-moving pulse exhibit a strong response to increasing taper angle and pump current combinations. A similar behavior has already been observed for the sole increase of the pump current at a constant taper angle as shown in Fig. 4.21. However, unlike that case, the increasing taper angle also produces a quantitatively stronger pulse shortening in the other gain sections, which leads to an overall reduction of the pulse width as shown in Fig. 4.35.

In summary, changing the taper angle does not qualitatively affect the previously described pulse-shaping mechanism (see Sect. 4.3.3). The increased pulse power at larger taper angles, however, boosts the pulse-shaping contributions of the individual stages, such that much stronger modulations of the pulse along its propagation can be observed.

4.7.3 Taper Angle Dependence of the Timing Jitter

Lastly, a detailed timing jitter analysis of the fundamental mode-locking state across the already established taper angle and pump current scan is carried out. The investigated parameter sets are equidistantly located along the black dashed lines in Fig. 4.34. The corresponding pulse performance is shown in Fig. 4.35 and the pulse shaping in Fig. 4.36.

The results from the direct time-domain method are presented in Fig. 4.37, where (a) shows the long-term timing jitter σ_{lt} (blue line), the pulse-to-pulse timing jit-

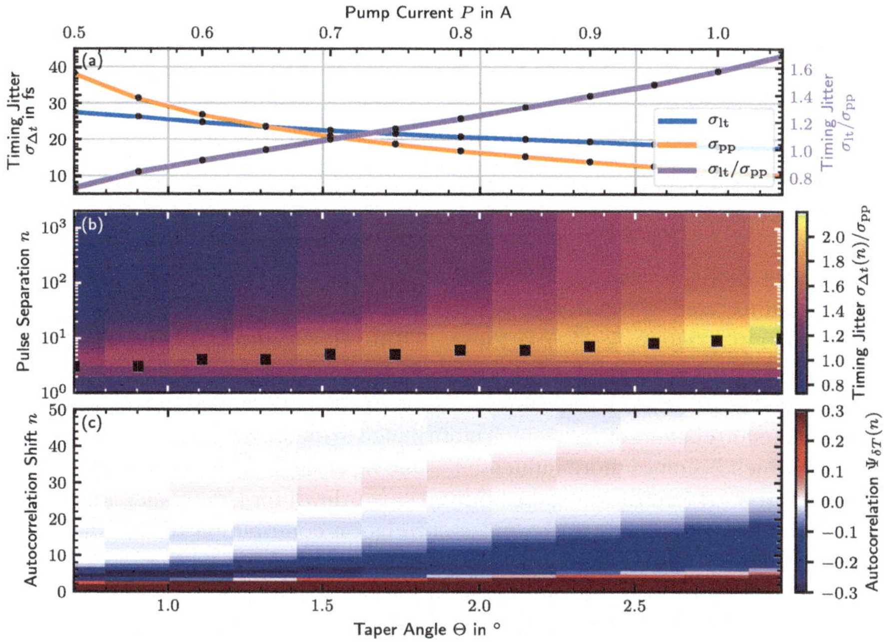

Fig. 4.37 Taper angle and pump current dependence of the timing jitter: **a** long-term timing jitter σ_{lt} (blue line, corresponding to the top row in (**b**)), pulse-to-pulse timing jitter σ_{pp} (orange line, corresponding to the bottom row in (b)), and their ratio σ_{lt}/σ_{pp} (purple line). **b** Color-coded pulse separation dependent timing jitter $\sigma_{\Delta t}(n)$ normalized to the pulse-to-pulse timing jitter σ_{pp}. **c** Color-coded pulse-period fluctuation autocorrelation functions $\Psi_{\delta T}$. Individual data points (denoted by black dots) are obtained from 100 realizations, each 5 μs long. The scanned parameter range is indicated by black dashed lines in Fig. 4.34. Laser parameters: $U = 6\,\text{V}$ and $D = 10^{-6}\,\text{cm ps}^{-1}[\eta]^{-1}$

ter σ_{pp} (orange line), and their ratio σ_{lt}/σ_{pp} (purple line). The latter measures the contributions of the pulse correlations to the long-term jitter σ_{lt}.

Both the long-term and the pulse-to-pulse timing jitter improve with an increasing taper angle. While the pulse-to-pulse timing jitter is approximately reduced by a factor of four from $\sigma_{pp} \approx 40\,\text{fs}$ to $\sigma_{pp} \approx 10\,\text{fs}$, the long-term timing jitter exhibits a weaker but still respectable reduction from $\sigma_{lt} \approx 28\,\text{fs}$ to $\sigma_{lt} \approx 18\,\text{fs}$. The decrease of pulse-to-pulse timing jitter with the increasing taper angle is explained by the increased pulse power and thus an improved signal-to-noise ratio. Along the reduction, both quantities cross each other: For taper angles $\Theta \lesssim 1.7°$ the pulse-to-pulse timing jitter is larger than the long-term timing jitter. For taper angles $\Theta \gtrsim 1.7°$ this relationship changes and the long-term timing jitter is larger than the pulse-to-pulse timing jitter. Correspondingly, their ratio starts of at $\sigma_{lt}/\sigma_{pp} \approx 0.8$ and increases approximately linearly to $\sigma_{lt}/\sigma_{pp} \approx 1.6$.

Figure 4.37b, furthermore, shows the color-coded pulse separation dependent timing jitter $\sigma_{\Delta t}(n)$, which has been normalized to the pulse-to-pulse timing jitter σ_{pp}. Black squares indicate the global timing jitter maximum. The crossing behavior of the pulse-to-pulse and long-term timing jitter can also be identified in this plot: At small taper angles, the timing jitter first increases to a global maximum and then relaxes towards its long-term equilibrium, which is smaller than the initial pulse-to-pulse jitter. For larger taper angles, on the other hand, the rise to the global maximum at a few pulse separations n is similar, but the final equilibrium value remains above the initial pulse-to-pulse jitter. Moreover, the global maximum in the timing jitter $\sigma_{\Delta t}(n)$, shifts to larger pulse-separations n with increasing taper angles and pump currents.

This characteristic behavior is a manifestation of the pulse-period correlations, which are illustrated in Fig. 4.37c, where the color-coded pulse-period fluctuation autocorrelation function $\Psi_{\delta T}(n)$ is shown. Using those autocorrelation functions, the increase of the ratio $\sigma_{\mathrm{lt}}/\sigma_{\mathrm{pp}}$ from $\approx$0.8 to $\approx$1.6 can be reconstructed. With an increasing taper angle and pump current, the observed autocorrelation features move to larger autocorrelation shifts n, which is attributed to the slower timescale of the gain sections, which becomes more influential as the taper angle is increased. Especially the initial drop-off (red region at the bottom) slows down for larger taper angles and pump currents, which adds to the pulse-correlation contribution of the long-term timing jitter. Moreover, the negative dip, which is associated with a reduction of the long-term timing jitter, becomes weaker for larger taper angles due to increased saturation by the stronger pulses. Thus, the self-healing property in the absorber section works worse if the pulses are too powerful. Additionally, for taper angles $\Theta \gtrsim 1.5$ a positive overshoot of the autocorrelation function appears at autocorrelation shifts between $n \approx 30$ and $n \approx 40$, which further causes an increase of the correlation contribution to the long-term timing jitter.

Nonetheless, the decrease of the pulse-to-pulse timing jitter, which is attributed to the improved signal-to-noise ratio, dominates the increase of pulse-period correlation contribution to the long-term timing jitter and yields an overall reduction of the long-term timing jitter σ_{lt} with an increased taper angle Θ. In conclusion, the presented timing jitter results suggest to increase the taper angle Θ as much as the transverse mode properties allow it, in order to achieve optimal temporal pulse train stability properties.

4.8 Shifting the Saturable Absorber Position

In this last section, the influence of the saturable absorber position within the three-section cavity is studied. For this purpose, the left side of the saturable absorber section is denoted as the absorber starting position z_{SA}. Using the numerical model, scans of the pump current for saturable absorber starting positions from $z^{\mathrm{s}}_{\mathrm{SA}} = 0.0\,\mathrm{mm}$ to $z^{\mathrm{s}}_{\mathrm{SA}} = 0.7\,\mathrm{mm}$ are performed, while the length of the absorber section is kept constant at 0.7 mm. The former configuration with $z_{\mathrm{SA}} = 0.0\,\mathrm{mm}$ corresponds to a traditional two-section design, with the saturable absorber at the highly reflecting end of the cavity, while the latter configuration with $z_{\mathrm{SA}} = 0.7\,\mathrm{mm}$ corresponds to the configuration that has been experimentally realized and studied in

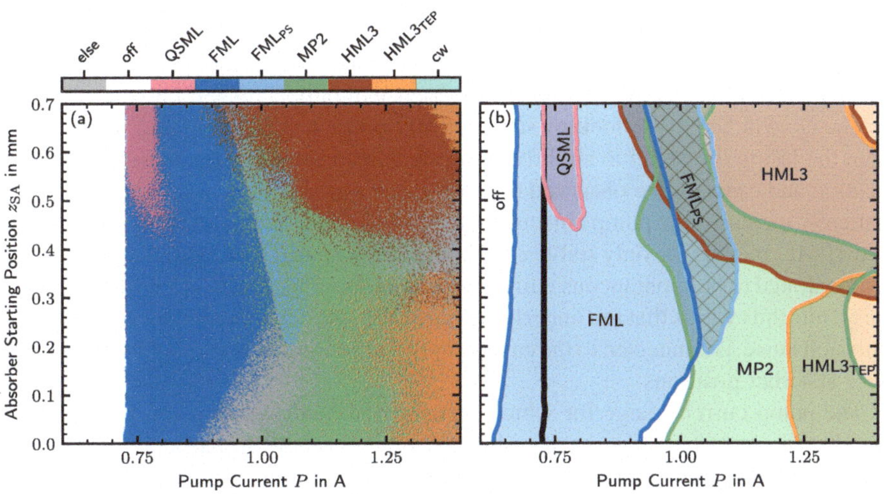

Fig. 4.38 Emission states observed in the pump current and absorber starting position (P, z_{SA}) parameter space: off, Q-switched mode-locking (QSML), fundamental mode-locking (FML), fundamental mode-locking with a pulse-switching instability (FML_{PS}), two non-identical pulses (MP2), third-order harmonic mode-locking (HML3), and third-order harmonic mode-locking with a strong trailing-edge plateau ($HML3_{TEP}$). **a** Laser (initial conditions) prepared in the off state. **b** Stable regions obtained from combined simulations with the laser prepared in the off, FML, HML3, and MP2 state, respectively. Hatches highlight the FML_{PS} state. Laser parameters: $U = 6.0\,V$ and $\Theta = 2.0°$

the previous sections. This procedure can be interpreted as performing pump current scans on an ensemble of lasers with different absorber positions z_{SA}. The absorber reverse bias and the taper angle are kept constant at $U = 6\,V$ and $\Theta = 2.0°$.

4.8.1 *Emission Dynamics and Pulse Performance*

The observed emission states are presented in Fig. 4.38 in a similar manner as in Figs. 4.17 and 4.33: Subfigure (a) presents the emission states, which have been obtained with the laser prepared in the off state for every data point. Subfigure (b) maps the regions of stability of the respective emission states, which have been extracted from multiple scans with the laser prepared in state specific initial conditions. The previously investigated and experimentally realized configuration $z_{SA} = 0.7$ mm is represented by the top rows of (a) and (b).[10]

As the results demonstrate, the saturable absorber position z_{SA}, i.e., the geometric cavity configuration, has a profound impact on the observable emission dynamics. For a decreasing absorber position, the lasing threshold (transition from off to lasing

[10] See Sect. 4.5 for detailed discussion.

in (a), black solid line in (b)) remains unaffected. This can be expected as the net gain (see Sect. 4.3.2), which induces the threshold once it becomes positive, does not depend on the cavity configuration in the absence of optical pulses, i.e., the off state.

The Q-switched mode-locking state (QSML, pink), on the other hand, shifts away from the lasing threshold to slightly larger pump currents for a decreasing absorber position and ceases to be observed for absorber positions $z_{SA} \lesssim 0.45$ mm. This dramatically increases the pump current range for which the fundamental mode-locking state (FML, blue) is the only stable emission state. Note that the QSML state has been demonstrated to be spontaneous emission noise induced (see Sect. 4.5.3). Combining the results thus yields that the underlying instability, which is excited by spontaneous emission noise, is connected to the cavity configuration and only exists for sufficiently large absorber positions.

The pump current range for which fundamental mode-locking can be observed initially increases for a decreasing absorber pump current. In particular, the lower pump current boundary monotonically shifts to lower pump currents, since the optical losses in the absorber section are reduced due to the shorter available recovery time between the forward and backward-traveling pulses. The upper pump current, on the other hand, shifts to larger pump currents until a maximum of $P \approx 1.04$ A at $z_{SA} \approx 0.22$ mm is reached. From there on, the upper stability boundary reduces again to $P \approx 0.91$ A and thus produces an overall reduction of the FML emission pump current range for smaller absorber positions. The maximum in the upper pump current stability boundary, furthermore, distinguishes the emission states, which can be observed beyond the stability boundary. For large absorber positions $z_{SA} \gtrsim 0.22$ mm, the pulse-switching unstable fundamental mode-locking can be observed (FML_{PS}, light blue, see Fig. 4.5c). For small absorber positions $z_{SA} \lesssim 0.22$ mm, complex irregular emission with a varying number of pulses can be observed. Those emission states are not further classified (else, gray in (a)). Similar to the QSML state, the pulse-switching unstable fundamental mode-locking state has been identified as a positive net-gain instability (see Fig. 4.8) that is excited by spontaneous emission noise (see Sect. 4.5.3). The results presented here further demonstrate that the pulse-switching instability is induced by the laser configuration, i.e., the absorber position z_{SA}.

Compared to the FML state, the third-order harmonic mode-locking state (HML3, red) shows a much greater dependence on the saturable absorber position z_{SA}. For decreasing absorber positions, the lower HML3 pump current stability boundary rapidly shifts to larger pump currents until no more HML3 emission can be observed within the scanned pump current range at $z_{SA} \approx 0.3$ mm. However, at smaller absorber positions $z_{SA} \lesssim 0.3$ mm, the third-order harmonic mode-locking state with a trailing-edge plateau ($HML3_{TEP}$, orange) appears independently from the HML3 for pump currents around $P \gtrsim 1.2$ A. Lastly, the multi-pulse emission state with two non-identical and non-equidistant pulses (MP2, green) can be observed for all absorber starting positions z_{SA}, but within a strongly varying range of pump currents. For absorber positions $z_{SA} \gtrsim 0.5$ mm, the MP2 state can only be observed for pump currents $0.95\text{ A} \gtrsim P \gtrsim 1.05$. For smaller absorber positions, on the other hand, the upper pump current stability boundary shifts to significantly larger values, such that the MP2 state can be found up to the largest scanned pump current $P = 1.4$ A.

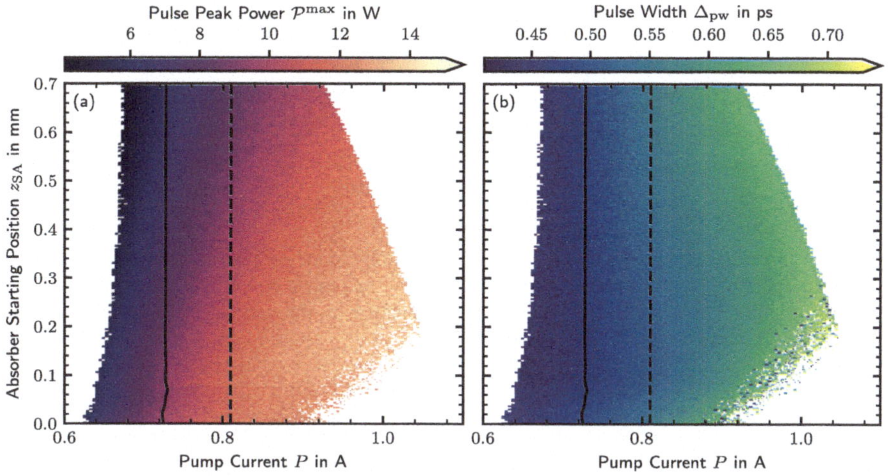

Fig. 4.39 Mode-locking performance characteristics in the pump current and absorber starting position (P, z_{SA}) parameter space: **a** pulse peak power and **b** pulse width of the fundamental mode-locking state. The black solid lines indicate the lasing threshold and the black dashed lines indicate the parameter scan presented in Fig. 4.40. Laser parameters: $U = 6.0\,\text{V}$ and $\Theta = 2.0°$

In conclusion, a large diversity of mode-locked emission states as well as their pronounced multistability, as discussed in Sect. 4.5, is predominately found for larger saturable absorber positions $z_{SA} \gtrsim 0.5\,\text{mm}$. At smaller absorber positions both the number of observable mode-locking states and their multistability is strongly reduced. As a result the maximal pump current range of monostable fundamental mode-locking is found at smaller absorber positions around $z_{SA} \approx 0.2\,\text{mm}$.

Pulse Performance Characteristics

Following the emission state dynamics, the pulse peak power and the pulse width of the fundamental mode-locking state are presented in Fig. 4.39a, b, respectively. The black solid lines in (a) and (b) denote the lasing threshold and the vertical black dashed lines the parameter scan presented in Fig. 4.40.

Along a given pump current scan, the shortest pulses are observed at the lower pump current stability boundary of the respective emission states as previously reported. The absorber position itself, however, also affects the pulse width, such that the shortest pulses are not found at the minimum achievable pump current at $z_{SA} = 0.0\,\text{mm}$ but at $z_{SA} \approx 0.15\,\text{mm}$ with a width of $\Delta_{pw} \approx 420\,\text{fs}$. The longest pulses, on the other hand, are indeed found with a width of $\Delta_{pw} \approx 700\,\text{fs}$ at the maximum achievable pump at $z_{SA} \approx 0.22\,\text{mm}$.

Contrary to the pulse width, the best pulse peak power is always found at the upper stability boundary. The strongest pulses with a peak power of $\mathcal{P}^{\max} \approx 15.0\,\text{W}$ are thus found at the maximum of the upper pump current stability boundary at $z_{SA} \approx 0.22\,\text{mm}$. The smallest peak power with a value of $\mathcal{P}^{\max} \approx 4.2\,\text{W}$ is, however, not found at the minimum achievable pump current, but at the largest absorber position

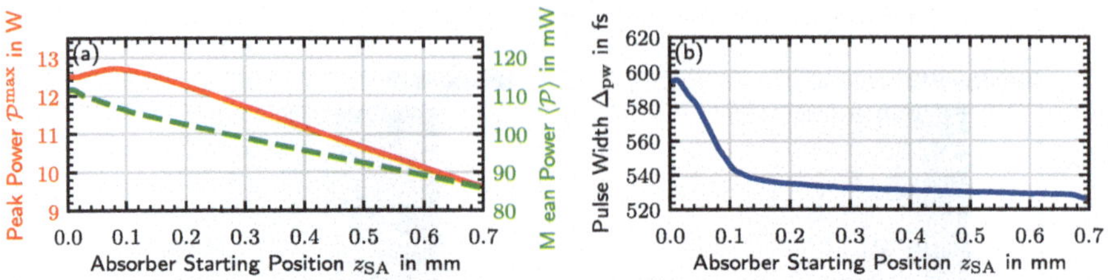

Fig. 4.40 Mode-locking performance in the fundamentally mode-locked state along the absorber starting position z_{SA}: **a** pulse peak power (red solid line) and mean power (green dashed line) and **b** pulse width. The presented parameter range is indicated by the vertical black dashed lines in Fig. 4.39. Laser parameters: $U = 6.0$ V and $\Theta = 2.0°$

$z_{SA} = 0.7$ mm. This behavior is attributed to a more effective absorber recovery at larger absorber positions due to the longer available recovery time between the forward and backward-traveling pulses.

To further examine the influence of the absorber starting position z_{SA} on the pulse performance in the fundamental mode-locking regime, a one-dimensional scan with a constant pump current $P = 0.81$ A is presented in Fig. 4.40. (a) Plots the pulse peak power (red line) and the mean power (green dashed line) and (b) plots the pulse width. The scanned parameter range is indicated by the vertical black dashed lines in Fig. 4.39. The mean optical power monotonically decreases from $\langle\mathcal{P}\rangle \approx 112$ mW to $\langle\mathcal{P}\rangle \approx 86$ mW for an increasing absorber starting position. The decreasing optical power is explained by a more efficient absorption of the forward-moving pulse due to the longer available recovery time between the backward and forward interaction of the pulse within the absorber section.

The pulse peak power, on the other hand, shows two distinct regimes: For small absorber starting positions $z_{SA} \lesssim 0.1$ mm, the peak power first increases to a maximum of $\mathcal{P}^{\max} \approx 12.7$ W and then decreases monotonically to $\mathcal{P}^{\max} \approx 9.6$ W for further increasing absorber positions. The two regimes can also be identified in the evolution of the pulse width, as shown in (b). For absorber starting positions up to $z_{SA} \approx 0.1$ mm, the pulse width quickly reduces from $\Delta_{pw} \approx 595$ fs to $\Delta_{pw} \approx 540$ fs. After that initial drop, the pulse width then only slowly decreases by an additional $\approx$10 fs to $\Delta_{pw} \approx 530$ fs at $z_{SA} = 0.7$ mm. Hence, it can be concluded that the initial increase of the pulse-peak power is caused by the reduction of the pulse width, i.e., a redistribution of the optical power, which overcompensates the decrease of the mean power. Both effects can be attributed to the short propagation time through the left gain section for small absorber starting positions, which does not allow for a sufficient recovery of the quantum-dot ground state occupation numbers in the left facing part of the absorber section.

Based on those results, an optimum performance can be predicted for a saturable absorber starting position at $z^{s}_{SA} \approx 0.1$ mm. This configuration increases the peak power by about 30% and avoids a significant pulse-width broadening compared to the experimental realized device with $z^{s}_{SA} \approx 0.7$ mm.

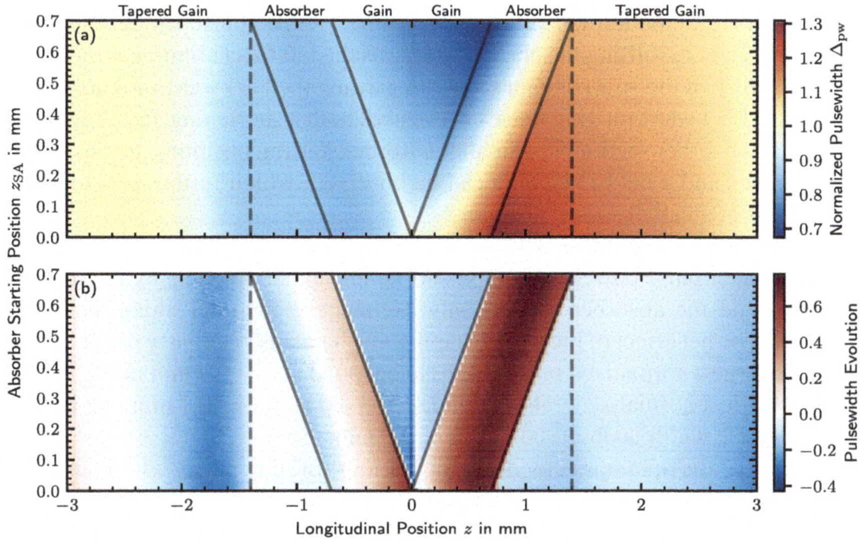

Fig. 4.41 Pulse shaping in the fundamental mode-locking state along the absorber starting position z_{SA}. The left side of both subplots shows the backward-moving field (denoted by negative positions z) and the right side shows the forward-moving field (denoted by positive positions z). Subfigure **a** presents the color-coded pulse width, which has been normalized to the respective out-coupled pulse width. Subfigure **b** plots the changes $\partial_z \Delta_{pw}$ of the normalized pulse width with respect to the longitudinal coordinate z. The absorber position is indicated by solid gray lines. The presented parameter range is indicated by the vertical dashed black line in Fig. 4.39. Laser parameters: $U = 6.0$ V and $\Theta = 2.0°$

4.8.2 Pulse Shaping Along the Absorber Position

To understand the influence of the saturable absorber starting position on the pulse-shaping mechanism in the fundamental mode-locking state, the spatial evolution of the pulse along the pulse propagation is studied. Figure 4.41 presents the pulse width Δ_{pw} (a) and the local pulse width changes $\partial_z \Delta_{pw}$ (b) as a function of the longitudinal position z within the unfolded laser cavity. Each row corresponds to an absorber starting position from the scan presented in Fig. 4.40, whose parameters are indicated by vertical black dashed lines in Fig. 4.39. The position of the absorber section is indicated by gray solid lines and the interface to the tapered gain section by gray dashed lines. The pulse width of each row has been normalized to the respective out-coupled pulse width.

Along the shift of the absorber starting positions z_{SA}, the pulse shaping qualitatively and quantitatively changes. Most evidently, the straight gain section, which encloses the absorber section for all intermediate absorber starting positions, changes its impact on the backward-moving pulse (left side of the plot). The part, which is located in between the tapered gain section and the absorber section neither shortens nor broadens the pulse width (white region in (b)). The part in between the absorber section and the left facet, however, always yields a pulse shortening (blue region in (b)). Hence, shifting the absorber towards the left facet eliminates the pulse shortening ability of the straight gain section regarding the backward-moving pulse. Moreover, the backward-moving pulse experiences both a shortening and a broadening within the absorber section. For smaller absorber starting positions, however, the latter dominates and a net broadening can be observed, which further acts towards broader pulses at small absorber positions.

The forward-traveling pulse is less affected by shifts of the absorber position. Due to the weaker pulse powers, the pulse shortening and broadening by the straight gain section and the absorber section only exchange their order along the pulse propagation. The net effect of the pulse-shaping mechanisms peculiarly results in the smallest and largest normalized pulse widths to be found at $z \approx 0.7$ mm in Fig. 4.41a. While the smallest normalized pulse width is observed at $z_{SA} = 0.7$ mm, the largest normalized pulse width is observed at $z_{SA} = 0.0$ mm.

Interestingly, the qualitatively different regions of the pulse width and the peak power evolution, which are separated approximately by $z_{SA} \approx 0.1$ mm (see Fig. 4.40), cannot be easily identified in the pulse-shaping mechanism. The only feature that is unique for the absorber starting position $z_{SA} \lesssim 0.1$ mm, is the initially weak pulse broadening of the forward-moving pulse in the absorber section. This effect, however, rather points towards shorter and not broader pulses. Hence, the steep increase of the pulse width is likely rather related to a global effect than to the local pulse-shaping dynamics.

For that reason, the spatial pulse peak power evolution is studied in Fig. 4.42. Subfigure (a) plots the pulse peak power, which has been normalized to the peak power at the out-coupling facet, as a function of the absorber starting position z_{SA} along the pulse propagation in the unfolded laser cavity. Subfigure (b) presents the corresponding differential gain g_P that the pulse peak experiences. This quantity has been directly calculated from the finite differences of the pulse peak power along the propagation.

The initial evolution of the backward-moving pulse through the tapered gain section appears to be very similar for all absorber starting positions. The role of the straight gain section, however, changes depending on its position: In the section that is located between the tapered gain section and the absorber section, the pulse peak power decreases (light blue triangle in (b)) due to the complete saturation of the gain by the leading edge as well as the non-resonant waveguide losses. In the straight gain section that is located between the absorber and the left facet, on the other hand, the pulses do not carry enough power to saturate the gain. Hence, the pulse peak power experiences only a weak net amplification (light red triangle in (b)). The results of those uneven amplification and absorption processes are pulses at the left facet, which are twice as powerful for the largest absorber positions (normalized peak power $\approx$0.27 at $z_{SA} = 0.7$ mm) compared to the smallest absorber position (normalized peak power $\approx$0.13 at $z_{SA} = 0.0$ mm).

The following forward propagation, however, peculiarly equalizes the pulse peak powers, such that for all absorber starting position z_{SA}, the normalized peak power upon leaving the absorber section in the forward direction amounts to $\approx$0.033. This

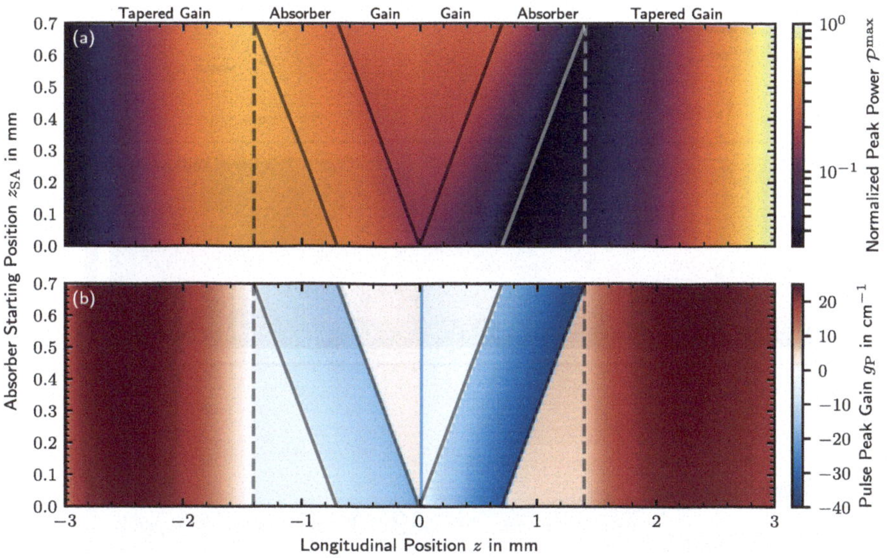

Fig. 4.42 Pulse peak power in the fundamental mode-locking state along the absorber starting position z_{SA}. The left side of both subplots shows the backward-moving field (denoted by negative positions z) and the right side shows the forward-moving field (denoted by positive positions z). Subfigure **a** presents the color-coded pulse peak power, which has been normalized to the respective out-coupled peak power. Subfigure **b** plots the differential pulse peak power gain g_P with respect to the longitudinal coordinate z. The absorber position is indicated by solid gray lines. The presented parameter range is represented by the vertical black dashed line in Fig. 4.39. Laser parameters: $U = 6.0\,V$ and $\Theta = 2.0°$

puts configurations with smaller absorber starting positions at a distinct advantage: The available length to amplify the pulse before reaching the out-coupling facet is longer. Even though, the differential pulse peak gain is relatively small in the straight gain section between the absorber and the tapered gain section, it is nonetheless critical to the final exponential amplification that yields the 30% increase of the pulse peak power, which has been observed in Fig. 4.40.

4.8.3 Absorber Starting Position Dependence of the Timing Jitter

In the next step, a detailed timing jitter analysis of the fundamental mode-locking state across the already established absorber starting position scan is performed. The investigated parameters are represented by the vertical black dashed lines in Fig. 4.39. The corresponding pulse performance is shown in Fig. 4.40.

The results from the direct time-domain method are presented in Fig. 4.43, where (a) shows the long-term timing jitter σ_{lt} (blue line), the pulse-to-pulse timing jitter

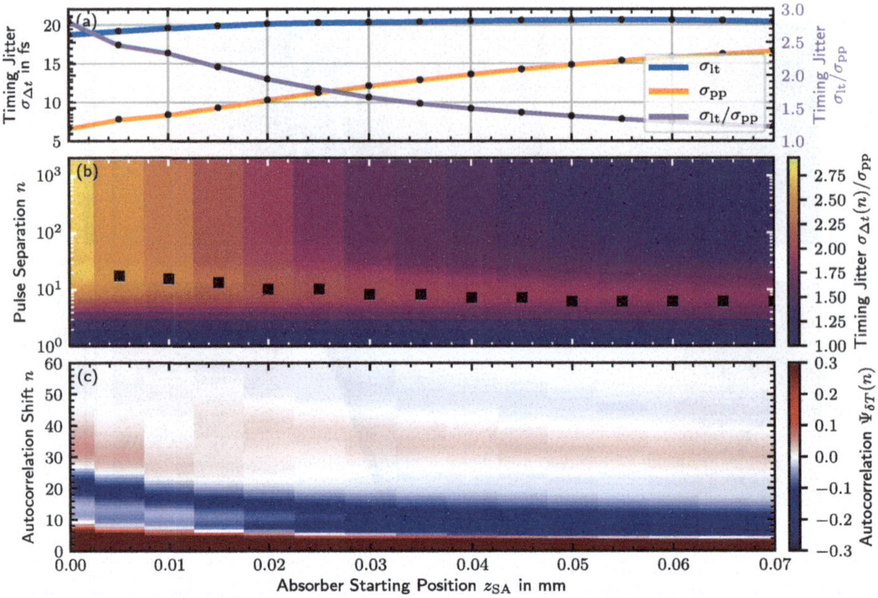

Fig. 4.43 Absorber starting position z_{SA} dependence of the timing jitter: **a** long-term timing jitter σ_{lt} (blue line, corresponding to the top row in (**b**)), pulse-to-pulse timing jitter σ_{pp} (orange line, corresponding to the bottom row in (a)), and their ratio σ_{lt}/σ_{pp} (purple line). **b** Color-coded pulse separation dependent timing jitter $\sigma_{\Delta t}(n)$ normalized to the pulse-to-pulse timing jitter σ_{pp}. **c** Color-coded pulse-period fluctuation autocorrelation functions $\Psi_{\delta T}$. Individual data points (denoted by black dots) are obtained from 100 realizations, each 5 µs long. The scanned parameter range is indicated by black dashed lines in Fig. 4.39. Laser parameters: $P = 0.81$ A, $U = 6$ V and $D = 10^{-6}$ cm ps$^{-1}[\eta]^{-1}$

σ_{pp} (orange line), and their corresponding ratio σ_{lt}/σ_{pp} (purple line). For an increasing absorber starting position, the pulse-to-pulse timing jitter significantly increases from $\sigma_{pp} \approx 6.9$ fs to $\sigma_{pp} \approx 16.7$ fs, while the long-term timing jitter only marginally increases from $\sigma_{lt} \approx 18.8$ fs to $\sigma_{lt} \approx 20.5$ fs. Moreover, the small increase of the long-term timing jitter only occurs for absorber starting positions $z_{SA} \lesssim 0.2$ mm and is otherwise approximately constant. Along the increasing absorber position, the ratio σ_{lt}/σ_{pp} decreases from ≈2.65 to ≈1.25.

The different evolution of the pulse-to-pulse and long-term timing jitter points towards changing pulse-period correlations, which further manifest in the pulse separation dependent timing jitter $\sigma_{\Delta t}(n)$, which is presented in Fig. 4.43b. The individual timing jitter functions are normalized to their respective pulse-to-pulse timing jitter σ_{pp} and black squares indicate their global maxima. Most notably, for $z_{SA} = 0.0$ mm, the timing jitter $\sigma_{\Delta t}(n)$ does not exhibit the otherwise characteristic local maximum and simply increases towards its long-term equilibrium. For absorber starting positions $z_{SA} > 0.0$ mm the local maximum reappears and moves towards smaller pulse separations n for an increasing absorber starting position.

The pulse-period correlations are further illustrated in Fig. 4.43, where (c) plots the color-coded pulse-period fluctuation autocorrelation function $\Psi_{\delta T}(n)$ as a function of the autocorrelation shift n. For an increasing absorber starting position z_{SA}, the autocorrelation function features, including the initial drop-off, generally shift to smaller autocorrelation shifts n. Moreover, the first dip below zero (blue colors) gains additional depth as the absorber position is increased. Both properties contribute the observed reduction of the ratio $\sigma_{\mathrm{lt}}/\sigma_{\mathrm{pp}}$ in (a).

A detailed examination of the autocorrelation features for small absorber starting positions further reveals additional nuances. The overshoot (red colors) of the autocorrelation function after the first dip below zero first moves to smaller autocorrelation shifts around $n \approx 30$ at $z_{\mathrm{SA}} \approx 0.1$ mm, but then jumps back to $n \approx 40$ at $z_{\mathrm{SA}} \approx 0.15$ mm. The position of this jump coincides with the local maximum in the pulse peak power and the fast change of the pulse width shown in Fig. 4.40b. Both effects are related to the recovery time of the quantum-dot ground state occupations in the saturable absorber: The propagation time through the very short gain section does not allow the absorber to be fully recovered. The jump in the overshoot position occurs simultaneously with a weaker dip below zero and therefore has little impact on the ratio $\sigma_{\mathrm{lt}}/\sigma_{\mathrm{pp}}$, but leads to a vanishing local maximum in the timing jitter $\sigma_{\Delta t}(n)$ at $z_{\mathrm{SA}} = 0.0$ mm.

The evolution of the timing jitter with the absorber starting position is explained by the changes of the gain and absorber recovery processes. For an increasing absorber starting position z_{SA}, the saturation, and thus pulse-shifting effects, increase in both the straight gain and absorber sections, where the former increases the pulse-to-pulse timing jitter and the latter decreases the pulse-period correlation contribution to the long-term timing jitter via the mechanisms discussed in Sect. 4.6.2. Both effects change by a factor of two for the considered absorber position shifts. However, the combination of those opposing effects results only in marginal changes of the long-term timing jitter σ_{lt}.

Hence, the long-term timing jitter does neither limit nor identify an optimal device configuration. The pulse-to-pulse timing jitter, on the other hand, clearly improves for small absorber starting positions, but only implies little relevance for applications, as discussed in Chap. 3. The optimization of the absorber starting position z_{SA} can therefore be almost exclusively based on the pulse performance figures as presented in Figs. 4.39 and 4.40.

4.9 Chapter Conclusions

In this chapter, the emission dynamics and the pulse performance characteristics of a passively mode-locked tapered three-section semiconductor quantum-dot laser have been studied. On that account, a computationally efficient numerical model has been developed, which couples the charge-carrier dynamics to the traveling-wave description of the electric field via the Maxwell-Bloch equations. A spatio-temporally self-consistent description of the laser dynamics is thereby achieved. When compared to the experimental laser characterization, the model has been demonstrated to excellently reproduce the results both qualitatively and quantitatively. This validates the modeling approach and opens up possibilities for further analyses and investigations via the numerical model.

To study the pulse-shaping mechanism of the laser, a co-moving spatio-temporal representation of the electric field intensity and the gain dynamics has been introduced. The investigation of the fundamental and third-order harmonic mode-locking states has revealed that the pulse-shaping is contrary to the mechanisms reported in the published literature [8, 71, 106–109]: Pulses broaden in the absorber section and shorten in the gain sections due to the interplay of the saturable and the non-saturable gain and absorption.

The simulated time series have furthermore been shown to allow for the precise identification and classification of any occurring instabilities. On that account, the experimentally observed unstable fundamental mode-locking state with pronounced timing and amplitude jitter has been identified as a pulse-switching instability. This instability is induced by a positive net-gain window, which is produced by the forward and backward-traveling pulses and is characteristic for the device geometry. Those dynamics highlight the need for the spatio-temporal modeling, as such instabilities can not be described by the popular ring-geometry delay-differential equation modeling approach [37, 59–61, 63, 142].

The numerical model has further been used to create detailed maps of the observable emission dynamics and the pulse performance figures of the fundamental mode-locking state. It has been shown that the pulse peak power and the pulse width can not be optimized simultaneously. However, a reasonable compromise between the two can be found at intermediate absorber reverse biases around $U \approx 6.0\,\mathrm{V}$, which corresponds to the experimentally utilized value. Motivated by this result, the implications of the absorber reverse bias have further been unraveled by separately studying the induced effects: The analysis has revealed that the observed emission states critically depend on the interplay of the absorber redshift and absorber recovery rate, which are both governed by the applied reverse bias. Fundamental mode-locking only appears if both are properly tuned. Moreover, this has demonstrated the value of the numerical model, since the effect of the absorber redshift can not be easily implemented in the aforementioned delay-differential equation modeling approach.

The timing jitter of the fundamental mode-locking state has furthermore been studied by utilizing the tools that have been introduced and developed in Chap. 3. A comprehensive map of the long-term timing jitter in the operation parameter space has revealed that the timing jitter exhibits the lowest values at the strongest driving conditions. The large parameter scan has been enabled by the newly developed pulse-period fluctuation autocorrelation method. In order to understand the underlying mechanism, the pulse-period fluctuations and their correlations along the pulse propagation through the cavity have been analyzed. This has allowed to associate the impact of the gain sections with a decrease of the pulse-to-pulse timing jitter and an increase of long-term timing jitter deteriorating correlations and the impact of the absorber section with the opposite. Tuning the operation parameters can thus be interpreted as giving either of those two mechanisms more or less influence. The resulting long-term timing jitter is then determined by the interplay of the pulse-to-pulse timing jitter and the pulse-period correlations.

Lastly, the numerical model has been employed to predict optimal laser design parameters regarding the angle of the long tapered gain section and the position of the saturable absorber section. The results have shown that all pulse performance characteristics in the fundamental mode-locking emission state improve with an increasing taper angle. A maximum taper angle would thus be limited by the transverse electric field properties, which are beyond the scope of this work. The absorber section, on the other hand, exhibits an optimal position within the laser cavity that maximizes both the pump current range for which fundamental mode-locking can be obtained as well as the achievable pulse peak power. It has further been demonstrated that these results can be interpreted and understood in terms of the pulse shaping and the gain and absorption recovery processes.

Open Questions and Outlook

The most significant shortcoming of the numerical model is likely the approximation of the inhomogeneously broadened quantum-dot ensemble by a single transition with an appropriate effective homogeneous linewidth. While this approach has enabled the efficient simulation of the spatio-temporal laser dynamics, it does not allow for the description of spectral hole burning effects. The extension, however, is straightforward via the introduction of quantum-dot subgroups, which are organized by their optical transition frequency [14, 28, 69, 73, 76, 77, 79, 81]. Such an approach would allow to adjust the number of quantum-dot subgroups, i.e., the simulation costs, and their respective effective homogeneous linewidth as needed. Large parameter scans or detailed timing jitter simulations could be performed with a low number of subgroups and detailed spatio-temporal pulse propagation studies with a large number. The former in particular would allow to quickly find suitable parameter sets for the latter.

Nevertheless, the derived numerical model can be used to perform further laser design explorations, since it has been proven to reproduce the experimental results. A large number of parameters lend themselves for such investigations: The reflectivities of the laser facets [13, 130] and the length [8, 11, 12] and position [10, 16, 17] of the absorber section have been popular optimization parameters and allow for comparisons with literature. Similarly, the semiconductor quantum-dot specific properties such as their density, confinement depth, and scattering rates [26, 28, 76, 79] represent suitable optimization parameters.

When investigating other mode-locked laser devices and active medium designs, it would be important to study the pulse-shaping mechanisms. This would reveal, whether the mechanism discovered in this work represents only a peculiarity of the specific laser design or transfers to a broader class of lasers. The co-moving spatio-temporal representation of the laser dynamics, which has been introduced in this work, would provide the necessary means. This open question, furthermore, demonstrates the importance of suitable visualization techniques, which are required in order to obtain insights from complex dynamical simulation data.

References

1. Guzmán R, Gordon C, Orbe L, Carpintero G (2017) 1 GHz InP on-chip monolithic extended cavity colliding-pulse mode-locked laser. Opt Lett 42(12):2318–2321. https://doi.org/10.1364/ol.42.002318
2. Lo MC, Guzmán R, Ali M, Santos R, Augustin L, Carpintero G (2017) 1.8-THz-wide optical frequency comb emitted from monolithic passively mode-locked semiconductor quantum-well laser. Opt Lett 42(19):3872–3875. https://doi.org/10.1364/ol.42.003872
3. Lo MC, Auth D, Weber C, Fiala P, Sauer P, Carpintero G, Breuer S (2019) Photonic integrated circuit extended cavity passively mode-locked dual absorber symmetric ring laser. Opt Lett 44(14):3566
4. Hoogland S, Dhanjal S, Tropper AC, Roberts JS, Häring R, Paschotta R, Morier-Gemoud F, Keller U (2000) Passively mode-locked diode-pumped surface-emitting semiconductor laser. IEEE Photon Technol Lett 12(9):1135–1137. ISSN 1941-0174. https://doi.org/10.1109/68.874213
5. Keller U, Tropper AC (2006) Passively modelocked surface-emitting semiconductor lasers. Phys Rep 429:67–120
6. Gaafar MA, Rahimi-Iman A, Fedorova KA, Stolz W, Rafailov EU, Koch M (2016) Mode-locked semiconductor disk lasers. Adv Opt Photonics 8(3):370–400. https://doi.org/10.1364/aop.8.000370
7. Waldburger D, Link SM, Mangold M, Alfieri CGE, Gini E, Golling M, Tilma BW, Keller U (2016) High-power 100 fs semiconductor disk lasers. Optica 3(8):844–852. https://doi.org/10.1364/optica.3.000844
8. Thompson MG, Rae AR, Xia M, Penty RV, White IH (2009) InGaAs quantum-dot mode-locked laser diodes. IEEE J Quantum Electron 15(3):661–672
9. Javaloyes J, Balle S (2010) Mode-locking in semiconductor Fabry-Pérot lasers. IEEE J Quantum Electron 46(7):1023–1030
10. Xu T, Montrosset I (2013) Quantum dot passively mode-locked lasers: relation between intra-cavity pulse evolution and mode locking performances. IEEE J Quantum Electron 49(1):65–71. https://doi.org/10.1109/jqe.2012.2229967
11. Rossetti M, Tianhong X, Bardella P, Montrosset I (2011) Impact of gain saturation on passive mode locking regimes in quantum dot lasers with straight and tapered waveguides. IEEE J Quantum Electron 47(11):1404. https://doi.org/10.1109/jqe.2011.2167131
12. Mee JK, Raghunathan R, Wright JB, Lester LF (2014) Device geometry considerations for ridge waveguide quantum dot mode-locked lasers. J Phys D 47:233001
13. Javaloyes J, Balle S (2011) Anticolliding design for monolithic passively mode-locked semiconductor lasers. Opt Lett 36(22):4407–4409
14. Simos H, Mesaritakis C, Xu T, Bardella P, Montrosset I, Syvridis D (2013) Numerical analysis of passively mode-locked quantum-dot lasers with absorber section at the low-reflectivity output facet. IEEE J Quantum Electron 49(1):3–10
15. Thompson MG, Marinelli C, Zhao X, Sellin RL, Penty RV, White IH, Kaiander IN, Bimberg D, Kang DJ, Blamire MG (2005) Colliding-pulse modelocked quantum dot lasers. Electron Lett 41(5):248–250
16. Xin YC, Li Y, Kovanis V, Gray AL, Zhang L, Lester LF (2007) Reconfigurable quantum dot monolithic multi-section passive mode-locked lasers. Opt Express 15(12):7623–7633. https://doi.org/10.1364/oe.15.007623
17. Li Y, Chiragh Furqan L, Xin YC, Lin CY, Kim J, Christodoulou CG, Lester LF (2010) Harmonic mode-locking using the double interval technique in quantum dot lasers. Opt Express 18(14):14637–14643. https://doi.org/10.1364/oe.18.014637
18. Mar A, Helkey RJ, Zou WX, Young WB, Bowers JE (1995) High-power mode-locked semiconductor lasers using flared waveguides. Appl Phys Lett 66(26):3558–3560. https://doi.org/10.1063/1.113786

19. Nikitichev DI, Ding Y, Ruiz M, Calligaro M, Michel AN, Krakowski M, Krestnikov I, Livshits D, Cataluna MA, Rafailov EU (2011) High-power passively mode-locked tapered InAs/GaAs quantum-dot lasers. Appl Phys B 103(3):609–613. https://doi.org/10.1007/s00340-010-4290-5
20. Thompson MG, Rae A, Sellin RL, Marinelli C, Penty RV, White IH, Kovsh AR, Mikhrin SS, Livshits DA, Krestnikov IL (2006) Subpicosecond high-power mode locking using flared waveguide monolithic quantum-dot lasers. Appl Phys Lett 88:133119
21. Rossetti M, Xu T, Bardella P, Montrosset I (2011) Modelling of passive mode-locking in InAs quantum-dot lasers with tapered gain section. Phys Status Solidi C 9(2):286–289. https://doi.org/10.1002/pssc.201100243
22. Weber C, Drzewietzki L, Rossetti M, Xu T, Bardella P, Simos H, Mesaritakis C, Ruiz M, Krestnikov I, Livshits D, Krakowski M, Syvridis D, Montrosset I, Rafailov EU, Elsäßer W, Breuer S (2015) Picosecond pulse amplification up to a peak power of 42 W by a quantum-dot tapered optical amplifier and a mode-locked laser emitting at 1.26 μm. Opt Lett 40(3):395–398. https://doi.org/10.1364/ol.40.000395
23. Bardella P, Drzewietzki L, Krakowski M, Krestnikov I, Breuer S (2018) Mode locking in a tapered two-section quantum dot laser: design and experiment. Opt Lett 43(12):2827–2830. https://doi.org/10.1364/ol.43.002827
24. Chow WW, Koch SW (2005) Theory of semiconductor quantum-dot laser dynamics. IEEE J Quantum Electron 41:495–505. https://doi.org/10.1109/jqe.2005.843948
25. Bimberg D, Fiol G, Kuntz M, Meuer C, Lämmlin M, Ledentsov NN, Kovsh AR (2006) High speed nanophotonic devices based on quantum dots. Phys Status Solidi A 203(14):3523–3532. https://doi.org/10.1002/pssa.200622488
26. Lüdge K (2011) Modeling quantum dot based laser devices. In: Nonlinear laser dynamics—from quantum dots to cryptography, vol 102, pp 3–34. ISBN 9783527411009 (Chapter 1)
27. Chow WW, Jahnke F (2013) On the physics of semiconductor quantum dots for applications in lasers and quantum optics. Prog Quantum Electron 37:109–184. s0079-6727(13)00009-8. https://doi.org/10.1016/j.pquantelec.2013.04.001
28. Lingnau B, Herzog B, Kolarczik M, Woggon U, Lüdge K, Owschimikow N (2017) Dynamic phase response and amplitude-phase coupling of self-assembled semiconductor quantum dots. Appl Phys Lett 110:241102. https://doi.org/10.1063/1.4985705
29. Lelarge F, Dagens B, Renaudier J, Brenot R, Accard A, van Dijk F, Make D, Gouezigou OL, Provost JG, Poingt F, Landreau J, Drisse O, Derouin E, Rousseau B, Pommereau F, Duan GH (2007) Recent advances on InAs/InP quantum dash based semiconductor lasers and optical amplifiers operating at 1.55 μm. IEEE J Sel Top Quantum Electron 13(1):111. ISSN 1077-260X. https://doi.org/10.1109/jstqe.2006.887154
30. Rafailov EU, Cataluna MA, Sibbett W (2007) Mode-locked quantum-dot lasers. Nat Photonics 1(7):395–401. https://doi.org/10.1038/nphoton.2007.120
31. Kuntz M, Fiol G, Lämmlin M, Meuer C, Bimberg D (2007) High-speed mode-locked quantum-dot lasers and optical amplifiers. Proc IEEE 95(9):1767–1778. ISSN 0018-9219. https://doi.org/10.1109/jproc.2007.900949
32. Norman JC, Jung D, Zhang Z, Wan Y, Liu S, Shang C, Herrick RW, Chow WW, Gossard AC, Bowers JE (2019) A review of high-performance quantum dot lasers on silicon. IEEE J Quantum Electron 55(2):1–11. ISSN 0018-9197. https://doi.org/10.1109/jqe.2019.2901508
33. Norman JC, Mirin RP, Bowers JE (2021) Quantum dot lasers-history and future prospects. J Vac Sci Technol A 39(2):020802. https://doi.org/10.1116/6.0000768
34. von der Linde D (1986) Characterization of the noise in continuously operating mode-locked lasers. Appl Phys B 39(4):201. https://doi.org/10.1007/bf00697487
35. Solgaard O, Lau KY (1993) Optical feedback stabilization of the intensity oscillations in ultrahigh-frequency passively modelocked monolithic quantum-well lasers. IEEE Photon Technol Lett 5(11):1264
36. Paschotta R, Schlatter A, Zeller SC, Telle HR, Keller U (2006) Optical phase noise and carrier-envelope offset noise of mode-locked lasers. Appl Phys B 82:265–273. https://doi.org/10.1007/s00340-005-2041-9

37. Otto C, Lüdge K, Vladimirov AG, Wolfrum M, Schöll E (2012) Delay induced dynamics and jitter reduction of passively mode-locked semiconductor laser subject to optical feedback. New J Phys 14:113033. https://doi.org/10.1088/1367-2630/14/11/113033
38. Drzewietzki L, Breuer S, Elsäßer W (2013) Timing jitter reduction of passively mode-locked semiconductor lasers by self- and external-injection: numerical description and experiments. Opt Express 21(13):16142–16161. https://doi.org/10.1364/oe.21.016142
39. Jiang LA, Abedin KS, Grein ME, Ippen EP (2002) Timing jitter reduction in modelocked semiconductor lasers with photon seeding. Appl Phys Lett 80(10):1707–1709. https://doi.org/10.1063/1.1459112
40. Valling S, Krauskopf B, Fordell T, Lindberg AM (2006) Experimental bifurcation diagram of a solid state laser with optical injection. Opt Commun 271:532–542
41. Bajek D, Cataluna MA (2021) Fast optical sampling by electronic repetition-rate tuning using a single mode-locked laser diode. Opt Express 29(5):6890–6902. https://doi.org/10.1364/oe.413045
42. Ahmad FR, Rana F (2008) Fundamental and subharmonic hybrid mode-locking of a high-power (220 mw) monolithic semiconductor laser. IEEE Photon Technol Lett 20(15):1308–1310. ISSN 1041-1135. https://doi.org/10.1109/lpt.2008.926911
43. Heck MJR, Salumbides EJ, Renault B, Bente EAJM, Oei YS, Smit MK, van Veldhoven R, Nötzel R, Eikema KSE, Ubachs W (2009) Analysis of hybrid mode-locking of two-section quantum dot lasers operating at 1.5 μm. Opt Express 17(20):18063–18075. https://doi.org/10.1364/oe.17.018063
44. Fiol G, Arsenijević D, Bimberg D, Vladimirov AG, Wolfrum M, Viktorov EA, Mandel P (2010) Hybrid mode-locking in a 40 GHz monolithic quantum dot laser. Appl Phys Lett 96(1):011104. https://doi.org/10.1063/1.3279136
45. Habruseva T, Rebrova N, Hegarty SP, Huyet G (2010) Quantum-dot mode-locked lasers with dual mode optical injection. Proc SPIE 7720:1–8. https://doi.org/10.1117/12.854338
46. Rebrova N, Huyet G, Rachinskii D, Vladimirov AG (2011) Optically injected mode-locked laser. Phys Rev E 83(6):066202. https://doi.org/10.1103/physreve.83.066202
47. Fiol G, Kleinert M, Arsenijević D, Bimberg D (2011) 1.3 μm range 40 GHz quantum-dot mode-locked laser under external continuous wave light injection or optical feedback. Semicond Sci Technol 26(1):014006. https://doi.org/10.1088/0268-1242/26/1/014006
48. Habruseva T, Arsenijević D, Kleinert M, Bimberg D, Huyet G, Hegarty SP (2014) Optimum phase noise reduction and repetition rate tuning in quantum-dot mode-locked lasers. Appl Phys Lett 104(021112):1–4
49. Arkhipov RM, Habruseva T, Pimenov AS, Radziunas M, Hegarty SP, Huyet G, Vladimirov AG (2016) Semiconductor mode-locked lasers with coherent dual-mode optical injection: simulations, analysis, and experiment. J Opt Soc Am B 33(3):351–359. https://doi.org/10.1364/josab.33.000351
50. Helkey RJ, Derickson DJ, Mar A, Wasserbauer JG, Bowers JE, Thornton RL (1992) Repetition frequency stabilisation of passively mode-locked semiconductor lasers. Electron Lett 28(20):1920–1922. ISSN 0013-5194. https://doi.org/10.1049/el:19921229
51. Merghem K, Rosales R, Azouigui S, Akrout A, Martinez A, Lelarge F, Duan GH, Aubin G, Ramdane A (2009) Low noise performance of passively mode locked quantum-dash-based lasers under external optical feedback. Appl Phys Lett 95(13):131111. https://doi.org/10.1063/1.3238324
52. Breuer S, Elsäßer W, McInerney JG, Yvind K, Pozo J, Bente EAJM, Yousefi M, Villafranca A, Vogiatzis N, Rorison J (2010) Investigations of repetition rate stability of a mode-locked quantum dot semiconductor laser in an auxiliary optical fiber cavity. IEEE J Quantum Electron 46(2):150. ISSN 0018-9197. https://doi.org/10.1109/jqe.2009.2033255
53. Lin CY, Grillot F, Li Y (2011) Microwave characterization and stabilization of timing jitter in a quantum dot passively mode-locked laser via external optical feedback. IEEE J Sel Top Quantum Electron 17(5):1311. https://doi.org/10.1109/jstqe.2011.2118745
54. Haji M, Hou L, Kelly AE, Akbar J, Marsh JH, Arnold JM, Ironside CN (2012) High frequency optoelectronic oscillators based on the optical feedback of semiconductor mode-locked laser diodes. Opt Express 20(3):3268–3274. https://doi.org/10.1364/oe.20.003268

55. Nikiforov O, Jaurigue LC, Drzewietzki L, Lüdge K, Breuer S (2016) Experimental demonstration of change of dynamical properties of a passively mode-locked semiconductor laser subject to dual optical feedback by dual full delay-range tuning. Opt Express 24(13):14301–14310. https://doi.org/10.1364/oe.24.014301
56. Meinecke S, Drzewietzki L, Weber C, Lingnau B, Breuer S, Lüdge K (2019) Ultra-short pulse generation in a three section tapered passively mode-locked quantum-dot semiconductor laser. Sci Rep 9:1783. https://doi.org/10.1038/s41598-018-38183-1
57. Kolarczik M, Böhm F, Woggon U, Owschimikow N, Pimenov AS, Wolfrum M, Vladimirov AG, Meinecke S, Lingnau B, Jaurigue LC, Lüdge K (2020) Coherent and incoherent dynamics in quantum dots and nanophotonic devices. In: Semiconductor nanophotonics. Springer series in solid-state sciences, vol 194, p 91. Springer. ISBN 978-3-030-35655-2 (Chapter 4). https://doi.org/10.1007/978-3-030-35656-9_4
58. Meinecke S, Lüdge K (2021) Efficient timing jitter simulation for passively mode-locked semiconductor lasers. Appl Phys Lett 118(1):011104. https://doi.org/10.1063/5.0036928
59. Vladimirov AG, Turaev DV (2005) Model for passive mode locking in semiconductor lasers. Phys Rev A 72(3):033808. https://doi.org/10.1103/physreva.72.033808
60. Viktorov EA, Mandel P, Vladimirov AG, Bandelow U (2006) Model for mode locking of quantum dot lasers. Appl Phys Lett 88(20):201102. https://doi.org/10.1063/1.2203937
61. Vladimirov AG, Bandelow U, Fiol G, Arsenijević D, Kleinert M, Bimberg D, Pimenov AS, Rachinskii D (2010) Dynamical regimes in a monolithic passively mode-locked quantum dot laser. J Opt Soc Am B 27(10):2102. https://doi.org/10.1364/josab.27.002102
62. Vladimirov AG, Rachinskii D, Wolfrum M (2011) Modeling of passively mode-locked semiconductor lasers. In: Lüdge K (ed) Nonlinear laser dynamics—from quantum dots to cryptography. Reviews in nonlinear dynamics and complexity. Wiley-VCH, Weinheim, pp 183–213. ISBN 978-3-527-41100-9 (Chapter 8)
63. Rossetti M, Bardella P, Montrosset I (2011) Modeling passive mode-locking in quantum dot lasers: a comparison between a finite-difference traveling-wave model and a delayed differential equation approach. IEEE J Quantum Electron 47(5):569
64. Jaurigue LC, Krauskopf B, Lüdge K (2017) Multipulse dynamics of a passively mode-locked semiconductor laser with delayed optical feedback. Chaos 27(11):114301. https://doi.org/10.1063/1.5006743
65. Bandelow U, Radziunas M, Vladimirov AG, Hüttl B, Kaiser R (2006) 40 GHz mode locked semiconductor lasers: theory, simulation and experiments. Opt Quantum Electron 38:495. https://doi.org/10.1007/s11082-006-0045-2
66. Mulet J, Mørk J (2006) Analysis of timing jitter in external-cavity mode-locked semiconductor lasers. IEEE J Quantum Electron 42(3):249. https://doi.org/10.1109/jqe.2006.869808
67. Avrutin EA, Russell BM (2009) Dynamics and spectra of monolithic mode-locked laser diodes under external optical feedback. IEEE J Quantum Electron 45(11):1456
68. Radziunas M, Vladimirov AG, Viktorov EA, Fiol G, Schmeckebier H, Bimberg D (2011) Pulse broadening in quantum-dot mode-locked semiconductor lasers: simulation, analysis, and experiments. IEEE J Quantum Electron 47(7):935–943
69. Rossetti M, Bardella P, Montrosset I (2011) Time-domain travelling-wave model for quantum dot passively mode-locked lasers. IEEE J Quantum Electron 47(2):139
70. Bimberg D, Grundmann M, Ledentsov NN (1999) Quantum dot heterostructures. Wiley, New York. 978-0-471-97388-1
71. Rafailov EU, Cataluna MA, Avrutin EA (2011) Ultrafast lasers based on quantum dot structures. Wiley-VCH, Weinheim. 978-3-527-40928-0
72. Bayer M, Forchel A (2002) Temperature dependence of the exciton homogeneous linewidth in $In_{0.60}Ga_{0.40}As$/GaAs self-assembled quantum dots. Phys Rev B 65(4):041308
73. Lingnau B, Lüdge K, Herzog B, Kolarczik M, Kaptan Y, Woggon U, Owschimikow N (2016) Ultrafast gain recovery and large nonlinear optical response in submonolayer quantum dots. Phys Rev B 94(1):014305. https://doi.org/10.1103/physrevb.94.014305
74. Nambu Y, Tomita A, Saito H, Nishi K (1999) Effects of spectral broadening and cross relaxation on the gain saturation characteristics of quantum dot laser amplifiers. Jpn J Appl Phys 38:5087

75. Lei W, Offer M, Lorke A, Notthoff C, Meier C, Wibbelhoff O, Wieck AD (2008) Probing the band structure of InAs/GaAs quantum dots by capacitance-voltage and photoluminescence spectroscopy. Appl Phys Lett 92(19):193111. https://doi.org/10.1063/1.2920439
76. Lingnau B (2015) Nonlinear and nonequilibrium dynamics of quantum-dot optoelectronic devices. Springer theses. Springer International Publishing, Switzerland. 978-3-319-25803-4. https://doi.org/10.1007/978-3-319-25805-8
77. Kolarczik M, Owschimikow N, Korn J, Lingnau B, Kaptan Y, Bimberg D, Schöll E, Lüdge K, Woggon U (2013) Quantum coherence induces pulse shape modification in a semiconductor optical amplifier at room temperature. Nat Commun 4:2953. https://doi.org/10.1038/ncomms3953
78. Nielsen TR, Gartner P, Jahnke F (2004) Many-body theory of carrier capture and relaxation in semiconductor quantum-dot lasers. Phys Rev B 69:235314. https://doi.org/10.1103/physrevb.69.235314
79. Majer N, Lüdge K, Schöll E (2010) Cascading enables ultrafast gain recovery dynamics of quantum dot semiconductor optical amplifiers. Phys Rev B 82:235301. https://doi.org/10.1103/physrevb.82.235301
80. Lüdge K, Schöll E (2009) Quantum-dot lasers—desynchronized nonlinear dynamics of electrons and holes. IEEE J Quantum Electron 45(11):1396–1403. https://doi.org/10.1109/jqe.2009.2028159
81. Bardella P, Columbo L, Gioannini M (2017) Self-generation of optical frequency comb in single section quantum dot Fabry-Perot lasers: a theoretical study. Opt Express 25(21):26234–26252
82. Dong M, Mangan NM, Kutz JN, Cundiff ST, Winful HG (2017) Traveling wave model for frequency comb generation in single-section quantum well diode lasers. IEEE J Quantum Electron 53(6):2500311. https://doi.org/10.1109/jqe.2017.2756641
83. Lingnau B, Dillane M, O'Callaghan J, Corbett B, Kelleher B (2019) Multimode dynamics and modelling of free-running and optically injected Fabry-Pérot quantum-dot lasers. Phys Rev A 100(6):063837
84. Javaloyes J, Balle S (2012) Multimode dynamics in bidirectional laser cavities by folding space into time delay. Opt Express 20(8):8496–8502. https://doi.org/10.1364/oe.20.008496
85. Lingnau B, Lüdge K (2018) Rabi-oscillation-enhanced frequency conversion in quantum-dot semiconductor optical amplifiers. Opt Quantum Electron 50(2):111. ISSN 1572-817X. https://doi.org/10.1007/s11082-018-1380-9
86. Pérez-Serrano A, Vilera M, Tijero JMG, Balle S, Esquivias I (2018) A voltage driven traveling-wave model for the simulation of an integrated three-section MOPA under static and modulated operation. IEEE J Quantum Electron 54(2):1–10. ISSN 0018-9197. https://doi.org/10.1109/jqe.2018.2797888
87. Lingnau B, Chow WW, Schöll E, Lüdge K (2013) Feedback and injection locking instabilities in quantum-dot lasers: a microscopically based bifurcation analysis. New J Phys 15:093031. https://doi.org/10.1088/1367-2630/15/9/093031
88. Lingnau B, Chow WW, Lüdge K (2014) Amplitude-phase coupling and chirp in quantum-dot lasers: influence of charge carrier scattering dynamics. Opt Express 22(5):4867–4879. https://doi.org/10.1364/oe.22.004867
89. Petermann K (1988) Laser diode modulation and noise. KTK Scientific Publishers, Tokyo
90. Coldren LA, Corzine SW, Mashanovitch M (2012) Diode lasers and photonic integrated circuits. In: Wiley series in microwave and optical enginieering, 2nd edn. Wiley
91. Haug H, Koch SW (2004) Quantum theory of the optical and electronic properties of semiconductors, 2nd edn. World Scientific, Singapore
92. de Melo AM, Petermann K (2008) On the amplified spontaneous emission noise modeling of semiconductor optical amplifiers. Opt Commun 281:4598–4605
93. Gardiner CW (2002) Handbook of stochastic methods for physics, chemistry and the natural sciences. Springer, Berlin
94. Malins DB, Gomez-Iglesias A, White SJ, Sibbett W, Miller A, Rafailov EU (2006) Ultrafast electroabsorption dynamics in an InAs quantum dot saturable absorber at 1.3 μm. Appl Phys Lett 89(17):171111. ISSN 0003-6951. https://doi.org/10.1063/1.2369818

95. Breuer S, Rossetti M, Drzewietzki L, Montrosset I, Krakowski M, Hopkinson M, Elsäßer W (2013) Dual-state absorber-photocurrent characteristics and bistability of two-section quantum-dot lasers. IEEE J Sel Top Quantum Electron 19(5):1901609. https://doi.org/10.1109/jstqe.2013.2255264
96. Viktorov EA, Erneux T, Mandel P, Piwonski T, Madden G, Pulka J, Huyet G, Houlihan J (2009) Recovery time scales in a reversed-biased quantum dot absorber. Appl Phys Lett 94(26):263–502. https://doi.org/10.1063/1.3159838
97. Wegert M, Schwochert D, Schöll E, Lüdge K (2014) Integrated quantum-dot laser devices: modulation stability with electro-optic modulator. Opt Quantum Electron 46:1337–1344. https://doi.org/10.1007/s11082-014-9878-2
98. Lorke M, Nielsen TR, Seebeck J, Gartner P, Jahnke F (2006) Influence of carrier-carrier and carrier-phonon correlations on optical absorption and gain in quantum-dot systems. Phys Rev B 73:085324. https://doi.org/10.1103/physrevb.73.085324
99. Xu T, Bardella P, Rossetti M, Montrosset I (2012) Beam propagation method simulation and analysis of quantum dot flared semiconductor optical amplifiers in continuous wave high-saturation regime. IET Optoelectron 6(2):110–116. https://doi.org/10.1049/iet-opt.2011.0056
100. Press WH, Flannery BP, Teukolsky SA, Vettering WT (2007) Numerical recipes, 3rd edn. Cambridge University Press, Cambridge
101. New G (1974) Pulse evolution in mode-locked quasi-continuous lasers. IEEE J Quantum Electron 10(2):115. ISSN 0018-9197. https://doi.org/10.1109/jqe.1974.1145781
102. Lüdge K (2012) Nonlinear laser dynamics—from quantum dots to cryptography. Wiley-VCH, Weinheim. 978-3-527-41100-9
103. Herzog B, Lingnau B, Kolarczik M, Kaptan Y, Bimberg D, Maaßdorf A, Pohl UW, Rosales R, Schulze JH, Strittmatter A, Weyers M, Woggon U, Lüdge K, Owschimikow N (2016) Strong amplitude-phase coupling in submonolayer quantum dots. Appl Phys Lett 109:201102. https://doi.org/10.1063/1.4967833
104. Otto C (2014) Dynamics of quantum dot lasers—effects of optical feedback and external optical injection. Springer theses. Springer, Heidelberg. 978-3-319-03785-1. https://doi.org/10.1007/978-3-319-03786-8
105. Lu ZG, Poole PJ, Liu JR, Barrios PJ, Jiao ZJ, Pakulski G, Poitras D, Goodchild D, Rioux B, Springthorpe AJ (2011) High-performance 1.52 μm InAs/InP quantum dot distributed feedback laser. Electron Lett 47(14):818–819. ISSN 0013-5194. https://doi.org/10.1049/el.2011.0946
106. Haus HA (1975) Theory of mode locking with a fast saturable absorber. J Appl Phys 46(7):3049–3058. https://doi.org/10.1063/1.321997
107. Derickson DJ, Helkey RJ, Mar A, Karin JR, Wasserbauer JG, Bowers JE (1992) Short pulse generation using multisegment mode-locked semconductor lasers. IEEE J Quantum Electron 28(10):2186–2202
108. Ippen EP (1994) Principles of passive mode locking. Appl Phys B 58:159–170. https://doi.org/10.1007/bf01081309
109. Haus HA (2000) Mode-locking of lasers. IEEE J Sel Top Quantum Electron 6(6):1173–1185. https://doi.org/10.1109/2944.902165
110. Avrutin EA, Marsh JH, Portnoi EL (2000) Monolithic and multi-gigahertz mode-locked semiconductor lasers: constructions, experiments, models and applications. IEE Proc Optoelectron 147(4):251. https://doi.org/10.1049/ip-opt:20000282
111. Fork R, Shank C, Yen R, Hirlimann C (1983) Femtosecond optical pulses. IEEE J Quantum Electron 19(4):500–506. ISSN 0018-9197. https://doi.org/10.1109/jqe.1983.1071898
112. Bischoff S, Sørensen MP, Mørk J, Brorson SD, Franck T, Nielsen JM, Møller-Larsen A (1995) Pulse-shaping mechanism in colliding-pulse mode-locked laser diodes. Appl Phys Lett 67(26):3877–3879. https://doi.org/10.1063/1.115303
113. Bischoff S, Mørk J, Franck T, Brorson SD, Hofmann M, Fröjdh K, Prip L, Sørensen MP (1997) Monolithic colliding pulse mode-locked semiconductor lasers. Quantum Semiclass Opt 9(5):655–674. https://doi.org/10.1088/1355-5111/9/5/002

114. Weber C, Klehr A, Knigge A, Breuer S (2018) Picosecond pulse generation and pulse train stability of a monolithic passively mode-locked semiconductor quantum-well laser at 1070 nm. IEEE J Quantum Electron 54(3):1–9. ISSN 0018-9197. https://doi.org/10.1109/jqe.2018.2832288
115. Kefelian F, O'Donoghue S, Todaro MT, McInerney JG, Huyet G (2008) RF linewidth in monolithic passively mode-locked semiconductor laser. IEEE Photon Technol Lett 20(16):1405. ISSN 1041-1135. https://doi.org/10.1109/lpt.2008.926834
116. Weber C, Drzewietzki L, Breuer S (2016) Amplitude jitter and timing jitter characterization of a monolithic high-power passively mode-locked tapered quantum dot laser. Proc SPIE 9892:9892–989. https://doi.org/10.1117/12.2229739
117. Huang X, Stintz A, Li H, Lester LF, Cheng J, Malloy KJ (2001) Passive mode-locking in 1.3 μm two-section InAs quantum dot lasers. Appl Phys Lett 78(19):2825–2827
118. Gubenko A, Livshits D, Krestnikov I, Mikhrin SS, Kozhukhov AV, Kovsh AR, Ledentsov N, Zhukov A, Portnoi EL (2005) High-power monolithic passively modelocked quantum-dot laser. Electron Lett 41(20):1124–1125
119. Zhang L, Cheng L, Gray AL, Huang H, Kutty SPG, Li H, Nagyvary J, Nabulsi F, Olona L, Pease E (2006) High-power low-jitter quantum-dot passively mode-locked lasers. In: Physics and simulation of optoelectronic devices XIV, vol 6115. International Society for Optics and Photonics, SPIE, p 611502
120. Carpintero G, Thompson MG, Penty RV, White IH (2009) Low noise performance of passively mode-locked 10-GHz quantum-dot laser diode. IEEE Photon Technol Lett 21(6):389–391. https://doi.org/10.1109/lpt.2008.2011918
121. Lin CY, Grillot F, Li Y, Raghunathan R, Lester LF (2010) Characterization of timing jitter in a 5 GHz quantum dot passively mode-locked laser. Opt Express 18(21):21932. https://doi.org/10.1364/oe.18.021932
122. Liu S, Norman JC, Jung D, Kennedy MJ, Gossard AC, Bowers JE (2018) Monolithic 9 GHz passively mode locked quantum dot lasers directly grown on on-axis (001) Si. Appl Phys Lett 113(4):041108. https://doi.org/10.1063/1.5043200
123. Liu S, Wu X, Jung D, Norman JC, Kennedy MJ, Tsang HK, Gossard AC, Bowers JE (2019) High-channel-count 20 GHz passively mode-locked quantum dot laser directly grown on Si with 4.1 Tbit/s transmission capacity. Optica 6(2):128–134. https://doi.org/10.1364/optica.6.000128
124. Auth D, Liu S, Norman J, Bowers JE, Breuer S (2019) Passively mode-locked semiconductor quantum dot on silicon laser with 400 Hz RF line width. Opt Express 27(19):27256–27266. https://doi.org/10.1364/oe.27.027256
125. Krakowski M, Resneau P, Garcia M, Vinet E, Robert Y, Parillaud O, Gérard B, Kundermann S, Torcheboeuf N, Boiko DL (2019) Stabilized high pulse energy passively mode-locked monolithic and external cavity tapered lasers for space applications. IEEE J Sel Top Quantum Electron 25(6):1100615. ISSN 1077-260X. https://doi.org/10.1109/jstqe.2019.2927577
126. Hausen J, Meinecke S, Lingnau B, Lüdge K (2019) Pulse cluster dynamics in passively mode-locked semiconductor vertical-external-cavity surface-emitting lasers. Phys Rev Appl 11:044055. https://doi.org/10.1103/physrevapplied.11.044055
127. Grillot F, Lin CY, Naderi NA, Pochet M, Lester LF (2009) Optical feedback instabilities in a monolithic InAs/GaAs quantum dot passively mode-locked laser. Appl Phys Lett 94(15):153503. https://doi.org/10.1063/1.3114409
128. Kuntz M, Fiol G, Lämmlin M, Bimberg D, Thompson MG, Tan KT, Marinelli C, Wonfor A, Sellin RL, Penty RV, White IH, Ustinov VM, Zhukov AE, Shernyakov YM, Kovsh AR, Ledentsov NN, Schubert C, Marembert V (2004) Direct modulation and mode locking of 1.3 μm quantum dot lasers. New J Phys 6:181
129. Schmeckebier H, Fiol G, Meuer C, Arsenijević D, Bimberg D (2010) Complete pulse characterization of quantum dot mode-locked lasers suitable for optical communication up to 160 Gbit/s. Opt Express 18(4):3415. https://doi.org/10.1364/oe.18.003415
130. Simos H, Simos C, Mesaritakis C, Syvridis D (2015) Amplitude and timing noise in a noncolliding passively mode-locked quantum dot laser. IEEE Photon Technol Lett 27(5):506–509

131. Marconi M, Javaloyes J, Balle S, Giudici M (2014) How lasing localized structures evolve out of passive mode locking. Phys Rev Lett 112:223901. https://doi.org/10.1103/physrevlett.112.223901
132. Marconi M, Javaloyes J, Balle S, Giudici M (2015) Passive mode-locking and tilted waves in broad-area vertical-cavity surface-emitting lasers. IEEE J Sel Top Quantum Electron 21(1):85–93. ISSN 1077-260X. https://doi.org/10.1109/jstqe.2014.2346614
133. Camelin P, Javaloyes J, Marconi M, Giudici M (2016) Electrical addressing and temporal tweezing of localized pulses in passively-mode-locked semiconductor lasers. Phys Rev A 94:063854. https://doi.org/10.1103/physreva.94.063854
134. Hausen J, Meinecke S, Javaloyes J, Gurevich SV, Lüdge K (2020) Phase-incoherent photonic molecules in V-shaped mode-locked vertical-external-cavity surface-emitting semiconductor lasers. Phys Rev Appl 14:044059. https://doi.org/10.1103/physrevapplied.14.044059
135. Schelte C, Hessel D, Javaloyes J, Gurevich SV (2020) Dispersive instabilities in passively mode-locked integrated external-cavity surface-emitting lasers. Phys Rev Appl 13:054050. https://doi.org/10.1103/physrevapplied.13.054050
136. Auth D, Weber C, Breuer S, Bardella P, Columbo L (2019) Bistability in a monolithic multi-section quantum dot semiconductor laser. In: Novel in-plane semiconductor lasers XVIII, vol 10939. SPIE, pp 205–210. https://doi.org/10.1117/12.2510221
137. Viktorov EA, Mandel P, Kuntz M, Fiol G, Bimberg D, Vladimirov AG, Wolfrum M (2007) Stability of the mode-locked regime in quantum dot lasers. Appl Phys Lett 91(23):231116. https://doi.org/10.1063/1.2822808
138. Pimenov AS, Habruseva T, Rachinskii D, Hegarty SP, Huyet G, Vladimirov AG (2014) The effect of dynamical instability on timing jitter in passively mode-locked quantum-dot lasers. Opt Lett 39(24):6815. https://doi.org/10.1364/ol.39.006815
139. Jaurigue LC, Pimenov AS, Rachinskii D, Schöll E, Lüdge K, Vladimirov AG (2015) Timing jitter of passively mode-locked semiconductor lasers subject to optical feedback; a semi-analytic approach. Phys Rev A 92(5):053807. https://doi.org/10.1103/physreva.92.053807
140. Haus HA, Mecozzi A (1993) Noise of mode-locked lasers. IEEE J Quantum Electron 29(3):983. https://doi.org/10.1109/3.206583
141. Jiang LA, Grein ME, Haus HA, Ippen EP (2001) Noise of mode-locked semiconductor lasers. IEEE J Sel Top Quantum Electron 7(2):159. https://doi.org/10.1109/2944.954125
142. Jaurigue LC (2017) Passively mode-locked semiconductor lasers: dynamics and stochastic properties in the presence of optical feedback. Springer thesis edn. Springer, Cham. ISBN 978-3-319-58874-2. https://doi.org/10.1007/978-3-319-58874-2
143. Otto C, Jaurigue LC, Schöll E, Lüdge K (2014) Optimization of timing jitter reduction by optical feedback for a passively mode-locked laser. IEEE Photonics J 6(5):1501814. https://doi.org/10.1109/jphot.2014.2352934

Chapter 5
V-Shaped External Cavity Laser

5.1 Introduction

Passively mode-locked vertical-external-cavity surface-emitting lasers (VECSELs) based on semiconductor gain chips[1] are capable of generating ultra short and high power pulses [1–4]. They compete with solid-state lasers, e.g., ones based on the Ti:sapphire [5, 6] and Nd:YAG [7, 8] platform, and fiber lasers [9, 10], on the one hand, and with monolithically integrated semiconductor lasers [11, 12], e.g., as presented in Chap. 4, on the other hand. While they do not achieve the pulse performance of the solid-state based lasers, they offer more emission wavelength flexibility due to the diverse semiconductor band gap engineering technologies and are comparatively inexpensive due to mature semiconductor fabrication processes [3, 4]. Nevertheless, mode-locked VECSELs outperform edge-emitting mode-locked semiconductor lasers by achieving pulse peak powers on the order of a few kilowatts [13] instead of watts and pulse widths as low as 100 fs [14] instead of picoseconds (s. Table 4.2).

The external cavity setup inherently provides flexible and thus adaptable cavity designs, since the individual components can be chosen and configured independently. This allows for repetition rates ranging from 85 MHz [15] to 190 GHz [16, 17], the simple introduction of additional optical elements, such as a birefringent crystal [18], and the construction of elaborate multi-pass geometries [15, 19–21]. The latter is employed to achieve higher pulse peak powers by overcoming multi-pulse instability inducing upper state lifetime limitations [22]. This chapters' V-shaped geometry represents a basic and thus easy to configure multi-pass geometry.

Due to the described properties, mode-locked VECSELs find many applications in frequency comb generation for metrology and spectroscopy [2, 23–26], supercontinuum generation [4, 27–29], and two-photon microscopy [30, 31]. All applica-

[1] Also known as semiconductor disk lasers.

S. Meinecke, *Spatio-Temporal Modeling and Device Optimization of Passively Mode-Locked Semiconductor Lasers*, Springer Theses,
https://doi.org/10.1007/978-3-030-96248-7_5

tions benefit from ultra short and high peak power pulses. Hence, the optimization of mode-locked VECSELs is highly desirable and an ongoing quest in both science and the industry.

5.1.1 V-Shaped External Cavity Semiconductor Lasers

The considered V-shaped surface-emitting semiconductor laser is comprised of three optical components as depicted in Fig. 5.1: an output coupler (gray), a gain chip (green), and a saturable absorber chip (red). They are arranged such that the electric field (light blue), which is generated in between the components, geometrically forms the letter *V* and thus gives rise to the characteristic name. The passive free space in between the optical elements results in the field propagation time τ_1 between the output coupler and the gain chip and the field propagation time τ_2 between the gain chip and the absorber chip. Collectively, all components form a laser cavity, where the light-matter interaction is spatially strongly localized to the gain and absorber chip. The dynamics thus arise due to the interplay of propagation times between the different elements and the recovery times of the respective charge-carrier distributions.

Such configurations are popular with semiconductor gain materials and have therefore been realized experimentally many times with examples given in Refs. [13, 14, 32–34]. The device investigated in this chapter is based on the one presented in [35], which has been introduced and characterized in detail in [4]. The following paragraphs summarize the relevant features and additionally discuss the general design requirements of the three individual components.

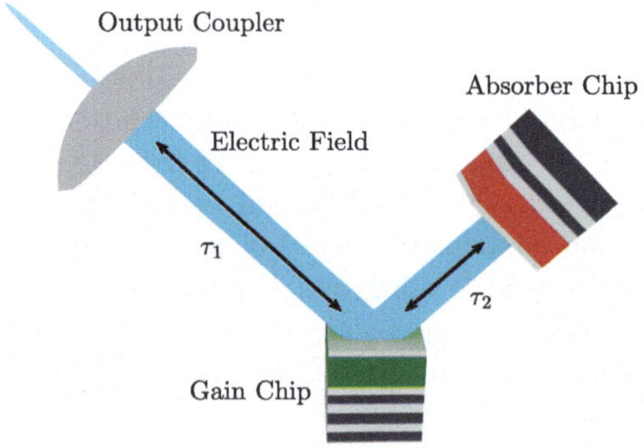

Fig. 5.1 Sketch of the V-shaped external cavity laser. The setup contains three optical components: a high reflectivity output coupler, a semiconductor gain chip and a semiconductor saturable absorber chip. The two latter are mounted on top of high reflectivity DBRs. The active regions are indicated by green and red colors, respectively. The length of the cavity arms is denoted by their propagation times τ_1 and τ_2. Based on artwork by Jan Hausen

Despite its name, the output coupler is a highly reflective mirror, which only outcouples approximately 1% of the light and thereby ensures high intra-cavity electric field intensities and long photon lifetimes.

The gain chip provides the optical amplification via a semiconductor active medium that is subject to an inversion generating pump mechanism. It is, furthermore, bundled with a high reflectivity distributed Bragg reflector (DBR) on its backside. In the case of the considered laser, the active medium contains ten surface oriented InGaAs quantum wells, which are embedded within layers of GaAs. They can be optically pumped by an additional diode laser, which generates electron-hole pairs far above the band gap. Semiconductor quantum wells are an ubiquitous choice for semiconductor gain chips due to their mature fabrication technology, which provides great flexibility and has lead to excellent results [2, 3, 17, 36]. Nonetheless, semiconductor gain chips have also been realized with Stranski-Krastanov quantum dots [37–40] and submonolayer quantum dots [41]. If the active region is further more grown into a *p-i-n* scheme and contacted, the gain chip can also be electrically pumped [42–44].

Lastly, the absorber chip introduces the nonlinear saturable losses, which are required for passive mode-locking, to the laser cavity. If, furthermore, bundled with a high reflectivity distributed Bragg reflector, such chips are commonly referred to as semiconductor saturable absorber mirrors (SESAMs) [45, 46]. In order to facilitate efficient passive mode-locking, such absorber chips have to exhibit a smaller saturation energy than the corresponding gain chip and a fast recovery process. The former can be easily achieved by stacking less quantum-well (or quantum-dot-in-a-well) layers and the latter by growing the absorber active region at low temperatures and introducing additional barriers [4, 40, 46]. In order to further reduce the recovery time, the absorber chip can be contacted and reverse biased [47, 48]. This also introduces the reverse bias as an additional control parameter to the experiment. Similar to the semiconductor gain chips, the absorber active medium can incorporate quantum wells [2, 4, 46], Stranski-Krastanov quantum dots [40], and submonolayer quantum dots [49]. The considered device, however, utilizes a simple absorber chip that contains a single quantum well, which is not subject to a reverse bias.

5.1.2 Experimentally Observed Emission States

This chapters' investigations are motivated by the experimental characterization of the passively mode-locked emission dynamics, which have been presented in Ref. [35] in detail. The goal, however, is not to quantitatively reproduce all measurements, but to derive a minimal model, which qualitatively replicates the different emission states and thus allows to understand their underlying mechanisms [50]. Hence, this subsection briefly summarizes the relevant experimental results, which characterize the observed distinct mode-locking states. For more details including quantitative pulse performance and pulse stability properties see Refs. [4, 35].

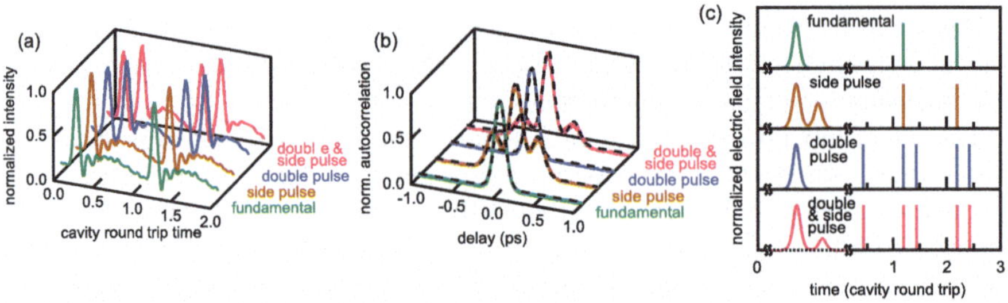

Fig. 5.2 Experimentally observed passively mode-locked emission states. For an increasing pump current, fundamental mode-locking (green), fundamental mode-locking with a side pulse (yellow), pulse cluster mode locking with two pulses (purple), and pulse cluster mode-locking with two pulses and side pulses (pink) are observed. Subfigure **a** plots sampling oscilloscope time traces, **b** plots normalized intensity autocorrelation traces, and **c** shows sketches of the time series. Adapted with permission from [35] © The Optical Society (Figs. 2, 3, and 7)

On that account, Fig. 5.2 presents sampling oscilloscope time traces (a), intensity autocorrelation traces (b), and sketched pulse pattern dynamics (c).[2] For an increasing pump current four distinct mode-locking states are observed: fundamental mode-locking (green), fundamental mode-locking with a side pulse (yellow), pulse-cluster mode locking with two pulses (purple, labeled with double pulse), and pulse-cluster mode-locking with two pulses and side pulses (pink, labeled with double pulse).

The normalized sampling oscilloscope time traces (a) are not capable of resolving the exact pulse shapes, but nonetheless reveal the temporal pulse positions. In the fundamental mode-locking and fundamental mode-locking with a side pulse state one pulse circulates within the laser cavity with a repetition rate close to the cold-cavity round-trip time T. The pulse cluster and pulse cluster with a side pulse state are characterized by two non-equidistantly spaced pulses, which are separated by approximately $T/4$ from each other.

The intensity autocorrelation traces, on the other hand, do not convey the full time-domain dynamics but provide access to the pulse shape. The fundamental and pulse-cluster mode-locking states exhibit sech^2-shaped pulses[3] with a deconvoluted FWHM of ≈ 100 fs. As the name suggests, the fundamental and pulse-cluster states with a side pulses show side pulses at both sides of the autocorrelation peak. The side pulses exhibit attenuated maxima and are located at an autocorrelation shift, which is on the order of the pulse width.

Lastly, Fig. 5.2c sketches time series that would produce the observed sampling oscilloscope and autocorrelation time series. The left side presents short slices of time, which exemplify possible deconvoluted pulse profiles and the right side presents larger time intervals, which highlight the pulse positions but obscure to pulse shapes.

[2] All presented experiments have been performed by Dominik Waldburger in the group of Ursula Keller at the ETH Zürich.

[3] Soliton mode-locking is characterized by sech^2-shaped pulses, which exhibit almost no chirp and are close to transform limited [46, 51].

The questions that are to be answered regard the mechanism that generates the double pulse states as well as the intensity autocorrelation side pulses. Since the experimental measurements are limited by the detection bandwidth of the electronics, a suitable numerical model is required, which provides access to the complete spatio-temporal electric field a the gain and absorber chip charge-carrier dynamics and thereby helps to unravel the underlying mode-locking mechanics.

5.1.3 Chapter Outline

This chapter studies the emission dynamics of a V-shaped external cavity mode-locked laser by the means of a relatively simple delay-differential equation system.[4] A special focus is put on the mechanisms that lead to structured multi-pulse emission states and their relation to asymmetrically configured laser cavities.

Following the introduction, which has presented the device as well as the experiment, this chapter continues in Sect. 5.2 by deriving the delay-differential equation model based on the semiconductor laser modeling fundamentals, which have been discussed in Chap. 2. The obtained model initially does not specify the gain medium and is adapted in the next step to the semiconductor quantum-well active media of the considered laser.

Section 5.3 presents the observed emission dynamics for a symmetric V-shaped cavity configuration. The obtained dynamics are discussed in terms of the pump current operation parameter as well as the unsaturated absorption and gain bandwidth device parameters. The gain and absorber chip dynamics are utilized to understand both the pulse-shaping mechanism and the occurrence of various multi-pulse emission states.

Lastly, the impact of geometrically asymmetric laser cavity configurations is investigated in Sect. 5.4. A focus is placed on the stability boundaries of the fundamental mode-locking state for which analytic approximations are derived. Those, furthermore, explain the different occurring instabilities and predict an optimal geometric cavity configuration. The findings, furthermore, allow for a discussion of the experimental results, which is presented in Sect. 5.4.2. Finally, the chapter conclusions are drawn in Sect. 5.5.

5.2 V-Shaped External Cavity Laser Model

This section presents the detailed derivation of the delay-differential equation model for the passively mode-locked V-shaped external cavity semiconductor laser. The simulation challenge is posed by large photon lifetimes up to microsecond

[4] Parts of this chapter have been published in Refs. [35, 50, 52].

timescales,[5] which lead to very long transient times for the laser to settle on a mode-locked solution. Hence, a model that is as simple as possible, while still maintaining the essential geometric features of the laser cavity, is required.

For that reason, the chosen modeling approach requires that the light-matter interaction is not mediated by a dynamical polarization but rather by a charge-carrier number dependent gain function. Coming from the Maxwell-Bloch equations, such an interaction is typically obtained by adiabatically eliminating the dynamical polarization equations as presented in Sect. 2.5. This approach has been shown to work well, as long as the considered electric field dynamics are significantly slower than the polarization dephasing time [53–55]. The disadvantage, however, is that the spectral features of the gain medium are lost and must be added retroactively to account for the laser's gain bandwidth. The advantage of the resulting interaction nonetheless is that it lends itself to an advanced integration scheme for the traveling-wave equation, which was first proposed in Ref. [56] for a linearized quantum-well gain function and has been derived in detail in Sect. 2.5 for an arbitrary gain function.

Based on this, the derivation of a minimal delay-differential equation model for the V-shaped external cavity laser is presented in the following subsections. The resulting equations are applicable to a wide range of active media, since the gain function has not been specified up to this point. This allows to easily implement different active media, by simply adapting the gain function and supplying the respective equations of motion for the charge-carrier dynamics.

For convenience, the starting point and the key results regarding the integration scheme for electric field and active medium dynamics are reproduced from Chap. 2 in the following: The traveling-wave equation for the forward (+) and backward (−) moving slowly-varying electric field amplitude (s. Eq. (2.87)) is written as

$$\left[\pm\partial_z + \frac{1}{v_g}\partial_t\right] E^{\pm}(z,t) = \left[-\frac{\alpha_{\text{int}}}{2} + \frac{g\left(\rho(z,t)\right)}{2}\right] E^{\pm}(z,t), \tag{5.1}$$

where g is a complex valued gain function that depends on the charge-carrier number $\rho(z,t)$. α_{int} describes linear non-resonant losses. The corresponding equation for the charge-carrier dynamics (s. Eq. (2.88)) is given by

$$\frac{\mathrm{d}}{\mathrm{d}t}\rho(z,t) = -\frac{\rho(z,t)}{\tau_\rho} + R - v_g \mathrm{Re}\left\{g\left(\rho(z,t)\right)\right\}\left[|E^+(z,t)|^2 + |E^-(z,t)|^2\right], \tag{5.2}$$

where τ_ρ denotes the charge-carrier lifetime, R summarizes all charge-carrier scattering processes and v_g denotes the group velocity. All parameters may have a spatial dependence, which is, however, not explicitly written in the equations above. Applying a suitable spatial discretization z_k and following Sect. 2.5 produces the algebraic propagation equations

[5] The long photon lifetimes result from the large cavities and the high-reflectivity components.

$$E^+(z_{k+1}, t) = E^+(z_k, t - \tau_k) \exp\left(-\frac{\alpha_{\text{int}}}{2}\Delta z_k + \frac{1}{2}G_k(t)\right) \tag{5.3}$$

$$E^-(z_k, t) = E^+(z_{k+1}, t - \tau_k) \exp\left(-\frac{\alpha_{\text{int}}}{2}\Delta z_k + \frac{1}{2}G_k(t)\right) \tag{5.4}$$

for the forward and backward-traveling electric field and the differential equation

$$\frac{\mathrm{d}}{\mathrm{d}t}\langle\rho\rangle_k(t) = -\frac{\langle\rho\rangle_k(t)}{\tau_\rho} + R - \frac{v_g}{\Delta z_k} e^{-\frac{\alpha_{\text{int}}}{2}\Delta z_k}\left(e^{\text{Re}\{G_k(t)\}} - 1\right) \times \left[\left|E^+(z_k, t - \tau_k))\right|^2 + \left|E^-(z_{k+1}, t - \tau_k))\right|^2\right] \tag{5.5}$$

for the spatially averaged carrier number in the kth section. Δz_k and τ_k denote the length of the kth section and the associated propagation time. Lastly, $G_k(t)$ denotes the integrated gain function, which is approximated by

$$G_k(t) = \int_{z_k}^{z_{k+1}} \mathrm{d}\zeta\, g\left(\rho\left(\zeta, t - \frac{\Delta t_k}{2}\right)\right) \approx \Delta z_k g\left(\langle\rho\rangle_k(t)\right). \tag{5.6}$$

All details regarding the derivation can be found in Sect. 2.5.

5.2.1 *V-Shaped External Cavity*

To construct a model for the V-shaped external cavity laser, a 'straightened' cavity geometry, which only contains four sections, is utilized: Two passive sections that correspond to the free space cavity arms with the propagation times τ_1 and τ_2 and one section each for the semiconductor gain chip and the semiconductor saturable absorber mirror. A sketch of the 'straightened' setup is shown in Fig. 5.3. The bottom row indicates the relevant positions z_k within the laser cavity, where OC denotes the output coupler, G and G′ the left and right sides of the gain section and Q and Q′ the left and right side of the absorber section. Contrary to the real geometry, the interaction with the gain chip is modeled by a single pass through the active region with the length ℓ_G. G_G and G_Q denote the integrated gain within the gain and absorber chip.

The use of only one numerical section for both the gain and absorber is justified by the slowly varying envelope approximation. The gain and the absorber chip typically only extend for a few wavelengths [4] along the optical axis of the laser cavity. Hence, as long as the traveling-wave description is valid, the electric field envelope is approximately constant in space along the gain and absorber chip. Thus, a more detailed spatial discretization is only reasonable if the full wave equation Eq. (2.8) is used.

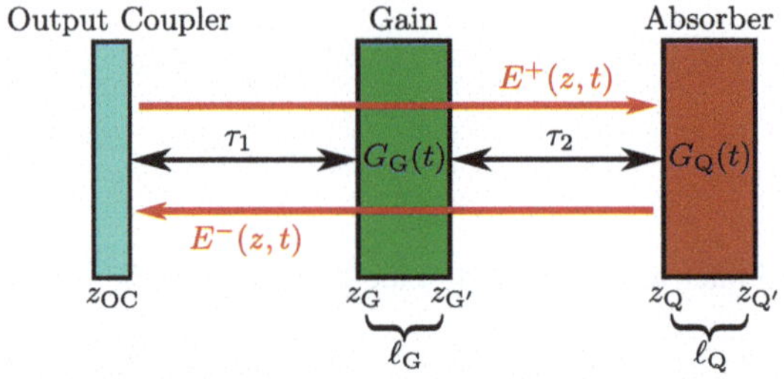

Fig. 5.3 Sketch of the straightened V-shaped cavity. The length of the cavity arms is encoded by the propagation times τ_1 and τ_2. The relevant cavity positions z_k are denoted on the bottom. The active region widths of the gain and absorber sections are drawn vastly out of proportion to highlight the widths $\ell_{G,Q}$. G_G and G_Q denote the integrated gain within the gain and absorber chip

Using suitable boundary conditions for the reflections at the optical elements and the electric field propagation equations (5.3) and (5.4), a closed system of delay-coupled algebraic equations for the electric field evolution at the relevant positions z_k can be formulated:

$$E^+(z_{OC},t) = \sqrt{r_{OC}}E^-(z_{OC},t) \tag{5.7}$$
$$E^+(z_G,t) = E^+(z_{OC},t-\tau_1) \tag{5.8}$$
$$E^+(z_{G'},t) = \sqrt{r_G}E^+(z_G,t-\ell_G/v_g)\,e^{-\frac{\alpha_{int}\ell_G}{2}+\frac{G_G(t)}{2}} \tag{5.9}$$
$$E^+(z_Q,t) = E^+(z_{G'},t-\tau_2) \tag{5.10}$$
$$E^+(z_{Q'},t) = E^+(z_Q,t-\ell_Q/v_g)\,e^{-\frac{\alpha_{int}\ell_Q}{2}+\frac{G_Q(t)}{2}} \tag{5.11}$$
$$E^-(z_{Q'},t) = \sqrt{r_Q}E^+(z_{Q'},t) \tag{5.12}$$
$$E^-(z_Q,t) = E^-(z_{Q'},t-\ell_Q/v_g)\,e^{-\frac{\alpha_{int}\ell_Q}{2}+\frac{G_Q(t)}{2}} \tag{5.13}$$
$$E^-(z_{G'},t) = E^-(z_{Q'},t-\tau_2) \tag{5.14}$$
$$E^-(z_G,t) = \sqrt{r_G}E^-(z_{G'},t-\ell_G/v_g)\,e^{-\frac{\alpha_{int}\ell_G}{2}+\frac{G_G(t)}{2}} \tag{5.15}$$
$$E^-(z_{OC},t) = E^-(z_G,t-\tau_1). \tag{5.16}$$

Equations (5.7) and (5.11) represent the reflections at the left and right side of the straightened laser cavity with the intensity reflectivity coefficients r_{PC} and r_Q. Similarly, the reflection coefficient r_G in Eqs. (5.9) and (5.15) accounts for the reflection at the gain-chip DBR that occurs in the real laser.

To proceed, reflection losses at the gain and absorber chip are neglected, such that the respective reflection coefficients are set to $r_G = r_Q = 1$. Moreover, the propagation times through the gain and absorber chip active regions, which are both only a few optical wavelengths long, are much shorter than all other timescales and are therefore neglected. Combining Eq. (5.7) to (5.16) and utilizing both approximations, the electric field $E^-(z_{OC},t)$ that arrives at the output coupler can be written as

$$E^-(z_{OC}, t) = E^+(z_{OC}, t - T)e^{\frac{1}{2}G_G(t-\tau_1)}e^{G_Q(t-\tau_1-\tau_2)}e^{\frac{1}{2}G_G(t-\tau_1-2\tau_2)}, \tag{5.17}$$

where $T = 2\tau_1 + 2\tau_2$ is the cold-cavity round-trip time. Thus, the electric field $E^-(z_{OC}, t)$ is constructed from the reflected electric field $E^+(z_{OC}, t - T)$, which is delayed by one cold-cavity round-trip time T and the respective interactions with the gain and absorber chip along the propagation. In this equation, the cavity geometry is encoded by the delay times T, τ_1 and τ_2. The interaction with the gain chip G_G appears twice and corresponds to the interaction with the forward and backward-moving field. The interaction with the absorber chip, on the other hand, only appears once due to the neglected propagation time through the absorber chip. The forward and backward interaction is, however, accounted for by the missing factor $1/2$.

Modeling the Gain Bandwidth

The elimination of the polarization dynamics, which was required for the integration scheme, has also lead to an infinitely broad gain spectrum. Hence, as mentioned before, the modeling of the gain spectrum has to be done retroactively in order to self-consistently describe the evolution of passively mode-locked pulses. For this reason, the boundary condition Eq. (5.7) at the output coupler is reformulated in frequency space and reads

$$\hat{E}^+(z_{OC}, \omega) = \sqrt{r_{OC}}\hat{f}(\omega)\hat{E}^-(z_{OC}, \omega), \tag{5.18}$$

where $\hat{E}^\pm(z, \omega)$ denotes the Fourier transform with respect to time of the electric field amplitude and $\hat{f}(\omega)$ represents a spectral filter that describes the combined gain bandwidth of the active media. This procedure corresponds to a lumped element approach [56–59], which localizes the spectral dynamics of the electric field to the output coupler. Note that the positioning of the spectral filter is arbitrary as it could be placed at any point within the laser cavity.[6] For mathematical simplicity, a Lorentzian filter

$$\hat{f}(\omega) = \frac{\gamma}{\gamma + i\omega} \tag{5.19}$$

is used. For practical, i.e., simulation purposes, Eq. (5.18) needs to be transformed to the time domain. This can be achieved by using the convolution theorem to yield the integral equation

$$E^+(z_{OC}, t) = \sqrt{r_{OC}} \int_{-\infty}^{t} d\tau f(t - \tau)E^-(z_{OC}, \tau) \tag{5.20}$$

[6] The placement of the spectral filter does impact the emission dynamics since it acts on the local electric field, which evolves along its propagation. In that sense, its position is not irrelevant but rather arbitrary by the nature of the lumped element approach, which does not dictate a specific position by any rules.

for the boundary condition at the output coupler. The time-domain transform of the Lorentzian filter can be calculated separately and reads [59]

$$f(t) = \gamma e^{-\gamma t} \quad \text{for } t > 0, \tag{5.21}$$

where $t > 0$ preserves causality. The integral equation Eq. (5.20) can be transformed into an ordinary differential equation by computing the total derivative with respect to time and using the Leibniz rule, which produces

$$\begin{aligned}\frac{\mathrm{d}}{\mathrm{d}t} E^+(z_{\mathrm{OC}}, t) &= \sqrt{r_{\mathrm{OC}}} \underbrace{f(0)}_{=\gamma} E^-(z_{\mathrm{OC}}, t) + \underbrace{\sqrt{r_{\mathrm{OC}}} \int_{-\infty}^{t} \mathrm{d}\tau \underbrace{\frac{\partial}{\partial t} f(t-\tau)}_{=-\gamma f(t-\tau)} E^-(z_{\mathrm{OC}}, \tau)}_{=-\gamma E^+(z_{\mathrm{OC}}, t)} \\ &= \gamma \left(\sqrt{r_{\mathrm{OC}}} E^-(z_{\mathrm{OC}}, t) - E^+(z_{\mathrm{OC}}, t) \right),\end{aligned} \tag{5.22}$$

where the underbraced term in the first line has been identified with Eq. (5.20). The resulting dynamical equation for $E^+(z_{\mathrm{OC}}, t)$ can be interpreted as a first-order low-pass filter with the time constant γ^{-1}, which is driven by the field $\sqrt{r_{\mathrm{OC}}} E^-(z_{\mathrm{OC}}, t)$.

Delay-Differential Equation System

Using the just derived differential equation boundary condition and Eq. (5.17), the time evolution of the forward-moving electric field at the output coupler can be expressed by the single delay-differential equation

$$\begin{aligned}\frac{\mathrm{d}}{\mathrm{d}t} E^+(z_{\mathrm{OC}}, t) = &-\gamma E^+(z_{\mathrm{OC}}, t) + \gamma \sqrt{r_{\mathrm{OC}}} E^+(z_{\mathrm{OC}}, t - T) \\ &\times e^{\frac{1}{2} G_{\mathrm{G}}(t-\tau_1)} e^{G_{\mathrm{Q}}(t-\tau_1-\tau_2)} e^{\frac{1}{2} G_{\mathrm{G}}(t-\tau_1-2\tau_2)}\end{aligned} \tag{5.23}$$

instead of the system of coupled Eqs. (5.7)–(5.16). Similar to Eq. (5.17), the incident electric fields at the gain and absorber chips can also be expressed in terms of $E^+(z_{\mathrm{OC}}, t)$

$$E^+(z_{\mathrm{G}}, t) = E^+(z_{\mathrm{OC}}, t - \tau_1) \tag{5.24}$$

$$E^+(z_{\mathrm{Q}}, t) = E^+(z_{\mathrm{OC}}, t - \tau_1 - \tau_2)\, e^{\frac{1}{2} G_{\mathrm{G}}(t-\tau_2)} \tag{5.25}$$

$$E^-(z_{\mathrm{G}}, t) = E^+(z_{\mathrm{OC}}, t - \tau_1 - 2\tau_2)\, e^{\frac{1}{2} G_{\mathrm{G}}(t-2\tau_2)} e^{G_{\mathrm{Q}}(t-\tau_2)}, \tag{5.26}$$

where the time delays encode the propagation time from the output coupler to the respective positions z_k. The exponentials account for the amplification and absorption, which are accumulated along the propagation. Plugging these expressions into Eq. (5.5) generates dynamical equations for the averaged carrier numbers in the gain and absorber chip, which only depend on the electric field $E^+(z_{\mathrm{OC}}, t)$ at the output coupler. Introducing the shorthand $E \equiv E^{+}(z_{\mathrm{OC}})$, the final system of delay-differential equations is formulated:

$$\frac{\mathrm{d}}{\mathrm{d}t}E(t) = -\gamma E(t) + \gamma\sqrt{r_{\mathrm{OC}}}E(t-T)e^{\frac{1}{2}G_{\mathrm{G}}(t-\tau_1)}e^{G_{\mathrm{Q}}(t-\tau_1-\tau_2)}e^{\frac{1}{2}G_{\mathrm{G}}(t-\tau_1-2\tau_2)} \tag{5.27}$$

$$\begin{aligned}\frac{\mathrm{d}}{\mathrm{d}t}\langle\rho\rangle_{\mathrm{G}}(t) = &-\frac{\langle\rho\rangle_{\mathrm{G}}(t)}{\tau_{\mathrm{G}}} + R_{\mathrm{G}} - \frac{v_{\mathrm{g}}}{\ell_{\mathrm{G}}}\left(e^{\mathrm{Re}\{G_{\mathrm{G}}(t)\}}-1\right)\\ &\times\left[|E(t-\tau_1)|^2 + |E(t-\tau_1-2\tau_2)|^2\,e^{2\mathrm{Re}\{G_{\mathrm{Q}}(t-\tau_2)\}}e^{\mathrm{Re}\{G_{\mathrm{G}}(t-2\tau_2)\}}\right]\end{aligned} \tag{5.28}$$

$$\begin{aligned}\frac{\mathrm{d}}{\mathrm{d}t}\langle\rho\rangle_{\mathrm{Q}}(t) = &-\frac{\langle\rho\rangle_{\mathrm{Q}}(t)}{\tau_{\mathrm{Q}}} + R_{\mathrm{Q}} - \frac{v_{\mathrm{g}}}{\ell_{\mathrm{Q}}}\left(e^{\mathrm{Re}\{G_{\mathrm{Q}}(t)\}}-1\right)\\ &\times|E(t-\tau_1-\tau_2)|^2\,e^{\mathrm{Re}\{G_{\mathrm{G}}(t-\tau_2)\}},\end{aligned} \tag{5.29}$$

where the integrated gain in the gain and absorber chip is calculated according to

$$G_{\mathrm{G}}(t) = \ell_{\mathrm{G}}g_{\mathrm{G}}\left(\langle\rho\rangle_{\mathrm{G}}(t)\right) \tag{5.30}$$

$$G_{\mathrm{Q}}(t) = \ell_{\mathrm{Q}}g_{\mathrm{Q}}\left(\langle\rho\rangle_{\mathrm{Q}}(t)\right). \tag{5.31}$$

To complete the model, gain functions for the gain and absorber chip must be specified. Depending on the charge-carrier model, additional equations to model the dynamics thereof must be introduced. In the most simple case with no additional charge carrier dynamics, a self-consistent description of the V-shaped external cavity laser is achieved with one complex (E) and two real-valued $\langle\rho\rangle_{\mathrm{G,Q}}$ delay-differential equations.

Numerical Implementation

The system of coupled delay-differential equations is numerically integrated using a fourth-order Runge-Kutta method [60]. The time-delayed terms are implemented via a history array, which is interpolated with a third-order cubic Hermite spline method if required. Empirically, the numerical integration has converged well with time step $h = (5\gamma)^{-1}$, where γ is the gain bandwidth, which is typically the fastest timescale in the system.

The integration is, furthermore, simplified by shifting the gain and absorption variables in time according to

$$G(t+\tau_1) \to G(t) \text{ and } Q(t+\tau_1+\tau_2) \to Q(t), \tag{5.32}$$

which reduces the number of time delays that have to be evaluated per time step.[7]

If spontaneous emission noise is, furthermore, added phenomenologically to the electric field equation, the corresponding stochastic term is solved alongside the

[7] See Ref. [50] for more details.

delay-differential equation with a simple Euler method. To ensure proper convergence, the combined system is spot checked and integrated with the Euler method and a much smaller time step.

5.2.2 *Quantum-Well Active Medium*

This last subsection implements a semiconductor quantum-well active medium to obtain the model used in this chapter. Semiconductor quantum wells provide charge-carrier confinement within one spatial dimension, which effectively produces a two-dimensional crystal lattice with a constant density of states for the in-plane localized charge-carrier states [54, 61]. The resulting gain function nonetheless depends non-trivially on the charge-carrier density due to the involved quasi Fermi-Dirac statistics, which produce an asymmetric population of the bands. To avoid the microscopic calculation of the gain [54, 62–64] logarithmic parameterizations have been shown to work very well [21, 65–68]. However, linear approximations of the gain function turn out to be sufficient in many cases and have therefore found a widespread use in the modeling of semiconductor quantum-well opto-electronic devices [53, 59, 68–74]. Hence, such a linear gain approximation will be used in this chapter. The gain function is then given by

$$g(n(z,t)) = (1 - i\alpha(z))\tilde{g}(z)\Gamma(z)\left[n(z,t) - n^{\mathrm{tr}}(z)\right], \tag{5.33}$$

where $\tilde{g}$ denotes the differential gain coefficient, Γ the optical confinement factor, n_{tr} the carrier density at transparency, and α the amplitude-phase coupling factor. In order to follow conventions, the charge-carrier density is represented by the variable $n(z,t)$ instead of the previously used variable $\rho(z,t)$. The semiconductor gain function may be linearized at different charge-carrier densities to reflect different operating conditions, e.g., as they are found in the gain and absorber chip [56, 59, 75]. Thus, both the gain coefficient $\tilde{g}$ and the transparency charge-carrier density n^{tr} depend on the linearization point.

According to Eqs. (5.30) and (5.31), the integrated gain for the gain and absorber chip reads

$$G_{\mathrm{G}}(t) = (1 - i\alpha_{\mathrm{G}})\underbrace{\ell_{\mathrm{G}}\tilde{g}_{\mathrm{G}}\Gamma_{\mathrm{G}}\left[\langle n(z,t)\rangle_{\mathrm{G}} - n^{\mathrm{tr}}_{\mathrm{G}}\right]}_{=G(t)} \tag{5.34}$$

$$G_{\mathrm{Q}}(t) = (1 - i\alpha_{\mathrm{Q}})\underbrace{\ell_{\mathrm{Q}}\tilde{g}_{\mathrm{Q}}\Gamma_{\mathrm{Q}}\left[\langle n(z,t)\rangle_{\mathrm{Q}} - n^{\mathrm{tr}}_{\mathrm{Q}}\right]}_{=Q(t)}, \tag{5.35}$$

respectively. The underbraced parts of both equations define the new variables G and Q. Their respective dynamical equations are found by multiplying Eq. (5.28) and Eq. (5.29) by $\ell_{\mathrm{G,Q}}\tilde{g}_{\mathrm{G,Q}}\Gamma_{\mathrm{G,Q}}$ and defining $J_{\mathrm{G}} = \ell_{\mathrm{G}}\tilde{g}_{\mathrm{G}}\Gamma_{\mathrm{G}}(R_{\mathrm{G}} - n^{\mathrm{tr}}_{\mathrm{G}}/\tau_{\mathrm{G}})$ and $Q_0 = \ell_{\mathrm{Q}}\tilde{g}_{\mathrm{Q}}\Gamma_{\mathrm{Q}}(\tau_{\mathrm{Q}}R_{\mathrm{Q}} - n^{\mathrm{tr}}_{\mathrm{Q}})$. Allocating all terms, the resulting equations read

$$\frac{\mathrm{d}}{\mathrm{d}t}G(t) = -\frac{G(t)}{\tau_{\mathrm{G}}} + J_{\mathrm{G}} - v_{\mathrm{g}}\tilde{g}_{\mathrm{G}}\Gamma_{\mathrm{G}}\left(e^{G(t)} - 1\right) \times \left[|E(t-\tau_1)|^2 + |E(t-\tau_1-2\tau_2)|^2 e^{2Q(t-\tau_2)} e^{G(t-2\tau_2)}\right] \tag{5.36}$$

$$\frac{\mathrm{d}}{\mathrm{d}t}Q(t) = \frac{Q_0 - Q(t)}{\tau_{\mathrm{Q}}} - v_{\mathrm{g}}\tilde{g}_{\mathrm{Q}}\Gamma_{\mathrm{Q}}\left(e^{Q(t)} - 1\right)|E(t-\tau_1-\tau_2)|^2 e^{G(t-\tau_2)}, \tag{5.37}$$

where J_{G} is identified as the pump current and Q_0 describes the unsaturated absorption. Note that this transform was enabled by the linear gain function. Any nonlinearity in the gain function would lead to additional terms, which arise when differentiating the gain function. To obtain a final dimensionless formulation of the full system of delay-differential equations, time is rescaled by the cold-cavity round-trip time $t \Rightarrow t/T$ and the electric field is rescaled according to $E \Rightarrow E/\sqrt{v_{\mathrm{g}}\tilde{g}_{\mathrm{G}}\Gamma_{\mathrm{G}}T}$. The complete set of equations is then given by

$$\frac{\mathrm{d}}{\mathrm{d}t}E(t) = -\gamma E(t) + \gamma\sqrt{r_{\mathrm{OC}}}E(t-T) \times e^{\frac{1-i\alpha_{\mathrm{G}}}{2}[G(t-\tau_1)+G(t-\tau_1-2\tau_2)]} e^{(1-i\alpha_{\mathrm{Q}})Q(t-\tau_1-\tau_2)} + \sqrt{D_{\mathrm{sp}}}\xi(t) \tag{5.38}$$

$$\frac{\mathrm{d}}{\mathrm{d}t}G(t) = -\gamma_{\mathrm{G}}G(t) + J_{\mathrm{G}} - \left(e^{G(t)} - 1\right) \times \left[|E(t-\tau_1)|^2 + |E(t-\tau_1-2\tau_2)|^2 e^{2Q(t-\tau_2)} e^{G(t-2\tau_2)}\right] \tag{5.39}$$

$$\frac{\mathrm{d}}{\mathrm{d}t}Q(t) = \gamma_{\mathrm{Q}}(Q_0 - Q(t)) - s\left(e^{Q(t)} - 1\right)|E(t-\tau_1-\tau_2)|^2 e^{G(t-\tau_2)}, \tag{5.40}$$

where $s = \tilde{g}_{\mathrm{Q}}\Gamma_{\mathrm{Q}}/(\tilde{g}_{\mathrm{G}}\Gamma_{\mathrm{G}})$ denotes the ratio of the differential gain coefficients and optical confinement factors and γ_{G} and γ_{Q} denote the inverse lifetimes $1/\tau_{\mathrm{G}}$ and $1/\tau_{\mathrm{Q}}$ in the gain and absorber chip. Additionally, spontaneous emission noise has been added phenomenologically to the electric field equation, where $\sqrt{D_{\mathrm{sp}}}$ denotes the noise strength and $\xi(t)$ a delta-correlated complex Gaussian white noise term.

Laser Parameters

To complete the quantum-well based V-shaped external cavity laser model, the system parameters are introduced and briefly discussed. A complete list of all parameters is presented in Table 5.1, where both their dimensionful and dimensionless values and a brief explanation are listed. Among them, a number of parameters are varied in the subsequent sections. The quoted values thus represent the set of default parameters.

The presented parameter values are chosen to describe the experimentally characterized laser, which is presented in Refs. [4, 35]. A subset of the parameters are unambiguously imposed by the device setup: The cavity geometry directly determines the propagation times T, τ_1 and τ_2. The output-coupler reflectivity r_{OC} is determined by the mirror specifications. The remaining parameters, on the other hand, must be

Table 5.1 Default parameters used in the V-shaped external cavity semiconductor quantum-well laser model

Symbol	Value	Dimensionless	Meaning
T	625 ps	1	Cold-cavity round-trip time
τ_1	156.25 ps	0.25	Output coupler to gain chip propagation time
τ_2	156.25 ps	0.25	Gain chip to absorber chip propagation time
γ	3200 ns^{-1}	2000	Gain bandwidth
r_{OC}	0.99	0.99	Output coupler intensity reflectivity
α_{G}	0.0	0.0	Gain-chip amplitude-phase coupling
α_{Q}	0.0	0.0	Absorber-chip amplitude-phase coupling
D_{sp}	0.16 ns^{-1}	0.1	Spontaneous emission noise strength
J_{G}	1.6 ns^{-1}	1.0	Gain chip pump current
γ_{G}	1.0 ns^{-1}	0.625	Gain chip recovery rate
Q_0	−0.58	−0.58	Unsaturated absorption
γ_{Q}	180.0 ns^{-1}	112.5	Absorber chip recovery rate
s	2	2	Differential gain ratio

The last column briefly indicates the meaning of each parameter. Parameters are chosen to describe the device presented in [4, 35]

tuned within reasonable margins to accurately reproduce the observed dynamics. On that account, the gain and absorber-chip recovery times have been determined by pump-probe measurements by the authors of Refs. [4, 35]. The reported values fall well within the range that can be found in the literature [4, 43, 76–79]. The ratio s is estimated based in the difference of the differential gain coefficients and assumes a conservative factor of two due the different operating conditions in the gain and absorber chips. The unsaturated absorption Q_0 is tuned to facilitate an effective pulse shaping and short pulses.

While typical amplitude-phase coupling coefficients in semiconductor quantum-well active media are found to be between $\alpha = 2$ and $\alpha = 10$ [4, 80–82], the modeling of mode-locked lasers has nonetheless been shown to be successful with the use of a vanishing amplitude-phase coupling [83–85]. Moreover, it has been demonstrated that the usage of a constant amplitude-phase coupling coefficient breaks down at fast and high-power excitations [86–88]. Hence, this chapters' investigations are performed with no amplitude-phase coupling.

Lastly, the gain bandwidth γ is adapted to obtain optical pulses as short as possible, while maintaining reasonable computational demands. Thus, the experimentally observed 100 fs pulse width is not achieved. The obtained $\sim$ 1 ps pulses are, nevertheless, much shorter than all other processes, i.e., the gain and absorber recovery, and thus the dynamics remain qualitatively identical for further increased gain bandwidths.

5.3 Symmetric Cavity Dynamics Characterization

This section introduces and discusses the emission states that can be obtained for the V-shaped external cavity laser with symmetric, i.e., equally long, cavity arms. In that context, the natural operation parameter is the pump current J_G, which controls the energy supply of the gain chip. The absorber chip properties are fixed with the fabrication and do not require an additional reverse bias to achieve mode-locking. Similarly, the occurring recovery times and gain spectra are also considered to be determined by the fabrication of the gain and absorber chips.[8] Hence, the observed states are discussed in terms of the pump current J_G, which is also easily tunable in experiments.

Lasing Threshold

The analysis starts off with the lasing threshold, i.e., the pump current that is required for the laser to transition from the off state to a continuous-wave (cw) lasing state. For the considered model, this problem is still in the realm of analytics: In the linear stability analysis framework, the threshold corresponds to the bifurcation in which the off state looses its stability and the cw state is created or becomes stable [74]. The cw state is defined by a constant intensity in time, which allows for an electric field with a constant amplitude and an arbitrary rotation in the complex plane. The lasing threshold is thus characterized by the transition

$$(0, G^*, Q^*) \rightarrow (A^* e^{i\omega t}, G^*, Q^*), \tag{5.41}$$

where the left side corresponds to the off state and the right side to the cw state. G^*, Q^* and A^* denote the steady state values and ω the mean rotation frequency in the complex plane. Physically, ω relates to the optical frequency via the chosen rotating frame (s. Sect. 2.2) and corresponds to the maximum gain mode. At the bifurcation point, the electric field exhibits an infinitesimal amplitude A^*, which allows to plug both states simultaneously into the dynamical equations (5.38)–(5.40) to produce the equations

$$i\omega = -\gamma + \gamma\sqrt{r_{\mathrm{OC}}} e^{(1-i\alpha_{\mathrm{G}})G^* + (1-i\alpha_{\mathrm{Q}})Q^* - i\omega T} \tag{5.42}$$
$$0 = J_{\mathrm{Q}} - \gamma_{\mathrm{G}} G^* \tag{5.43}$$
$$0 = Q_0 - Q^*. \tag{5.44}$$

Further splitting the first equation (5.42) into its real and imaginary part yields

$$\omega = -\gamma\sqrt{r_{\mathrm{OC}}} e^{G^*+Q^*} \sin\left(\alpha_{\mathrm{G}} G^* + \alpha_{\mathrm{Q}} Q^* + \omega T\right) \tag{5.45}$$
$$\gamma = \gamma\sqrt{r_{\mathrm{OC}}} e^{G^*+Q^*} \cos\left(\alpha_{\mathrm{G}} G^* + \alpha_{\mathrm{Q}} Q^* + \omega T\right). \tag{5.46}$$

---

[8] A more involved modeling approach would have to include a dependence of the mentioned quantities on the device (lattice) and carrier temperatures. The latter in particular would dynamically depend on the lasers' operation state.

These conditions can be reformulated by firstly dividing Eq. (5.45) by Eq. (5.46) and secondly by adding the squared equations to produce:

$$\omega^2 + \gamma^2 = \gamma^2 r_{\mathrm{OC}} e^{2G^* + 2Q^*} \tag{5.47}$$

$$\frac{\omega}{\gamma} = -\tan\left(\alpha_{\mathrm{G}} G^* + \alpha_{\mathrm{Q}} Q^* + \omega T\right). \tag{5.48}$$

These two transcendental equations determine the maximum gain mode ω and thus the optical frequency of the generated cw state. Plugging the steady-state gain G^* and absorption Q^*, obtained from Eqs. (5.43) and (5.44), respectively, into the former equation (5.47) finally yields the analytic equation for the threshold pump current:

$$J_{\mathrm{th}} = \gamma_{\mathrm{G}} \left[\frac{1}{2} \ln\left(\frac{\omega^2}{\gamma^2} + 1 \right) - \frac{1}{2} \ln\left(r_{\mathrm{OC}} \right) - Q_0 \right], \tag{5.49}$$

where ω is the previously determined maximum gain mode. If the amplitude-phase coupling coefficients α_{G} and α_{Q} are negligible or $\gamma \gg \omega$ holds, the first term in the parenthesis can be neglected. If the out-coupling losses are small, i.e., $r_{\mathrm{OC}} \approx 1$, the threshold pump current density is dominated by the unsaturated absorption and is given by

$$J_{\mathrm{th}} \approx -\gamma_{\mathrm{G}} Q_0. \tag{5.50}$$

This approximation holds well for the chosen laser parameters and thus for all cases that are investigated within this chapter.

Analysis of the Mode-Locking Dynamics

The cw emission state that is generated at the lasing threshold, however, turns out to be only stable within a very small region close to the lasing threshold. The further bifurcation analysis is beyond the scope of analytics and must be carried out by numerical path-continuation tools like DDE-Biftool [89, 90]. Unfortunately, the large gain bandwidth parameter γ renders the system numerically untraceable by such tools. Moreover, the stochastic spontaneous emission noise has profound consequences for the observed emission states.[9] Hence, the analysis of mode-locking dynamics must be conducted by the means of direct numerical integration.

Nevertheless, the bifurcation analysis of the deterministic equations can be performed for smaller gain bandwidth values and has been presented in [50]. The results show that the lasing threshold corresponds to an Andronov-Hopf bifurcation, which reduces to a pitchfork bifurcation in the proper rotating frame, which cancels the rotation in the complex plane. In this rotating frame, the such generated cw-emission state looses its stability in an Andronov-Hopf bifurcation, which then gives rise to the

[9] The observed mode-locking states are mostly destabilized by noise-induced perturbations that grow due to positive net-gain windows. In the absence of spontaneous emission noise, mode-locking states can be observed, which can be destabilized even by infinitesimal noise strengths and are therefore considered to be unphysical.

fundamental mode-locking solution. This behavior has also been obtained for ring-shaped laser models [59, 91, 92] that describe edge-emitting semiconductor lasers. It can therefore be assumed that the bifurcation structure of the V-shaped laser with a broad gain spectrum follows the same pattern.

5.3.1 Basic Mode-Locking States

The numerical characterization of the mode-locking dynamics starts with an illustration of the basic emission states that are observed for symmetric cavity arms. On that account, Fig. 5.4 presents pseudo space-time plots of the normalized optical intensity (left column), power spectral densities of the optical intensity (middle column), and normalized intensity autocorrelation traces (right column). The pseudo space-time plots are created by slicing a time-series into sections with the length T and stacking them on top of each other, such that the vertical axis corresponds to the round-trip number. The optical intensity is encoded by a white to red color code. The six rows (a)–(f) represent the following emission states, which are also denoted to the right of the autocorrelation traces: fundamental mode-locking (FML, $J_G = 1.0$), leading-edge unstable second order harmonic mode-locking (HML2$_{\mathrm{LEI}}$, $J_G = 2.0$), pulse-cluster mode-locking with two pulses (PC2, $J_G = 2.2$), leading-edge unstable fourth-order harmonic mode-locking (HML4$_{\mathrm{LEI}}$, $J_G = 3.5$), pulse-cluster mode-locking with three pulses (PC3, $J_G = 3.0$), and third-order harmonic mode-locking (HML3, $J_G = 3.2$).

Fundamental mode-locking (FML) is characterized by a single stable pulse, which circulates within the laser cavity. The nonlinear interaction with the gain and absorber chip and the spectral filtering element slows down the effective group velocity of the pulse and thus tilts the pulse train to the right in the space-time plot Fig. 5.4a1. The corresponding power spectrum (a2) exhibits well defined characteristic peaks at the fundamental and harmonic repetition rate frequencies. The autocorrelation trace only shows a well defined peak close to the cold-cavity round-trip time within the plotted time lag range, which corresponds to the mode-locking period.

Second-order leading-edge unstable harmonic mode-locking (HML2$_{\mathrm{LEI}}$) produces two equidistant unstable pulse trains, as can be seen in Fig. 5.4b1. The leading-edge instability continuously produces a new pulse in front of an existing pulse, which consequently takes away the gain from the main pulse and thus destabilizes the pulse trains. This invokes a leftward movement to earlier times in the space-time plot. The reoccurring process manifests in the power spectrum (b2) as broad and noisy repetition-rate line shapes at all harmonic frequencies. The second-order mode-locking nature, however, can still be identified by well defined sharp peaks that only sit on top of the line shapes with an even harmonic number. The autocorrelation trace gains additional small pedestals at the time lags $\Delta t = 0$ and $\Delta t \approx 1$ as well as a weak double peak at $\Delta t \approx 0.5$.

Pulse-cluster mode-locking (PC2) with two pulses as shown in Fig. 5.4c1 is characterized by two non-equidistant but approximately identical pulses that circulate within the laser resonator. In the case at hand, the pulse separations are found to be

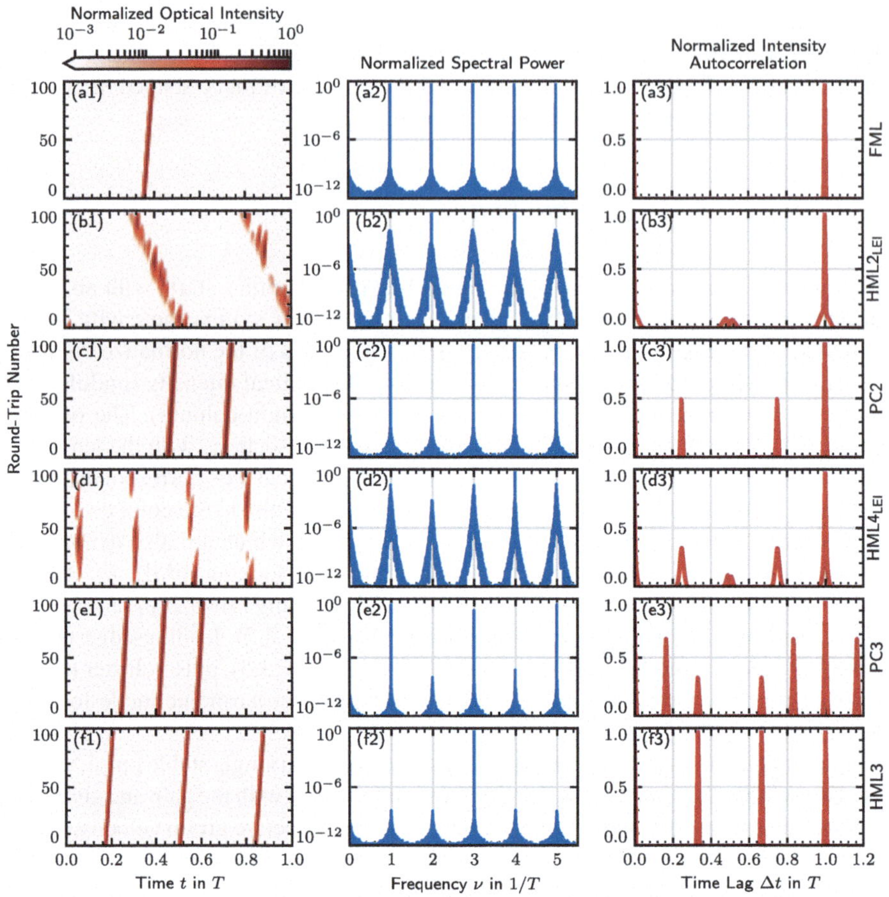

Fig. 5.4 Illustration of the basic mode-locking states. From left to right: pseudo space-time plots normalized to the maximum intensity, normalized power spectral densities, and normalized autocorrelation functions. From top to bottom: fundamental mode-locking (FML, $J_G = 1.0$), leading-edge unstable second order harmonic mode-locking ($HML2_{LEI}$, $J_G = 2.0$), pulse-cluster mode-locking with two pulses (PC2, $J_G = 2.2$), leading-edge unstable fourth-order harmonic mode-locking ($HML4_{LEI}$, $J_G = 3.5$), pulse-cluster mode-locking with three pulses (PC3, $J_G = 3.0$), and third-order harmonic mode-locking (HML3, $J_G = 3.2$). Laser parameters: $Q_0 = -0.58$ and $\tau_1 = 0.25$

≈ 0.25 and ≈ 0.75, where the former coincides with the length of the cavity arms τ_1 and τ_2. The corresponding power spectrum (c2) exhibits a strongly attenuated peak at the second harmonic frequency, when compared to the FML spectrum. The autocorrelation trace (c3) exhibits additional peaks at $\Delta t \approx 0.25$ and $\Delta t \approx 0.75$, which reflect the pulse separations observed in (c1). The height of ≈ 0.5 indicates that the pulse shapes of the two pulses within the cluster are indeed only approximately identical.

Fourth-order leading-edge unstable mode-locking ($HML4_{LEI}$) is qualitatively similar to the $HML2_{LEI}$ state, except for four instead of two pulse trains, which circulate within the laser cavity. However, the space-time representation Fig. 5.4d1 reveals that the leading-edge instability is much less pronounced compared to Fig. 5.4b1 and thus produces a much weaker leftwards temporal drift of the pulse trains. The power spectrum (d2), again, exhibits broad line shapes at all harmonic frequencies and a well defined peak at harmonic numbers that are multiples of four, which reflects the underlying fourth-order harmonic mode-locking state. The autocorrelation trace (d3) exhibits intermediate peaks at the time lags $\Delta t \approx 0.25$ and $\Delta t \approx 0.75$ as well as a small double peak at $\Delta t \approx 0.5$, which reflect the pulse separations as they are observed in (d1). The difference in peak strength suggests that, although unstable, neighboring pulses are more strongly correlated, i.e., more similar, than next-neighbor pulses.

Next, pulse-cluster mode-locking with three pulses is constituted by three non-equidistant but approximately identical stable pulse trains, which circulate within the laser cavity. As can be seen in Fig. 5.4e1, the pulse separation within the pulse cluster amounts to $\approx 1/6$ and the pulse separation between the cluster to $\approx 4/6$. The corresponding power spectrum (e2) exhibits well defined repetition-rate line shapes at the odd harmonics and strongly attenuated line shapes at the even harmonics within the plotted spectral range. The autocorrelation trace (e3) shows characteristic peaks at the time lags $\Delta t \approx 1/6$, $\Delta t \approx 2/6$, $\Delta t \approx 4/6$ and $\Delta t \approx 5/6$. Similar to the PC2 state, the autocorrelation peaks are also attenuated, which suggests that the pulses within the cluster are only approximately identical.

Lastly, third-order harmonic mode-locking (HML3, $J_G = 3.2$) is characterized by three identical and equidistant pulses that circulate within the laser cavity as shown in Fig. 5.4f1. The power spectrum (f2) exhibits strongly damped line shapes at the fundamental and second harmonic frequency and a characteristic high power and narrow line shape at the third harmonic. This pattern continues for the respective higher harmonics. The autocorrelation trace (f3) exhibits characteristic peaks at time lags that are an integer multiple of $T/3$. All peaks are narrow and close to unity, which indicates that the three pulses are identical.

In conclusion, the presented mode-locked emission states have been identified and characterized by their respective pseudo space-time plots. Importantly, the complementary power spectra and intensity autocorrelation traces also allow to reliably identify the various emission states via their distinct features. This facilitates a simple experimental characterization, where the power spectra and autocorrelation traces are easily available.

Gain and Net-Gain Dynamics

In order to understand the mechanisms that lead to the observed mode-locking states, the gain G, absorber Q, and net-gain $\mathcal{G}$ dynamics are studied in the following. The net gain, as introduced by New [57], is a small signal measure, which allows to assess the stability of a mode-locked emission state against perturbations.[10]

[10] A perturbation in the sense of the net gain refers to a small perturbation of an existing mode-locked state.

In the case of the V-shaped external cavity laser model, the net gain can be computed by integrating the intensity gain over one complete round trip within a co-moving frame, which yields

$$\mathcal{G}(t) = G(t - \tau_1) + G(t - \tau_1 - 2\tau_2) + 2Q(t - \tau_1 - \tau_2) + \ln(r_{OC}). \qquad (5.51)$$

The time delays, which appear in the arguments, result from the propagation times through the cavity arms. Positive net-gain values correspond to an amplification and negative values to an attenuation of a small perturbation. Mode-locking is labeled as stable if a window of positive net gain only exists alongside the desired mode-locked pulses.

However, since the net gain is a small-signal measure, mode-locking can still be stable despite positive net-gain windows in front of or behind the mode-locked pulses [59, 93]. In those cases the group velocity dispersion causes the mode-locked pulses to travel either faster or slower than the small perturbation, which leads to a merging of both and thereby protects the mode-locked state.

The analysis first focuses on the stable fundamental and pulse-cluster mode-locking states, which have been presented in Fig. 5.4a, c and e. Figure 5.5 presents the intensity $|E|^2$ (blue), gain G (green), and absorber Q (red) dynamics of the aforementioned mode-locking states in the first row and the corresponding net-gain $\mathcal{G}$ dynamics in the second row. The intensity $|E|^2$, the gain G, and the absorption Q have been normalized to 0.1. The zero baselines of the gain G and absorption Q have been shifted upwards proportional to their cavity position relative to the output coupler

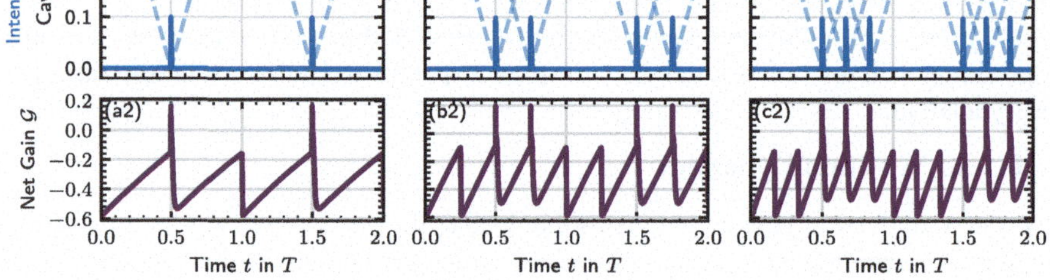

Fig. 5.5 Intensity and carrier dynamics of the FML ($J_G = 1.0$), PC2 ($J_G = 2.2$), and PC3 ($J_G = 3.0$) state. Subfigures **a1** to **c1**, respectively, show the temporal evolution of the intensity (blue lines), the gain (green lines) and the absorption (red lines). The intensity, gain, and absorption have been normalized to 0.1 and the respective zero positions (black dotted lines) have been shifted proportional to their cavity positions. The propagation of the pulses is indicated by blue dashed lines. **a2–c2** Plot the respective net gain $\mathcal{G}$. Laser parameters: $Q_0 = -0.58$ and $\tau_1 = 0.25$

and are indicated by horizontal black lines at 0.25 and 0.5. The propagation through the laser cavity is illustrated by blue dashed lines, such that the intersection with the black dotted lines, represents the pulse interactions with the gain and absorber chip.

Fundamental mode-locking, which is shown in the left column in Fig. 5.5, equates to one pulse circulating within the laser cavity. This implies one interaction with the absorber chip and two interactions with the gain chip per round trip. The gain-chip interactions are equidistant due to the symmetric cavity configuration. The cavity configuration and the gain dynamics also manifest in the net gain (a2), which apart from the short positive window that coincides with the pulse emission, shows a characteristic negative-valued local maximum right in between the pulse emissions. The reduction of the net gain after half a round-trip, which creates this local maximum, results from the pulse, which saturates the gain on its backward journey towards the output coupler. This, furthermore, illustrates the advantage of the V-shaped geometry: The second interaction prevents the net gain from growing too much and, consequently, creating a positive net-gain window in between the fundamental pulse emissions.

Pulse-cluster mode-locking with two pulses, which is presented in the middle column in Fig. 5.5, corresponds to two identical pulses with a separation of ≈ 0.25 that circulate within the symmetric cavity. This leads to two interactions with the absorber chip with the temporal separation of the pulse-cluster and four temporally-equidistant interactions with the gain chip. Regarding the latter, the neighboring gain interaction is always caused by the respective other pulse. The resulting net gain, as shown in (b2), exhibits two short positive windows that correspond to the pulse emissions and two negative-valued local maxima in between. Similar to fundamental mode-locking, the negative-valued local maxima are created by the gain interaction with the backward-moving pulses.

Lastly, pulse-cluster mode-locking with three pulses, as presented in the right column of Fig. 5.5, is characterized by three identical pulses with a separation of $\approx 1/6$ within their cluster. This produces three interactions with the absorber chip and six equidistant interactions with the gain chip. Continuing the previous pattern, the corresponding net gain (c2) exhibits three short positive windows, which coincide with the pulse emission and three negative-valued local maxima in between the pulse clusters, which are caused by the backward-traveling pulses.

From the presented examples, it can be concluded that the pulse cluster mode-locking states are among the natural successors of the fundamental mode-locking state at larger pump currents. Their critical feature is the equidistant interaction with the gain chip, which is caused by the symmetric V-shaped cavity configuration. The advantage of the equidistant gain interaction is that the gain recovery is rather kept at the most efficient initial part of the exponential relaxation and the associated non-radiative losses are minimized. Moreover, the twofold interaction of a single pulse with the gain section prevents destabilizing positive net-gain windows, as discussed in the case of fundamental mode-locking.

In the next step, the two unstable emission states that have been presented in Fig. 5.4, namely leading-edge unstable second and fourth-order mode-locking, are discussed. Both are not subject to any periodicity due to the stochastic nature of the instability. Hence, Fig. 5.6 shows pseudo space-time plots of the normalized intensity

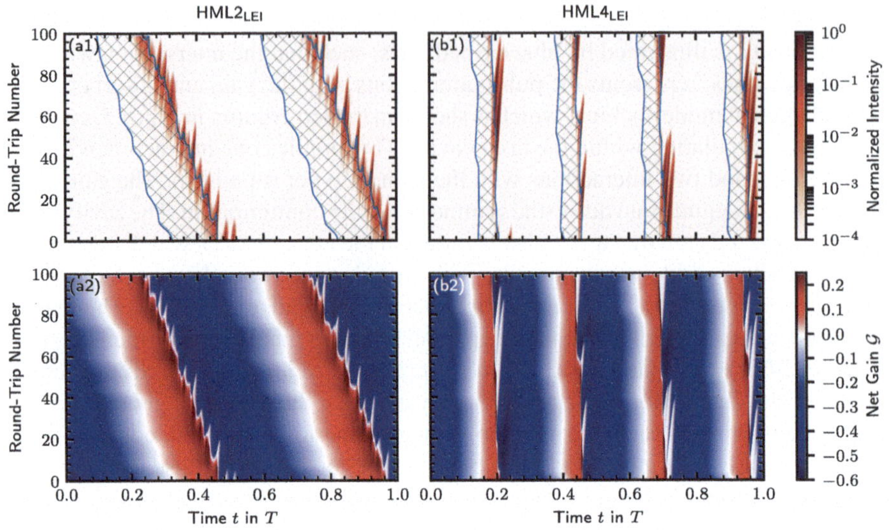

Fig. 5.6 Pseudo space-time plots of the pulse and net-gain dynamics: The top row shows the normalized intensity $|E|^2$ and the bottom the net gain $\mathcal{G}$. The columns correspond to second and fourth-order leading-edge unstable harmonic mode-locking ($J_G = 2.0$ and $J_G = 3.5$, respectively). Hatched regions in the top row indicate positive net gain $\mathcal{G} > 0$ and blue contour lines indicate $\mathcal{G} = 0$. Laser parameters: $Q_0 = -0.58$ and $\tau_1 = 0.25$

$|E|^2$ (top row) and the net gain $\mathcal{G}$ (bottom row) over a range of 100 round trips. The net gain is color coded and regions of positive net gain (red colors) are indicated in the intensity plots by hatched regions and blue contour lines. The left column, represents leading-edge unstable second-order harmonic mode-locking (HML2_{LEI}, $J_G = 2.0$) and the right column leading-edge unstable fourth-order mode-locking (HML4_{LEI}, $J_G = 3.5$).

Both states show a similar behavior: A large positive net-gain window is permanently open at the leading edge of all pulses, which continuously allows spontaneous emission noise induced perturbations to grow and destabilize the pulse train. In the case of the second-order state, both net-gain windows exhibit a temporal extend, which amounts to almost 15% of the round-trip time. In the case of the fourth-order state each net-gain window extends for about 6% of the round-trip time.

On top of the extended positive net-gain windows, both states suffer from another additional destabilizing mechanism: The equidistant pulse separation leads to a colliding pulse effect in the gain chip. Hence, two pulses always have to share the gain and are thereby tightly coupled. Thus any instability in one pulse train will immediately transfer to the respective other pulse train.

Lastly, the dynamics of third and fifth-order harmonic mode-locking is presented in Fig. 5.7. Similar to Fig. 5.5, the top row ((a1) and (b1)) plots the electric field, gain and absorption dynamics, where each quantity has been normalized to 0.1. The gain and absorption are shifted upwards proportional to their cavity position with respect

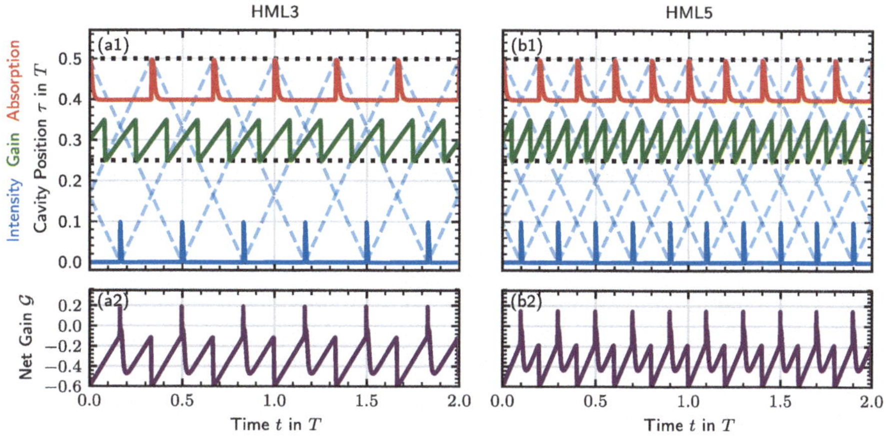

Fig. 5.7 Intensity and carrier dynamics of the HML3 ($J_G = 3.2$) and HML5 ($J_G = 4.0$) state. Subfigures **a1** and **b1**, respectively, show the temporal evolution of the intensity (blue lines), the gain (green lines), and the absorption (red lines). The intensity, gain, and absorption have been normalized to 0.1 and their respective zero positions (black dotted lines) have been shifted proportional to their cavity positions. The propagation of the pulses is indicated by blue dashed lines. **a2–b2** Plot the respective net gain $\mathcal{G}$. Laser parameters: $Q_0 = -0.58$ and $\tau_1 = 0.25$

to the output coupler. The pulse propagation is indicated by blue dashed lines. The bottom row ((a2) and (b2)) plots the corresponding net gain.

Both emission states produce equidistant pulses with three and five pulses circulating within the cavity, respectively. The odd number of pulses leads to equidistant interactions with the gain chip and thereby optimizes the gain recovery and minimizes non-radiative losses similar to the pulse cluster states. The net-gain dynamics in both cases show short positive windows that coincide with the pulse emission. In between pulse emissions, negative-valued local net-gain maxima can be observed. Comparing the PC3 and HML3 states, the observed gain dynamics are identical, due to the equidistant interactions of the pulses with the gain chip and thus the pulses are approximately identical in terms of the peak intensity and pulse width. The same mechanism applies to the PC5 state (not shown) and the HML5 state.

Summary

To conclude, the symmetric V-shaped cavity geometry has given rise to the pulse-cluster mode-locking states. These emission states achieve an equidistant pulse interaction with the gain chip and thereby optimize the gain recovery and minimize the non-radiative losses. This condition is also met by odd-order harmonic mode-locking states, where the odd-number of pulses within the cavity causes them to never collide within the gain chip. Stable even-order harmonic mode-locking, on the other hand, is counteracted by the collision of the pulses within the gain chip and the resulting positive net-gain instabilities. The entirety of stable and unstable emission states can be considered to be the characteristic feature of symmetric V-shaped external cavity lasers.

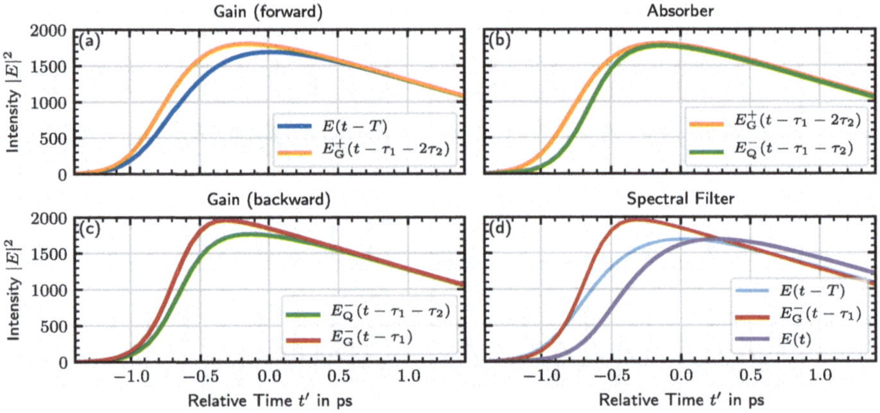

Fig. 5.8 Spatio-temporal pulse evolution of the fundamental mode-locking state. Subfigures **a**–**d** show the electric field along the pulse propagation through the four optical elements for one complete round trip. The pulse evolution is illustrated by plotting the field before and after the respective interactions. The colors throughout all subfigures denote the forward-moving field at the output coupler (delayed by one round trip, blue) and at the gain chip (orange), the backward-moving field at the absorber chip (green) and the gain chip (red), and the forward-moving field at the current time (purple). Laser parameters: $J_G = 1.0$, $Q_0 = -0.58$ and $\tau_1 = 0.25$

Pulse Shaping in the Fundamental Mode-Locking State

The operation state that is most valuable to applications is the fundamental state, due to its optimal use of the pump current, which does not need to be distributed among multiple pulses. Hence, a closer look at the pulse-shaping mechanism in the fundamental mode-locking state is taken in the following. On that account, the spatio-temporal pulse evolution along one complete round trip of the fundamental mode-locking state is presented in Fig. 5.8. Subfigures (a)–(d) present the pulse evolution along the propagation throughout the four interactions with the optical elements: The gain chip in the forward direction (a), the absorber chip (b), the gain chip in the backward direction (c), and the output coupler with the lumped element spectral filter (d). Although the electric field is only explicitly modeled at the output-coupler by the system of delay-differential equations, the respective electric fields at the various cavity positions can be easily calculated via the algebraic propagation equations (5.7) to (5.16).

The interaction with the optical elements is illustrated by plotting the electric field intensity before and after the respective optical element. The pulses are plotted in the co-moving frame, while the respective legends indicate the time in the lab frame. The colors throughout all subfigures denote the forward-moving field at the output coupler (delayed by one round trip, blue) and at the gain chip (orange), the backward-moving field at the absorber chip (green) and the gain chip (red), and the forward-moving field at the current time (purple).

Upon being reflected at the output coupler, the electric field $|E(t-T)|^2$ (blue line in Fig. 5.8a) passes the gain chip in the forward direction. The amplified field intensity $|E(t - \tau_1 - 2\tau_2)|^2$ (orange line in (a)) exhibits a steeper leading edge and a slightly increased pulse maximum. The trailing edge of the pulse, however, experiences almost no amplification, which indicates that the gain chip has already been completely saturated. The pulse width increases in this stage and the pulse position shifts to earlier times.

The following interaction with the absorber chip produces an intensity pulse profile (green line in (b)) that exhibits a reduced leading edge and hardly any other changes. This indicates that the leading edge of the pulse is sufficient to fully bleach the absorber chip. The pulse width reduces in this stage and the pulse position shifts very slightly to later times.

Subsequent to the reflection at the absorber chip, the backward-traveling pulse passes the gain chip a second time, where, again, the leading edge as well as the pulse maximum are amplified before the gain saturates and the trailing edge passes without modifications. The leading edge of the resulting pulse shape (red line in (c)) again steepens, the pulse width increases, and the pulse positions shifts to earlier times.

Lastly, the pulse arrives at the output coupler, where the spectral filter and the non-resonant losses act on the pulse profile. The final pulse shape after one round trip (purple line in (d)) is identical to the initial pulse shape (blue line in (a) and light blue line in (d)). It is, however, shifted by approximately 250 fs to later times, when compared to the cold-cavity round-trip time. The shift results both from the nonlinear interaction with the gain and absorber chips as well as the spectral filter. Both contributions can be distinguished by comparing the incoming (red line in (d)) and the resulting (purple line in (d)) pulse shape. The former pulse has been shifted by approximately 300 fs to earlier times due to the interactions with the gain chip. The spectral filtering element then eliminates large frequencies, which have been generated in the steepening of the leading edge, and thus smooths out the pulse shape. The resulting pulse maximum is shifted by approximately 550 fs to later times and produces the net shift of 250 fs compared to the cold-cavity round-trip time.

In conclusion, the pulse-shaping mechanism follows the pattern, which was described in the introduction and can be found throughout the mode-locking literature [11, 92, 94–98]: Pulses shorten in the absorber section and broaden in the gain section.[11]

It is, furthermore, important to note that the quantitative effect of the spectral filter is stronger than all three interactions with the gain and absorber chips. This points towards the quantitative limitations of the employed lumped-element filtering approach. Hence, if exact quantitative predictions are to be made, it is advisable to modify the model and implement a frequency dependent light-matter interaction for the gain and absorber chip. Nonetheless, qualitative predictions of the pulse

[11] It is important to note that the opposite pulse-shaping mechanism, which is reported in Chap. 4, is generated by the local interplay of the spectrally-resolved resonant gain and absorption, and the non-resonant losses. Such dynamics, however, can not be produced by the utilized V-shape model, which treats the optical losses and the spectral evolution of the electric field by a lumped element approach. Nevertheless, the impact of the interplay with the non-resonant losses in the gain and absorber chips can be assumed to be negligible, since those losses are typically very small [4, 35].

performance characteristic can be made if the changes in the pulse width and peak intensity surpass the typical effect of the gain bandwidth filter element. For the case at hand, this would imply peak intensity and pulse width variations $\gtrsim 15\%$.

5.3.2 *Pump Current Dependence of the Emission Dynamics*

To investigate the relationship between the various mode-locking emission states, their stability along the pump current J_G is probed. This 'poor mans' numerical continuation is performed by calculating multiple scans along the pump current, which utilize the different emission states (s. Fig. 5.4) as respective initial conditions for each pump current value. By combining the results, the respective states' regions of stability can be extracted.

The results are presented in Fig. 5.9, where the different emission states are distinguished by colors and their dynamics are illustrated by their time series local intensity maxima $|E|^2_{max}$. For the determination of each data point, a time series, which extends for 1000 cold-cavity round-trip times T, has been calculated. Stable emission, on the one hand, corresponds to a single unique maximum, whereas unstable emission, on the other hand, fills out a range of values. Regions of multistability correspond to the simultaneous existence of two or more emission states for a given pump current.

For an increasing pump current, lasing emission in terms of the fundamental mode-locking state (FML, blue line) is first observed at $J_G \approx 0.37$, which matches the previously derived analytic threshold. Within the pump current range of stable FML emission, the local intensity maxima, i.e., the pulse peak power steadily increases up to a maximum of $|E|^2_{max} \approx 2200$ at $J_G \approx 1.52$, which also represents the upper stability boundary.

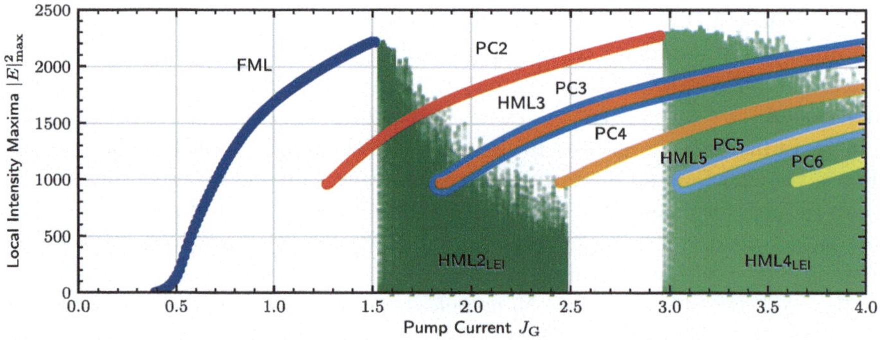

Fig. 5.9 Numerical bifurcation diagram along the pump current J_G. The dynamics are illustrated by plotting all the local maxima of the intensity $|E|^2$ recorded within 1000 T. The different observed emission states are plotted in different colors and labeled within the figure. The presented data has been assembled from different scans with custom build initial conditions to address the individual emission states. Laser parameters: $Q_0 = -0.58$ and $\tau_1 = 0.25$

Seemingly connected, the emission then transitions to the leading-edge unstable second-order harmonic mode-locking state ($HML2_{LEI}$, dark green dots), which extends to pump currents up to $J_G \approx 2.48$. For an increasing pump current, the local maxima of the $HML2_{LEI}$ decrease, which indicates that the instability is amplified and thus prevents the formation of short and high power pulses. The connection to the FML state is attributed to a positive net-gain instability of the FML state, which generates the $HML2_{LEI}$ state. The local net-gain maximum of the FML state that occurs in between the pulse emission (s. Fig. 5.5b2) grows with the pump current until it becomes positive at $J_G \approx 1.52$ and thereby destabilizes the FML state, giving rise to the $HML2_{LEI}$ state.

Within the just discussed pump current range, the PC2 (red line), PC3 (orange line), PC4 (light orange) PC5 (yellow), and PC6 emission state (light yellow) can also be observed. The PC2 state exhibits its lower stability boundary at $J_G \approx 1.27$, where it is bistable with the FML state and extends to the upper stability boundary at $J_G \approx 2.96$, where it is bistable with the PC3 and PC4 state. At the upper boundary, the PC2 states transitions to the $HML4_{LEI}$ state via the same mechanism that transforms the FML into the $HML2_{LEI}$ state. The two negative-valued local net-gain maxima (s. Fig. 5.5b) grow with the pump current and become positive at the transition point. Hence, two additional pulse trains are born and destabilize the existing ones via their mutual collision in the gain chip.

Similarly, the PC3 emission state stabilizes at $J_G \approx 1.84$, where it is multistable with the PC2 and $HML2_{LEI}$ state. It extends up to $J_G \approx 4.7$, where it is bistable with the PC4 state and transitions into the $HML6_{LEI}$ emission state (not shown). While the latter has not been previously presented, it continues the previous pattern of the HML_{LEI} states with six equidistant leading-edge unstable pulse trains. The destabilizing mechanism is identical to the one of the FML and PC2 emission states.

Along the PC3 state, the HML3 emission state (blue) can be observed with exactly the same stability boundaries and intensity extrema. Moreover, the HML3 also transitions to $HML6_{LEI}$ emission beyond its upper stability boundary at $J_G \approx 4.7$ (not shown). The identical pulse intensity extrema further highlight that the pulse properties are dominated by the gain interactions, which are identical among the HML3 and PC3 state (s. Figs. 5.5 and 5.7).

The PC4 state (orange line), which is characterized by four approximately identical pulses in a cluster with an internal pulse separation of $\approx 1/8$, has its lower stability boundary at $J_G \approx 2.43$, where it is multistable with the PC2, PC3, and $HML2_{LEI}$ state. Its upper stability boundary, which is outside the plotted pump current range, promotes the transition to a $HML8_{LEI}$ emission state in the same manner as the lower pulse cluster states.

The appearance of further higher-order pulse cluster states at increasing pump currents continues in the already established pattern and produces pulse clusters with up to 16 pulses for the given laser parameters. Figure 5.9, however, only further includes the PC5 and PC6 emission states within the presented pump current range. Note that the largest portrayed pump current $J_G = 4.0$ already corresponds to $\approx 11 \times J_{th}$, which likely exceeds the pump powers that an actual gain chip would be exposed to. Similarly, higher odd-order harmonic mode-locking states appear alongside the odd-order pulse cluster states for larger pump currents. The presented pump current range, furthermore, includes the HML5 state (light blue).

The transition to the even-order leading-edge unstable harmonic mode-locking states either occurs from a pulse-cluster or odd-order harmonic mode-locking state with half the number of pulses. In both cases, it is facilitated by a positive net-gain window that amplifies perturbations, which are induced by the spontaneous emission noise.

The presented results may suggest that the pulse cluster states directly bifurcate from the harmonic mode-locking states. However, the path-continuation at a lower gain bandwidths γ [50] reveals that all pulse cluster states are generated by a continuous deformation of the fundamental mode-locking emission state along the solution branch. The branch folds multiple times, i.e., is subject to saddle-node bifurcations, and each pair of left and right folds adds an additional pulse to the pulse cluster. The stability of the pulse cluster states is either limited by the mentioned saddle-node bifurcations or by additional torus bifurcations. The harmonic mode-locking states, on the other hand, are born in further independent Andronov-Hopf bifurcations of the cw state.

Furthermore, for sufficiently long cold-cavity round trip times T, which is fulfilled for the laser at hand, the fundamental mode-locking state detaches from the solution branch that is created in the first Andronov-Hopf bifurcation of the cw state. While this changes the bifurcation structure qualitatively, it has little impact on the observed stable emission states [50].

Pulse Characteristics in the Fundamental Mode-Locking Regime

After having discussed the emission state structure of the laser along the pump current J_G, the focus is now shifted to the pulse characteristics of the fundamental mode-locking state. On that account, Fig. 5.10 presents the pulse performance figures along the pump current J_G. Subfigure (a) plots the pulse peak intensity (red line) and the mean intensity (green dashed line) and subfigure (b) plots the pulse width as determined by the FWHM of the pulse profile.

The peak intensity first increases rather slowly up to $|E|^2_{\text{max}} \approx 150$ at $J_G \approx 0.5$, but then experiences a much stronger increase for intermediate pump currents. At larger pump currents, the increase of the peak intensity is subject to a rollover towards

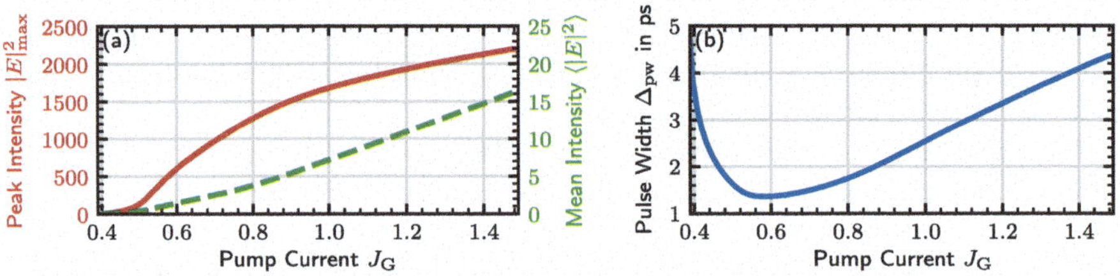

Fig. 5.10 Mode-locking performance in the fundamental mode-locking state along J_G. **a** Pulse peak intensity (red solid line) and mean intensity (green dashed line) and **b** pulse width. Laser parameters: $Q_0 = -0.58$ and $\tau_1 = 0.25$

the upper stability boundary. The mean intensity, on the other hand, steadily increases slightly superlinearly and does not exhibit the features of the pulse peak intensity.

Unlike the peak and mean intensity, the pulse width exhibits a non-monotonous behavior. While exceeding $\Delta_{\mathrm{pw}} = 4.0\,\mathrm{ps}$ close to the lasing threshold, the pulse width quickly drops to a global minimum of $\Delta_{\mathrm{pw}} \approx 1.37\,\mathrm{ps}$ at $J_G \approx 0.58$ and then linearly increases again to $\Delta_{\mathrm{pw}} \approx 4.4\,\mathrm{ps}$ at the upper stability boundary. The pulse width changes along the pump current are thus partly responsible for the non-trivial evolution of the pulse peak intensity. Shorter pulses redistribute the available mean intensity such that larger peak intensities are achieved.

Important for potential applications, the shortest pulses and largest peak intensity are obtained at very different operating points. The shortest pulses carry only $\approx$ 20% of the maximal peak intensity and the largest peak intensities are observed for pulses that are about three times longer than the shortest. Qualitatively, such behavior has been observed experimentally for multiple semiconductor quantum-well based external cavity laser [4, 13, 14, 99]. Combining the optimal pulse characteristics in one unique operation point, however, remains an unsolved problem.

Pulse-Shaping in the Fundamental Mode-Locking Regime

In order to understand the pulse width and power evolution along the pump current, the pulse shape as well as the gain and absorber-chip dynamics are investigated in the following. For this reason, Fig. 5.11a plots the normalized and color-coded pulse intensity profile along the pump current J_G within the fundamental mode-locking regime. Black contour lines indicate the leading and trailing half maximum of the pulse profile with respect to the maximum, which is centered at $t = 0.0\,\mathrm{ps}$. The

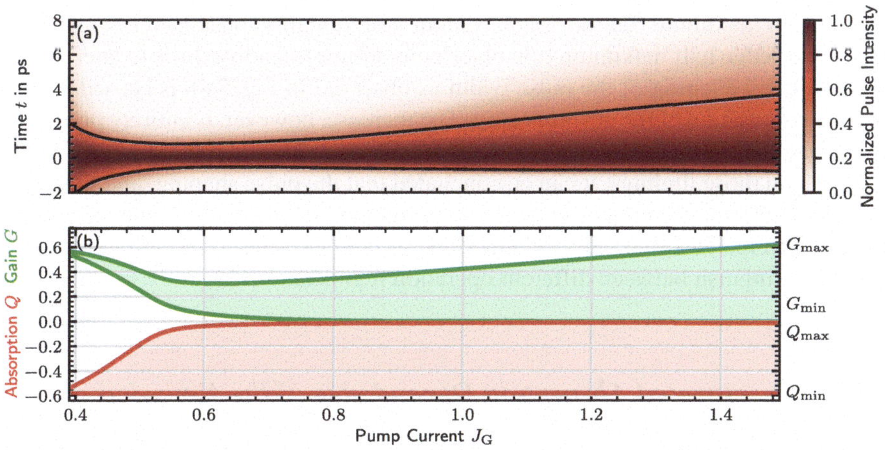

Fig. 5.11 Pulse shaping in the fundamental mode-locking state along the pump current J_G. **a** Illustrates the pulse shape via the color-coded normalized intensity $|E|^2$. The black contour lines indicate the leading and trailing half maxima and their vertical difference corresponds to the time-domain pulse width (FWHM). **b** Plots the global minimum and maximum of the integrated gain G (green) and absorption Q (red). Laser parameters: $Q_0 = -0.58$ and $\tau_1 = 0.25$

vertical distance between the two lines thus corresponds to the time-domain FWHM pulse width. Figure 5.11b, furthermore, plots the global maxima and minima, which the gain G (green lines) and absorption Q (red lines) reach within one round trip.

The examination of both plots reveals two qualitatively different operation regimes: For small pump currents $J_G \lesssim 0.55$, the pulse shape is approximately symmetrical with respect to the pulse maximum and neither the gain G nor the absorption Q are fully saturated (≈ 0). Close to the lower stability boundary in particular, the gain and absorption are both only subject to small modulations, i.e., their minima and maxima are almost identical. This suggest that the fundamental mode-locking state is indeed created in an Andronov-Hopf bifurcation with an infinitesimal modulation amplitude, which then grows with the pump current. Within the lower pump current regime, the pulse width reduces and the modulation depth of both the gain and absorption expand for increasing pump currents.

At the pump current $J_G \approx 0.55$, the saturation of the gain and absorption approaches zero and slows down considerably. This coincides with the increasing slope of the pulse peak intensity (s. Fig. 5.10a) and is close to the global pulse width minimum. Both are attributed to a more effective use of the gain, i.e., the pulse manages to make use of almost all of the provided inversion, and thus achieves an optimal pulse amplification. Simultaneously, the pulse also achieves the complete saturation of the absorber chip, which favors shorter pulse profiles.

For further increasing pump currents J_G, the maximum available gain (top green line in (b)) increases and the pulse thus improves its ability to fully saturate the absorber chip. As a consequence, the leading edge of an incoming pulse already manages to bleach the absorber, which allows the trailing edge to pass through with almost no losses (s. Fig. 5.8). The resulting pulse profile thus becomes increasingly asymmetric with an increasing pump current as can be seen in Fig. 5.11a: While the leading-half maximum (bottom black contour line) remains constant at $t \approx -0.7$ ps, the trailing-edge half maximum (top black contour line) steadily shifts to later times. Hence, the broadening of the pulse width as observed in Fig. 5.8b is caused by the complete saturation of the absorber chip. The increase, however, is entirely produced by the gain chip at the trailing edge of the pulse.

Based on those findings, it can be concluded that the pulse shaping, and thus also the pulse performance figures, critically depend upon the saturation behavior of both the gain and the absorber chip. Whether one or both become fully saturated can be used to distinguish between different operation regimes.

5.3.3 Unsaturated Absorption Dependence of the Emission Dynamics

Motivated by the just presented results, the dependence of the emission dynamics on the unsaturated absorption parameter Q_0 is investigated. Since the absorption always completely recovers after the interaction with a mode-locked pulse (s. Fig. 5.11), this

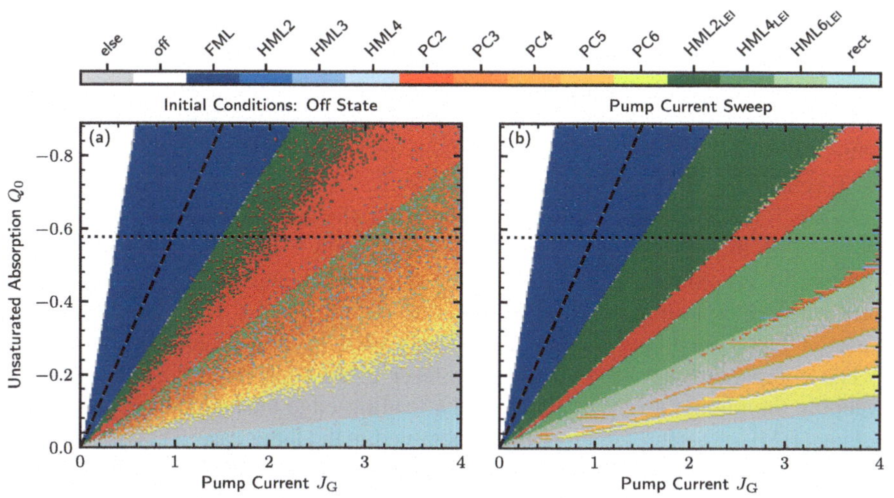

Fig. 5.12 Emission states observed in the pump current and unsaturated absorption parameter space (J_G, Q_0): off, fundamental/harmonic mode-locking (FML/HML), pulse cluster mode-locking (PC), leading-edge unstable harmonic mode-locking HML_{LEI}, and rectangular pulse (rect). The labels in the top row indicate the respective color codes. **a** Laser (initial conditions) prepared in the off state for each pixel. **b** Pump current up-sweep. The horizontal black dotted lines indicate the default unsaturated absorption. The diagonal black dashed lines indicate the scan presented in Fig. 5.14. Laser parameter: $\tau_l = 0.25$

parameter determines the 'depth' of the absorber chip, and thus also the amount of light that is required to saturate it. Within the model derivation, the unsaturated absorption has been defined by

$$Q_0 = \ell_Q \tilde{g}_Q \Gamma_Q (\tau_Q R_Q - n_Q^{tr}), \tag{5.52}$$

where the pump processes R_Q are set to zero and n_Q^{tr} represents the transparency charge-carrier density that results from the considered gain function linearization. Especially for absorber chips that are not subject to an additional reverse bias, the unsaturated absorption is not a parameter that can be easily controlled in the experiment,[12] but is rather fixed by the absorber chip design. For example, the most obvious design parameter is the length ℓ_Q of the absorber chip that is determined by the number of stacked quantum wells.

Hence, the following investigation presents two-dimensional scans in the pump current and unsaturated absorption parameter space (J_G, Q_0), which are to be interpreted as pump current scans performed on an ensemble of lasers with increasing unsaturated absorption parameters. Figure 5.12 presents the observed emission states within that parameter space, where (a) shows the results obtained with the laser started

[12] The device temperature does impact the carrier distribution and thereby the optical gain and the respective threshold density, and thus could be used to control the unsaturated absorption.

in the off state for every parameter combination and (b) shows the results for a continuously increased pump current (pump current up-sweep). The comparison of the two provides insights into the multistability among the various states. The recorded emission states are color coded and the codes are denoted on top of the plots. The horizontal black dotted lines indicate the default unsaturated absorption $Q_0 = -0.58$, which is used for all other simulations.

As predicted by the analytic expression (5.49), the lasing threshold J_{th} increases linearly with the unsaturated absorption as more gain is required to overcome the saturable losses. Similarly, the appearance of all other emission states shifts proportionally to the lasing threshold J_{th} to larger pump currents. Hence, the observed emission dynamics do not change qualitatively with the unsaturated absorption Q_0.

The comparison between starting the laser every time from the off state and sweeping the pump current reveals that the leading-edge unstable harmonic mode-locking (HML_{LEI}) states are more likely obtained in the latter case. This is attributed to their previously discussed generation mechanism from the FML and pulse-cluster states. Hence, if the laser operates in any of those states and is driven beyond the upper stability, the emission transitions to the corresponding HML_{LEI} state rather than directly to the next pulse cluster state. The pronounced multistability among the various pulse cluster states (s. Fig. 5.9) results in the noisy and speckled pattern (orange to yellow colors in (b)), which results if the laser is started from the off state.

Note that no even-order harmonic mode-locking is observed and odd-order harmonic mode-locking (HML3 in the presented case) is only found rarely when the laser is started from the off state and never when the pump current is swept. This suggests that the basin of attraction of the odd-order harmonic mode-locking states is rather small. Hence, in order to reliably achieve odd-order harmonic mode-locking, the laser must be seeded with a suitable pulse pattern, i.e., started with suitable initial conditions.

The absolute scale on which the pump current J_G is scanned, furthermore, results in large multiples of the threshold pump current J_{th} at smaller unsaturated absorption parameters Q_0. This produces further emission states that have not appeared in the previously shown one-dimensional scan presented in Fig. 5.9. As mentioned before, further pulse cluster states with an increasing number of pulses within the cluster are generated at larger pump currents. To preserve distinguishability among the smaller pulse clusters, these states are not explicitly color coded in Fig. 5.12 and therefore belong to the 'else' category. The creation of larger pulse cluster states ceases as soon as the absorption can not fully recover in between the increasingly closer packed pulse-cluster states. The collapse of the pulse clusters finally produces a rectangular shaped emission state (labeled with *rect* in Fig. 5.12), whose width amounts to approximately half the round-trip time.

Figure 5.13 illustrates the gain and emission dynamics of both the pulse cluster state with ten pulses (PC10, left column) and the rectangular-pulse state (right column) with a representation similar to Fig. 5.5: The top row shows the intensity $|E|^2$ (blue), gain G (green), and absorption Q (red) and the bottom row the net gain $\mathcal{G}$ (purple). The intensity, gain, and absorption have been normalized to 0.1 and their

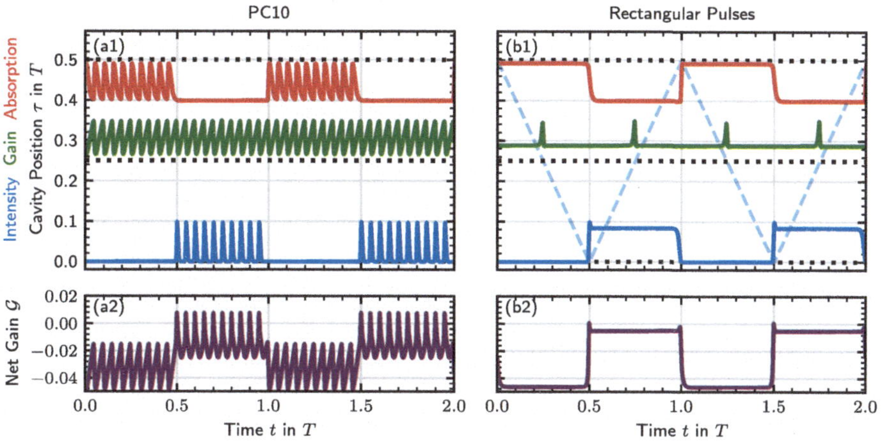

Fig. 5.13 Intensity and carrier dynamics of the PC10 ($J_G = 1.0$, $Q_0 = -0.04$) and rectangular pulse emission ($J_G = 3.0$, $Q_0 = -0.04$) states. Subfigures (a1) and (b1), respectively, show the temporal evolution of the intensity (blue lines), the gain (green lines), and the absorption (red lines). The intensity, gain, and absorption have been been normalized to 0.1 and their respective zero positions (black dotted lines) have been shifted proportional to their cavity positions. (a2) and (b2) plot the respective net gain $\mathcal{G}$. Laser parameter: $\tau_1 = 0.25$

zero baselines (horizontal black dotted lines) of the gain and absorption have been shifted upwards proportional to their cavity positions.

The PC10 state is characterized by a pulse cluster that contains ten approximately identical pulses with an intra-pulse spacing of $\approx 0.05\,T$. The identical spacing is imprinted on the absorber chip, where the absorption Q just manages to fully recover in between pulses. The gain, on the other hand, is subject to 20 equidistant pulse interactions, which are caused by the forward and backward-traveling pulses within the symmetric cavity. Hence, the pulse-cluster state is able to optimize the charge-carrier inversion to field conversion, by keeping the gain as close to its transparency density as possible. The resulting net gain (a2) exhibits ten short positive net-gain windows that correspond to the pulse emissions and ten additional characteristic negative-valued local maxima, which are caused by the gain-chip interaction with the backward-traveling pulse. Similar to the smaller pulse clusters, the PC10 state becomes unstable if the pump current is increased to the point, where the latter local net-gain maxima become positive.

The rectangular pulse emissions state (right column in 5.13) can be interpreted as a pulse cluster state, where the constituting pulses have become too close and thus have merged to single rectangularly-shaped pulse, which spans the width of the entire former pulse cluster. The pulse shape (blue in (b1)) is characterized by a steep leading edge, which produces a small overshoot of the intensity and is followed by a long plateau of constant intensity. The trailing edge does not drop as quickly as the leading edge rises and thus provides the gain (green in (b2), e.g., at $t \approx 0.75$) with a short window to recover before the backward-traveling pulse reaches the gain chip.

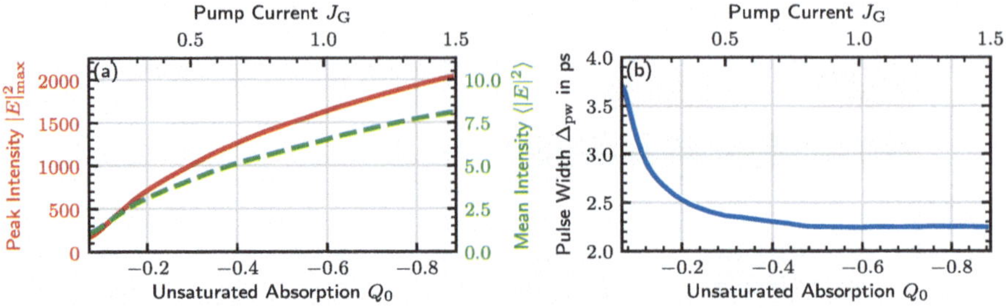

Fig. 5.14 Mode-locking performance in the fundamental mode-locking state across the pump current and unsaturated absorption parameter space (J_G, Q_0). The used parameter combinations are denoted at the top and bottom of the plots and are further indicated by the diagonal black dashed lines in Fig. 5.12. **a** Pulse peak intensity (red solid line) and mean intensity (green dashed line) and **b** pulse width. Laser parameter: $\tau_l = 0.25$

The absorption follows the pulse dynamics and is fully saturated for half a round trip and fully recovered for the other half. The resulting net gain exhibits only a short positive window at the leading edge of the rectangular pulse and remains constant at small values just below zero for the remainder of the pulse. This implies that the rectangular pulse shape itself is stable against small perturbations.

Pulse Performance Dependence on the Unsaturated Absorption

In the next step, the impact of the unsaturated absorption on the pulse performance figures is investigated. To create a fair comparison, i.e., to maintain a similar interplay between amplification and absorption, the pump current J_G is adapted alongside the unsaturated absorption to keep the relative distance to the lasing threshold constant. The scanned parameter combinations (J_G, Q_0) are indicated by the diagonal black dashed lines in Fig. 5.12. Along that line, Fig. 5.14 presents the pulse performance, where (a) plots the pulse peak intensity (red line) and the mean intensity (green dashed line) and (b) plots the pulse width as determined by the FWHM of the pulse profile.

For an increasing unsaturated absorption, i.e., more negative values of Q_0, both the pulse peak intensity and the mean intensity sublinearly increase: The former from $|E|^2_{max} \approx 200$ to $|E|^2_{max} \approx 2050$ and the latter from $\langle|E|^2\rangle \approx 1.0$ to $\langle|E|^2\rangle \approx 8.0$. Thus, even though the ratio J_G/Q_0 is kept constant, the out-coupled peak and mean power increase, which implies that the mode-locking process itself becomes more efficient. This manifests in the pulse width, which decreases from $\Delta_{pw} \approx 3.7$ ps to $\Delta_{pw} \approx 2.25$ ps along the (J_G, Q_0) scan. This observation lends itself to the interpretation that shorter pulses minimize the losses in the absorber chip and are therefore favored. Note, however, that the decrease of the pulse width approximately saturates for $Q_0 \lesssim -0.5$, while the peak and mean intensity still increases. This suggests that FWHM does not fully capture the adaptation of the pulse to the increased absorption and pump (s. Figs. 5.8 and 5.11 for the occurring asymmetric pulse profiles). Nonetheless, the increase of the pulse peak and mean intensity occurs considerably slower in the regime of the approximately constant pulse width compared to the regime of the decreasing pulse width.

In conclusion, the unsaturated absorption Q_0 hardly changes the observed qualitative dynamics, except for a shift to larger pump currents. Quantitatively, however, larger (more negative) unsaturated absorption parameters facilitate shorter pulses and larger pulse peak powers. Hence, if the energy consumption of the mode-locked laser is not a limiting factor, it is beneficial to increase the unsaturated absorption in order to achieve optimal pulse performance.

5.3.4 Gain Bandwidth Dependence of the Emission Dynamics

Lastly, the impact of the gain bandwidth parameter γ on the emission dynamics is investigated. Similar to the previously discussed unsaturated absorption, the gain bandwidth is determined by the optical components of the laser and typically cannot be controlled directly in experimental setups.[13] On that account, it is recalled that the parameter γ describes the round-trip gain spectrum in a lumped-element approach with a Lorentzian filter. In addition to the gain spectrum of the gain-chip active medium itself, it therefore also includes the spectral response of the absorber chip and the backside DBR mirrors. In the case of broadband gain chips [4], the latter may actually represent the limiting element regarding the overall round-trip gain.

Nevertheless, the gain bandwidth parameter γ can be considered to be mostly determined by the gain chip for the sake of this investigation. In order to increase its bandwidth, the individual quantum wells, which are contained within the gain chip, can be fabricated with inhomogeneously broadened band gaps. This can be achieved via various band gap engineering techniques, such as tuning the localization length, adapting the material ratios and introducing strain [54] .

In order to conduct the investigation, an ensemble of lasers with bandwidth parameters ranging from values of $\gamma = 50$ to $\gamma = 10{,}000$ is considered. This corresponds to gain bandwidths ranging from ≈ 0.105 meV to ≈ 21.1 meV. Pump current scans are performed for each laser and the results are recorded in the two-dimensional parameter space spanned by the pump current and the gain bandwidth (J_G, γ). Figure 5.15 presents the observed emission states within that parameter space, where (a) shows the results obtained with the laser started in the off state for each parameter combination and (b) shows the results for an continuously increased pump current (pump current up-sweep). The comparison between the two highlights the multistability among the various states. The recorded emission states are color coded and the respective codes are denoted on top of the plots. The horizontal black dotted lines indicate

[13] The gain chip device (lattice) temperature does impact the charge-carrier quasi-equilibrium distribution and thereby the gain bandwidth. This effect is, however, much smaller than the scanned parameter range.

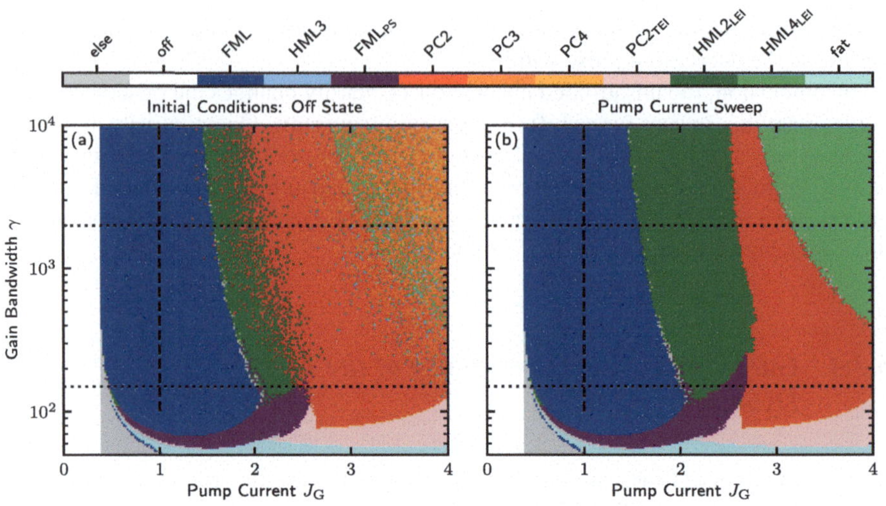

Fig. 5.15 Emission states observed in the pump current and gain bandwidth parameter space (J_G, γ): off, fundamental/harmonic mode-locking (FML/HML), pulse-cluster mode-locking (PC), leading-edge unstable mode-locking FML/HML$_{LEI}$, trailing-edge unstable pulse-cluster mode-locking PC2$_{TEI}$, and fat pulse mode-locking (fat). The labels in the top row indicate the respective color codes. **a** Laser (initial conditions) prepared in the off state. **b** Continuous increase of the pump current (up-sweep). The horizontal black dotted lines indicate the default bandwidth ($\gamma = 2000$, top) and the bandwidth used in [50] ($\gamma = 150$, bottom). The vertical black dashed lines indicate the scan in Fig. 5.17. Laser parameters: $Q_0 = -0.58$ and $\tau_1 = 0.25$

the default bandwidth ($\gamma = 2000$, top) and the bandwidth used in [50] ($\gamma = 150$, bottom). The vertical black dashed lines indicate the parameter scan used in Fig. 5.17.

As analytically predicted by Eq. (5.49), the lasing threshold J_{th} is not affected by the gain bandwidth γ due to the vanishing amplitude-phase coupling factors. The emission states that can be observed beyond the threshold, on the other hand, do critically change with the gain bandwidth.

At the lowest gain bandwidths $\gamma \lesssim 60$ the laser initially does not settle on any structured emission state (denoted with *else*), since noise induced perturbations continuously induce a sequence of irregular spikes. At larger pump currents, fat pulses (teal) emerge, which exhibit a pulse width $\Delta_{pw} \gtrsim 0.4T$. This emission state is related to the rectangular pulses, which have been presented in Figs. 5.12 and 5.13, however, with much smoother leading and trailing edges due to the smaller gain bandwidth.

At intermediate gain bandwidth parameters $60 \lesssim \gamma \lesssim 250$, two additional emission states appear: fundamental mode-locking subject to a pulse-switching instability (FML$_{PS}$, purple) at pump currents $J_G \lesssim 2.5$ and pulse-cluster mode-locking with two pulses and a trailing-edge instability (PC2$_{TEI}$, pink) at pump currents $J_G \gtrsim 2.5$. For increasing gain bandwidth parameters, the FML$_{PS}$ emission state transitions to either stable fundamental mode-locking (FML, blue) or leading-edge unstable second-order harmonic mode-locking (HML2$_{LEI}$, dark green), depending on the applied pump current. The PC2$_{TEI}$, on the other hand, transitions to stable PC2 emission (red).

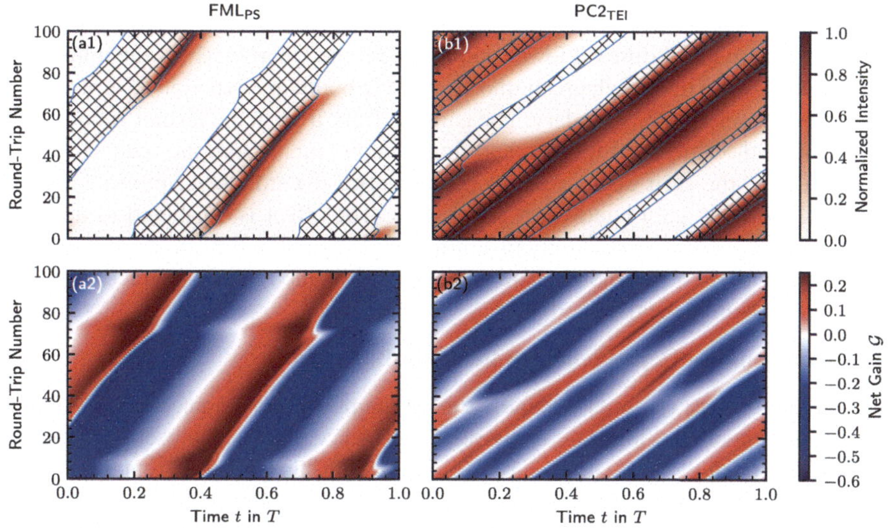

Fig. 5.16 Pseudo space-time plots of the pulse and net-gain dynamics: The top row shows the normalized intensity $|E|^2$ and the bottom row the net gain $\mathcal{G}$. The left column presents fundamental mode-locking subject to a pulse-switching instability ($\mathrm{FML_{PS}}$, $J_G = 2.0$, $\gamma = 60$) and the right column pulse-cluster mode-locking with two pulses and a trailing-edge instability ($\mathrm{PC2_{TEI}}$, $J_G = 3.0$, $\gamma = 60$). Hatched regions in the top row indicate positive net gain $\mathcal{G} > 0$ and blue contour lines indicate $\mathcal{G} = 0$. Laser parameters: $Q_0 = -0.58$ and $\tau_1 = 0.25$

The $\mathrm{FML_{PS}}$ and $\mathrm{PC2_{TEI}}$ emission states are, furthermore, illustrated in Fig. 5.16, where the top row shows pseudo space-time plots of the normalized intensity $|E|^2$ and the bottom row the corresponding net-gain dynamics. Hatched regions in the top row indicate positive net gain $\mathcal{G} > 0$ and blue contour lines indicate $\mathcal{G} = 0$. The $\mathrm{FML_{PS}}$ state (a1) is characterized by regularly switching pulse trains, where the new pulse train is always born half a round trip before/after the then dying pulse train. Peculiarly, as shown in (a2), two equidistant broad net-gain windows are open per round trip. This highlights that the net gain is only a small-signal stability measure, since the laser is not able to support two high-intensity pulse trains simultaneously in this case. The $\mathrm{PC2_{TEI}}$ state (b1) is characterized by a pulse cluster with two pulses, where periodically the leading pulse train dies, while a new pulse train is generated behind the trailing pulse train. This process shifts the $\mathrm{PC2_{TEI}}$ state to later times. The corresponding net gain (b2) exhibits four approximately equidistant, open, and relatively broad net-gain windows per round trip. Similar to the $\mathrm{FML_{PS}}$ state, the laser, however, can not support four high intensity pulse trains simultaneously.

For gain bandwidths $\gamma \gtrsim 250$, the pump current induced succession of emission states that has been presented in Figs. 5.9 and 5.12, is reestablished (FML, $\mathrm{HML2_{LEI}}$, PC2, and $\mathrm{HML4_{LEI}}$ for a continuously increasing pump current as shown in Fig. 5.15b). Within that regime, however, the transition boundaries between the emission states shift to lower pump currents. This effect is attributed to a decreasing pulse width, which reduces the interaction time with the gain chip and thus increases the available recovery time. A longer recovery time in between pulses then produces positive net-gain instabilities at lower pump currents and thereby shifts the transition boundaries. Due to this mechanism, the upper PC2 boundary experiences a stronger shift than the upper FML boundary. The gain recovery time in the former is only half as long as in the latter (s. Fig. 5.5), and thus benefits more from the reduced pulse width. Finally, the shift of the transition boundaries towards lower pump currents tapers off for large gain bandwidth parameters. The effect is attributed to a pulse width, which becomes increasingly insignificant compared to the round-trip time for large gain bandwidth parameters, and thereby reduces the relative interaction times of the pulses with the active media.

Pulse Performance Dependence on the Gain Bandwidth

Lastly, the influence of the gain bandwidth on the pulse performance figures in the fundamental mode-locking state is investigated. Hence, Fig. 5.17 presents the pulse performance from $\gamma = 100$ to $\gamma = 10{,}000$ for the constant pump current $J_G = 1$ as indicated by the vertical black dashed lines in Fig. 5.15. Subfigure (a) plots the pulse peak intensity (red line) and the mean intensity (green dashed line) and subfigure (b) plots the pulse width as determined by the FWHM of the pulse profile.

For an increasing gain bandwidth γ, both the pulse peak and mean intensity increase, which indicates a more efficient mode-locking processes with lower overall losses. Within the scanned parameter range, the peak intensity increases from $|E|^2_{\text{max}} \approx 17$ to $|E|^2_{\text{max}} \approx 10{,}000$ and the mean intensity from $\langle |E|^2 \rangle \approx 0.7$ to $\langle |E|^2 \rangle \approx 16$. The pulse width simultaneously decreases from $\Delta_{\text{pw}} \approx 26$ ps at $\gamma = 100$ to $\Delta_{\text{pw}} \approx 1.0$ ps at $\gamma = 10{,}000$. The shorter pulses do not only redistribute the available intensity to larger pulse peak values, but are also subject to smaller losses in the absorber chip, which leads to the increased mean intensity.

Furthermore, it has to be highlighted that the pulse width does not follow a simple γ^{-1} law as described in [59, 92, 98], but rather an approximate $\gamma^{-2/3}$ law (black dashed line in Fig. 5.17b), which is subject to slight local changes in the observed rate of decrease. This indicates that increasing gain bandwidth does not only directly affect the pulse properties via the filtering element, but also indirectly via the nonlinear

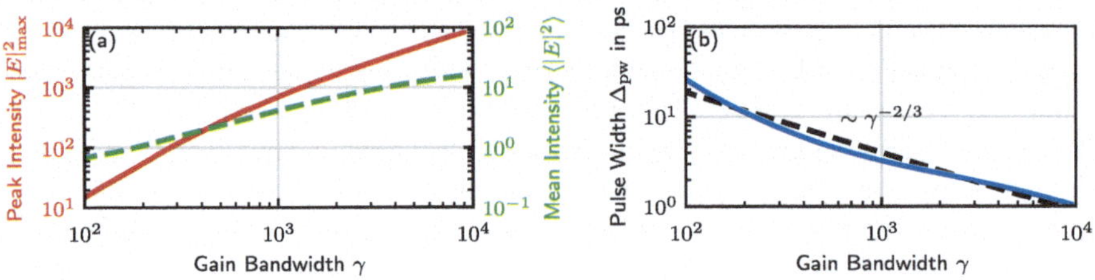

Fig. 5.17 Mode-locking performance in the fundamental mode-locking state along the increasing gain bandwidth γ. **a** Pulse peak intensity (red solid line) and mean intensity (green dashed line) and **b** pulse width. The black dashed line indicates a fitted power law. The scanned parameter range is indicated by the vertical black dashed lines in Fig. 5.15. Laser parameters: $J_G = 1.0$, $Q_0 = -0.58$ and $\tau_1 = 0.25$

interaction of the pulses with the gain and absorber chip. Hence, the shorter pulses produce a different response in the gain and absorber chip such that the pulse shaping mechanism is modified (s. Fig. 5.8), which effectively leads to the observed deviation from the γ^{-1} law.

In summary, a large gain bandwidth is highly desirable in order to achieve optimal pulses from the V-shaped laser, as it improves all performance characteristics simultaneously. The observed emission states, however, remain qualitatively identical for increasing gain bandwidths above $\gamma \gtrsim 250$. Based on this, the path continuation results regarding the fundamental and pulse-cluster mode-locking states, which have been presented in [50] for $\gamma = 150$, can be extrapolated to larger gain bandwidths. As the default gain bandwidth, this thesis chooses $\gamma = 2000$,[14] which produces pulses that are much shorter than the round-trip time, but also keep the computational integration costs reasonably low.

5.4 Asymmetric Cavity Configurations

This section investigates the impact of asymmetric laser cavities on the mode-locked emission dynamics. For that purpose, the overall cavity length, i.e., the cold-cavity round-trip time T, is kept constant while the ratio between the two cavity arms is changed. Note that asymmetry in this chapter only refers to the length of the cavity arms. In a broader sense, the laser configuration is always asymmetric, since the output coupler is located at one end and the absorber chip at the other end of the cavity. The asymmetry is then specified in terms of the propagation time τ_1 between the output coupler and the gain chip. The propagation time τ_2 between the gain and the absorber chip is then determined via the condition

$$2\tau_1 + 2\tau_2 = T, \tag{5.53}$$

where T is fixed as aforementioned. Within the dimensionless system, the propagation times thus take values $\tau_1, \tau_2 \in [0.0T, 0.5T]$. Hence, on the extremes, the gain chip is either located at the output coupler or at the absorber chip. These extreme cases are not directly experimentally realizable with a V-shaped external cavity since the optical components take up some space by themselves. The intermediate configurations, however, can be easily configured on an optical table by positioning the optical elements accordingly.

5.4.1 Emission States for Asymmetric Cavity Configurations

The analysis starts off by performing pump current scans for laser configurations with an asymmetry ranging from $\tau_1 = 0$ to $\tau_1 = 0.5$. The detected emission states

[14] Identical to [35].

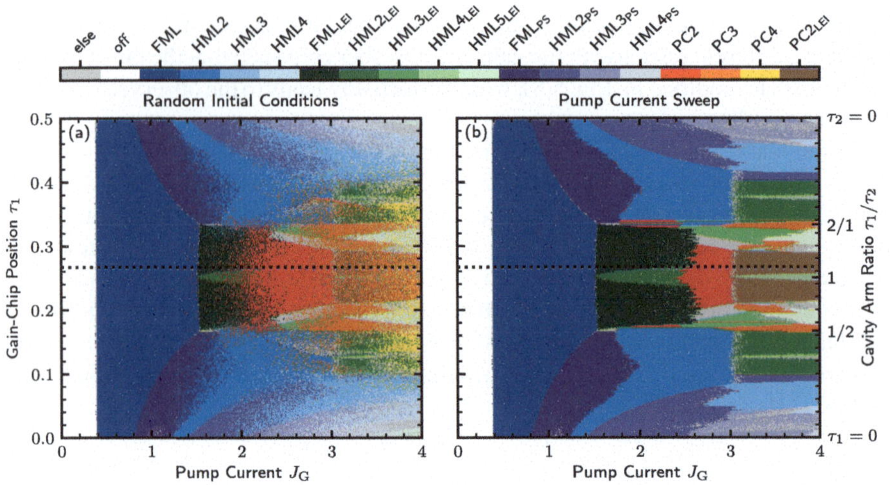

Fig. 5.18 Emission states observed in the pump current and gain-chip position parameter space (J_G, τ_1): off (white), fundamental/harmonic mode-locking (FML/HML, blue colors), pulse cluster mode-locking (PC, red to yellow colors), leading-edge unstable mode-locking (FML_{LEI}/HML_{LEI}, green colors), and pulse-switching unstable mode-locking (FML_{PS}/HML_{PS}, purple colors). The labels in the top row indicate the respective color codes. **a** Laser (initial conditions) prepared in the off state. **b** Continuous increase of the pump current (up-sweep). $\tau_1 = 0.0$ corresponds to the gain chip located at the output coupler and $\tau_1 = 0.5$ to the gain chip located at the absorber chip. Laser parameter: $Q_0 = -0.58$

are then recorded and collected in the two-dimensional parameter space spanned by the pump current and the gain-chip position (J_G, τ_1). The results are presented in Fig. 5.18, where (a) shows the emission states obtained with the laser started in the off state for each parameter combination and (b) shows the emission states for a continuously increased pump current (pump current up-sweep). The comparison between the two highlights the multistability among the various states. The recorded emission states are color coded and the respective codes are denoted on top of the plots. The horizontal black dotted lines indicate the experimentally realized cavity configuration [35] $\tau_1 = 0.267$, which is slightly off the symmetric configuration ($\tau_1 = 0.25$).

As Fig. 5.18 reveals, the cavity configuration has a profound impact on the observable emission states. Roughly speaking, the previously discussed pulse-cluster mode-locking states (red to yellow colors) only appear for approximately symmetric laser cavities. Stable even-order harmonic mode-locking states (blue colors), which have not been observed so far now appear for rather asymmetric configurations. With some minor exceptions at the boundaries, the observed emission states exhibit a reflection symmetry with respect to the symmetric cavity configuration $\tau_1 = 0.25$. This observed symmetry is attributed to an exchange symmetry of τ_1 and τ_2 in Eq. (5.53) with the assumption that the propagation effects dominate interaction with the absorber chip at one end and the output coupler at the other end of the cavity.

Fundamental Mode-Locking

The fundamental mode-locking state can be observed for all cavity configurations. The gain-chip position neither affects the lasing threshold (since Eq. (5.49) does not depend on the propagation times) nor the lower FML boundary. The upper pump current stability boundary, on the other hand, critically depends on the cavity configuration. With the gain-chip located at the output coupler, i.e., $\tau_1 = 0.0$, the upper FML stability boundary is found at $J_G \approx 0.8$ and pulse-switching unstable fundamental mode-locking (FML_{PS}, dark purple) is observed beyond the boundary. For an increasing gain-chip position τ_1, the upper FML stability boundary shifts towards larger pump currents until it reaches a local maximum at $J_G \approx 1.54$ for the configuration $\tau_1 = 1/6 \approx 0.167$. From there on, the upper FML stability boundary has reached an approximate plateau and only marginally reduces to the local minimum $J_G \approx 1.52$ at the symmetric configuration $\tau_1 = 0.25$. Within that plateau region, instead of the FML_{PS} state, leading-edge unstable fundamental mode-locking (FML_{LEI}, dark green) is observed beyond the stability boundary. That is, however, with the exception of the symmetric cavity configuration $\tau_1 = 0.25$, where, as previously reported in Sect. 5.3, leading-edge unstable second-order harmonic mode-locking ($HML2_{LEI}$, green) is found beyond the FML stability boundary. The remaining cavity configurations $\tau_1 \in]0.25, 0.5]$ follow the aforementioned reflection symmetry with respect to $\tau_1 = 0.25$.

The two new unstable emission states, namely leading-edge and pulse-switching unstable fundamental mode-locking (FML_{LEI} and FML_{PS}, respectively), to which the laser transitions after crossing the upper stability boundary of the fundamental mode-locking state, are further illustrated in the following. On that account, Fig. 5.19 shows pseudo space-time plots of the normalized intensity $|E|^2$ (top row) and the net gain $\mathcal{G}$ (bottom row) over a range of 100 round trips. The net gain is color coded and regions of positive net gain (red colors) are indicated in the intensity plots by blue contour lines and hatched regions. The left column plots leading-edge unstable fundamental mode-locking (FML_{LEI}, $J_G = 1.6$, $\tau_1 = 0.2$) and the right column plots pulse-switching unstable fundamental mode-locking (FML_{PS}, $J_G = 1.2$, $\tau_1 = 0.1$).

Leading-edge unstable fundamental mode-locking is characterized by a single pulse train that exhibits a large positive net-gain window at the leading edge of the pulse (hatched regions in (a1)). Hence, noise induced perturbations at the leading edge are amplified and thereby destabilize the pulse train, which produces large fluctuations of the pulse peak intensities and of the pulse positions. The net gain as shown in Fig. 5.19a2, furthermore, exhibits two additional characteristic negative valued local maxima at $t \approx 0.66$ and $t \approx 0.86$.

Pulse-switching unstable fundamental mode-locking,[15] on the other hand, is characterized by pulse trains which only live for a limited number of round trips before they die and a new pulse train is born at a different position within the cavity. As indicated in Fig. 5.19b1, the birth of a new pulse train is preceded by a broad positive

[15] Pulse-switching unstable fundamental mode-locking has also been observed for the monolithically integrated three-section laser in Chap. 4. Both the qualitative emission dynamics and the cavity configuration induced generation mechanism are identical.

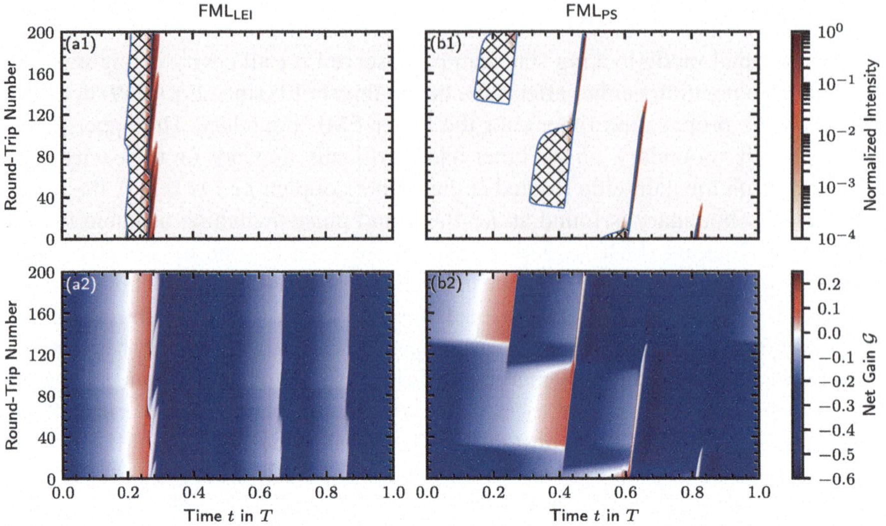

Fig. 5.19 Pseudo space-time plots of the pulse and net-gain dynamics: the top row shows the normalized intensity $|E|^2$ and the bottom the net gain $\mathcal{G}$. The left column shows leading-edge unstable fundamental mode-locking (FML$_{\text{LEI}}$, $J_G = 1.6$, $\tau_1 = 0.2$) and the right column shows pulse-switching unstable fundamental mode-locking (FML$_{\text{PS}}$, $J_G = 1.2$, $\tau_1 = 0.1$). Hatched regions in the top row indicate positive net gain $\mathcal{G} > 0$ and blue contour lines indicate $\mathcal{G} = 0$. Laser parameter: $Q_0 = -0.58$

net-gain window. The newly born pulse then emerges at the trailing edge of the net-gain window. As soon as the new pulse train has reached enough power to bleach the gain and absorber chip, it causes the positive net-gain window to shrink and only overlap with the pulse. However, shortly after the new pulse train is established, a new broad net-gain window at a different cavity position is reestablished as well. Hence, the switching process regularly repeats and thereby prohibits stable fundamental mode-locking emission. For the parameters at hand, a typical pulse trains lives for $\approx$ 150 round trips ($\approx$ 94 ns) and the switching processes, where two pulse trains coexist, last for $\approx$ 40 round-trips ($\approx$ 25 ns).

With the addition of the leading-edge unstable second-order harmonic mode-locking state (HML2$_{\text{LEI}}$), which is observed beyond the upper FML stability boundary for the symmetric cavity configuration ($\tau_1 = 0.25$), it can be reasoned that three different cavity configuration dependent destabilizing mechanisms are at work. A detailed analysis of those mechanisms is presented in Sect. 5.4.4.

Harmonic Mode-Locking States

Rather asymmetric cavity configurations lead to the stabilization of the even-order harmonic mode-locking states. For an increasing pump current, a cascade of increasing higher-order harmonic mode-locking states with intermediate pulse-switching instabilities is generated. Moreover, the harmonic mode-locking states at the rather asymmetric cavity configurations continue a pattern in their upper pump current stability boundary, which is similar to the FML state. For cavity configurations that become more symmetric, i.e., $\tau_l \rightarrow 0.25$, the respective upper stability boundary initially shifts to larger pump currents, as can be seen for the HML2 and HML3 states in Fig. 5.18. Beyond the stability boundary, pulse-switching unstable harmonic mode-locking of the same order, i.e., $\mathrm{HML2_{PS}}$ and $\mathrm{HML3_{PS}}$, respectively, is observed. Similar to the FML state, the increase of the upper stability boundary, however, reaches a local maximum, which is followed by a plateau as can be seen for HML2 emission in Fig. 5.18. The turning points towards the HML2 plateau are found at $\tau_l \approx 0.083$ and $\tau_l \approx 0.417$ instead of $\tau_l \approx 0.167$ and $\tau_l \approx 0.333$ for the FML emission case.

The observed emission states beyond the plateau are either leading-edge unstable second-order harmonic mode-locking ($\mathrm{HML2_{LEI}}$, green) or leading-edge unstable fourth-order harmonic mode-locking ($\mathrm{HML4_{LEI}}$, light green). The latter resembles the $\mathrm{FML_{LEI}}$ to $\mathrm{HML2_{LEI}}$ transition as discussed for the FML emission state at a symmetric cavity configuration (symmetry axes at $\tau_l = 0.125$ and $\tau_l = 0.375$ instead of $\tau_l = 0.25$).

Third-order harmonic mode-locking, which is, unlike second and fourth-order harmonic mode-locking, a stable emission state at rather symmetric configurations, can only be seen in very small pockets in Fig. 5.18, due to the large multistability among the diverse emission states (s. Fig. 5.9). It is therefore predominantly observed at strong cavity asymmetries.

The gain and net-gain dynamics of the harmonic mode-locking states, which can be observed for pronounced asymmetries, are presented in Fig. 5.20. Second-order harmonic mode-locking (HML2) is shown in the left column ((a1) and (a2), $J_G = 2.0$,

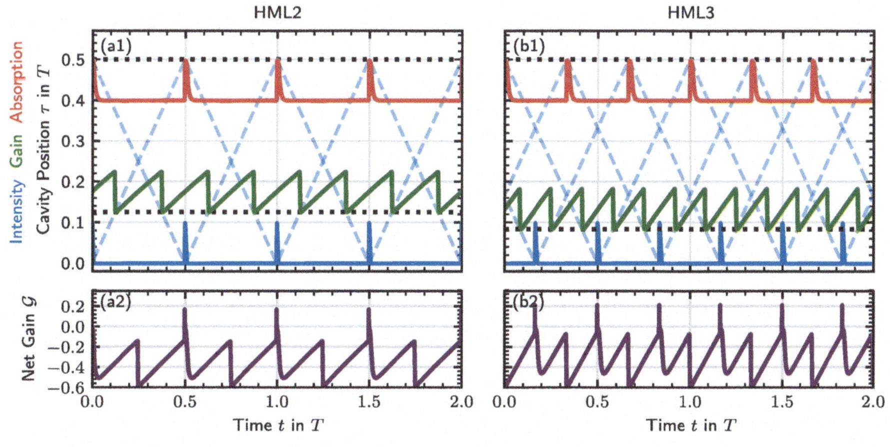

Fig. 5.20 Intensity and carrier dynamics of the HML2 ($J_G = 2.0$, $\tau_l = 0.25/2 = 0.125$) and HML3 ($J_G = 3.2$, $\tau_l = 0.25/3 \approx 0.0833$) states. Subfigures **a1** and **b1** show the temporal evolution of the intensity (blue lines), the gain (green lines), and the absorption (red lines), respectively. The intensity, gain and, absorption have been normalized to 0.1 and the respective zero positions (black dotted lines) have been shifted proportional to their cavity positions. The propagation of the pulses is indicated by blue dashed lines. **a2** and **b2** Plot the respective net gain $\mathcal{G}$. Laser parameter: $Q_0 = -0.58$

$\tau_1 = 0.25/2 = 0.125$) and third-order harmonic mode-locking (HML3) is shown in the right column ((b1) and (b2), $J_G = 3.2$, $\tau_1 = 0.25/3 \approx 0.0833$). The top row ((a1) and (b1)) presents the intensity $|E|^2$ (blue), gain G (green), and absorber Q (red) dynamics and the bottom row ((a2) and (b2)) the corresponding net-gain $\mathcal{G}$ dynamics. The intensity $|E|^2$, the gain G and the absorption Q have been normalized to 0.1. The zero baselines of the gain and absorption have been shifted proportional to their cavity position with respect to the output coupler and are indicated by horizontal black dotted lines. The propagation through the laser cavity is indicated by blue dashed lines. The intersection with the black dotted lines thus represents the interaction with the gain and absorber chips.

Both the second and third-order harmonic mode-locking dynamics are qualitatively identical. The respective cavity configurations ensure that the gain chip experiences equidistant interactions with the circulating pulse trains. The gain chip experiences four pulse interactions per round trip in the case of the HML2 state and six pulse interactions per round trip in the case of the HML3 state. Similar to fundamental mode-locking (s. Fig. 5.5), the corresponding net-gain dynamics exhibit short positive windows at the pulse emissions and characteristic negative-valued local maxima in between pulse emissions.

Note that the presented examples are chosen to be at resonant cavity configures, which enable the equidistant gain interactions with the pulses. At slightly off-resonant configurations, both the HML2 and HML3 emission states can be nonetheless observed with neither shifts of the pulse positions nor non-identical pulse shapes among the coexisting pulse trains. This behavior is contrary to the pulse cluster states, where the peak intensities within the cluster change for off-resonant configurations [52].

As reported above, the cavity configuration does affect the destabilizing mechanism, which determines the upper pump current boundary of the regions of harmonic mode-locking. The single negative-valued local net-gain maximum, as shown in Fig. 5.20a2 and b2, only occurs for the resonant configuration. For HML2 emission this leads to leading-edge unstable fourth-order harmonic mode-locking beyond the pump current stability boundary $J_G \approx 3$ ($\mathrm{HML4_{LEI}}$, green in Fig. 5.18). For off-resonant configurations, the negative-valued local net-gain maximum splits into two non-identical local maxima. As a consequence, the leading-edge unstable second-order harmonic mode-locking appears beyond the upper pump current stability boundary ($\mathrm{HML2_{LEI}}$, green in Fig. 5.18). Similarly, the HML3 emission state transitions into either $\mathrm{HML3_{LEI}}$ or $\mathrm{HML6_{LEI}}$ depending on the cavity configuration (not shown in Fig. 5.18).

Pulse Cluster Mode-Locking States

The pulse-cluster mode-locking states (PC, red to yellow colors) predominantly appear for symmetric configurations $\tau_1 \in [0.167, 0.333]$. For an increasing cavity asymmetry, i.e., $\tau_1 \neq 0.25$, the pulses within a cluster become increasingly non-identical. Typically, the leading pulse acquires additional energy and thus exhibits a larger peak intensity than the other pulses. Similarly, the temporal pulse spacing within the cluster adapts to optimize the gain-chip interactions with the pulses. On that account, the pulse clusters with three and four pulses (PC3 and PC4), which can be found for asymmetric cavity configurations up to $\tau_l \approx 0.1$ and $\tau_l \approx 0.4$, respectively, are strongly deformed.

As indicated in Fig. 5.18, the pulse-cluster states are subject to complex stability boundaries, which shall not be discussed in their entirety.[16] For that reason, this sections only focuses on the PC2 emission state. As presented in Fig. 5.18, the PC2 state shares some similarities with the FML and the HML emission states. For an increasing cavity symmetry, the upper stability boundary of the PC2 shifts to larger pump currents until a plateau is reached. Along the increasing stability boundary, the emission transitions to complex and irregular dynamics, which are not further classified and are thus labeled with 'else' (gray). For asymmetric configurations along the plateau, the PC2 state transitions to leading-edge unstable PC2 emission ($PC2_{LEI}$, brown). Only for the symmetric configuration, the laser emits leading-edge unstable fourth-order harmonic mode-locking ($HML4_{LEI}$, light green), as presented and discussed in Sect. 5.3. In that sense, the same pattern that has been observed for the FML and HML2 states is reproduced.

Furthermore, the plateau is reached at $J_G \approx 3.0$, which matches the HML2 plateau at asymmetric configurations and is not a coincidence. Both in the PC2 and the HML2 state, the gain is subject to four approximately equidistant pulse interactions per round trip (s. Figs. 5.5 and 5.20). In both cases, the resulting net gain exhibits two short and positive windows along the pulse emission and two characteristic negative-valued local maxima. The stability is either limited by a leading-edge instability of the current state or the transition to $HML4_{LEI}$ emission. Hence, the stability boundary is either determined by the pump current that causes the leading-edge instability ($HML2/PC2_{LEI}$) or by the negative-valued local net-gain maxima to become positive ($HML4_{LEI}$). Due to the similar operation conditions, this occurs at approximately the same pump currents. It can be expected that higher-order harmonic and pulse-cluster mode-locking states are subject to the same mechanism.

Summary

In conclusion, the gain-chip positions τ_l, i.e., the cavity asymmetry parameter, further reveals the rich diversity of mode-locking states in passively mode-locked external cavity lasers. With respect to the symmetric cavity, further emission states appear (even-order harmonic mode-locking) or become favorable compared to the pulse-cluster states (odd-order harmonic mode-locking). Additionally, the destabilizing mechanisms of the fundamental, harmonic, and pulse-cluster states diversify as well, which lead to qualitatively different irregular emission states (pulse-switching and leading-edge unstable mode-locking).

In that sense, the symmetric cavity configuration that was investigated in Sect. 5.3 represents a degenerate case, where both cavity arms are of the same length. This configuration uniquely gives rise to the $HML2_{LEI}$ instability of the fundamental mode-locking state and the $HML4_{LEI}$ instability of the pulse-cluster state with two pulses.

---

[16] See Refs. [50, 52] for a detailed analysis of the pulse cluster emission states and their stability.

The common feature that all stable emission states share is an approximately temporal equidistant interaction of the gain chip with the pulses. This optimizes the gain recovery and minimizes non-radiative losses in the gain chip. It, furthermore, prevents positive net-gain instabilities to occur at relatively low pump currents. For asymmetric cavity configurations, this condition is fulfilled by the harmonic mode-locking states. For the approximately symmetric configurations, on the other hand, the pulse-cluster states and the odd-order harmonic mode-locking states match this condition. Fundamental mode-locking can be observed for all cavity configurations. Nevertheless, a greatly improved upper pump stability boundary at rather symmetric configurations improves the operation flexibility.

5.4.2 Discussion of the Experimental Results

The presented results now allow for a discussion of the experimental results. On that account, the experimentally observed succession of emission states (s. Sect. 5.1 and Fig. 5.2) is reproduced for slightly but sufficiently asymmetric cavity configurations. As indicated by the black dotted lines in Fig. 5.18, the experimentally realized configuration is found just above the region of leading-edge unstable second-order harmonic mode-locking.

The experimentally observed side pulses in the intensity autocorrelation trace are, furthermore, associated with the simulated leading-edge unstable emission states. On that account, Fig. 5.21a1 and a2 present simulated autocorrelation traces at $J_G = 2.0$ and $\tau_l = 0.267$. The difference between the two is embodied in the integration time t_{int}. Figure 5.21a1 utilizes 100 round trips and (a2) 1000 round trips (similar to Fig. 5.4). Instead of a simple decay, Fig. 5.21a1 shows a characteristic peak at the time lag $\Delta t \approx 0.018$, which is on the order of the pulse width. The longer integration time (a2), on the other hand, smooths out the characteristic peak, but nonetheless exhibits two-stage decay of the autocorrelation trace, which is reminiscent of the peak seen in (a1).

The difference induced by the integration time t_{int} is attributed to the irregularity of the noise induced leading-edge instability. On sufficiently short time scales, the perturbations that are generated in front of the pulses (s. Fig. 5.19) remain sufficiently self-similar to create the characteristic peak in the autocorrelation trace. On longer time scales, however, the irregular noise induced perturbations average out and produce the observed slower second stage of the decay of the autocorrelation trace.

At this point, it is important to note that the simulated dynamics are able to qualitatively reproduce the experimentally observed side pulse in the autocorrelation trace. However, the simulations do not fully correspond to the experimentally observed dynamics, since the simulated leading-edge instabilities cause broad noisy harmonics in the power spectral densities (s. Fig. 5.4), which are not observed in the experiment (s. [35] for details). This suggest that reality includes a stabilizing mechanism, which regularizes the pulses produced in the leading-edge instability and thereby produces a periodic pulse train with a clean characteristic line shape in the power spectral density. Candidates for such a mechanism are likely found in the complex gain dynamics, which have been subject to strong simplifications in the employed modeling approach. For example, more quantitative models [64, 100, 101] include effects such as asymmetric gain spectra, kinetic hole burning, and carrier-distribution amplitude-phase coupling.

Nonetheless, the appearance of the side pulse in the autocorrelation trace can be attributed to the leading-edge instability, since the instability occurs due to the recovery of the gain chip in the absence of the optical pulses. This process would equally occur for more complex gain models. The stabilization of side pulses created in the leading-edge instability would further enable them to be detectable in the autocorrelation trace for arbitrary integration times.

Lastly, leading-edge unstable pulse-cluster mode-locking with two pulses ($PC2_{LEI}$, brown color in Fig. 5.18) is shown in a pseudo space-time representation in Fig. 5.21b. As the name suggests, the emission is characterized by a pulse cluster with two pulses (PC2, s. Fig. 5.4), where both pulse trains are subject to a leading-edge instability. This instability is induced by the large positive net-gain window in front of the pulse, as indicated by the hatched regions and blue contour lines. Additional positive net-gain windows appear in between the pulse cluster as well, but do not lead to the creation of high power pulses. The resulting autocorrelation traces of the $PC2_{LEI}$ are qualitatively identical to those of the FML_{LEI} state (Fig. 5.21a1 and a2) and are therefore not presented. Hence, following the discussion of the FML_{LEI} state, the simulated $PC2_{LEI}$ emission state corresponds to the experimentally observed PC2 state with intensity autocorrelation side pulses.

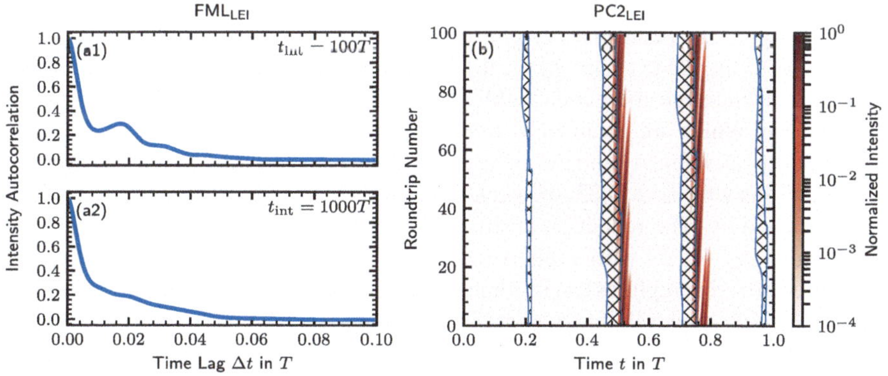

Fig. 5.21 Intensity autocorrelation trace of the leading-edge unstable fundamental mode-locking state (FML_{LEI}, $J_G = 2.0$ and $\tau_l = 0.267$) for an integration time $t_{int} = 100\,T$ (**a1**) and $t_{int} = 1000\,T$. **b** Pseudo space-time plot of the normalized intensity of the leading-edge unstable pulse-cluster state with two pulses ($PC2_{LEI}$, $J_G = 3.5$ and $\tau_l = 0.267$). Hatched regions in the top row indicate positive net gain $\mathcal{G} > 0$ and blue contour lines indicate $\mathcal{G} = 0$. Laser parameter: $Q_0 = -0.58$

5.4.3 Pulse Performance Dependence on the Cavity Asymmetry

In the next step, the pulse performance characteristics of the fundamental mode-locking state are investigated in the pump current and gain-chip position parameter space (J_G, τ_1). On that account, Fig. 5.22 presents the color-coded pulse peak intensity in (a) and the color-coded pulse width (time-domain FWHM) in (b). The depicted parameter space corresponds to the dark blue regions in Fig. 5.18. Note that the horizontal line scan at $\tau_1 = 0.25$ represents the symmetric cavity and has already been shown in Fig. 5.10 and discussed in Sect. 5.3.

As both Fig. 5.22a and b reveal, the pulse performance characteristics found for the symmetric cavity are representative for all other cavity configurations. The pulse peak intensity monotonically increases for an increasing pump current and the pulse width first decreases to a global minimum at $0.5 \lesssim J_G \lesssim 0.6$ and then monotonically grows again. In that sense, the cavity configuration, i.e., the gain-chip position τ_1, mostly affects the performance by determining the upper stability boundary.

This strongly affects the maximum achievable pulse peak intensity. For the strongly asymmetric configurations $\tau_1 = 0.0$ and $\tau_1 = 0.5$, the peak intensity only reaches $|E|^2_{max} \approx 1000$ before the FML state becomes unstable. At the local maxima of the upper stability boundary $\tau_1 = 0.167$ and $\tau_1 = 0.333$, on the other hand, peak intensities of $|E|^2_{max} \approx 2250$ and $|E|^2_{max} \approx 2150$, respectively, can be observed. Contrary to the peak intensity, the cavity configuration is irrelevant with respect to the minimal achievable pulse width as this occurs for all configurations below the upper stability boundary.

Taking a closer look, the pulse performance characteristics are not exactly symmetric with respect to the symmetric cavity configuration $\tau_1 = 0.25$. For that reason Fig. 5.22c and d, respectively, present the pulse peak intensity and the pulse width normalized to the value observed for the symmetric configuration for a given pump current J_G. Hence, the color code highlights relative changes, i.e., improvements and degradations, which are obtained by tuning the cavity geometry.

Both the peak intensity and the pulse width show a non-trivial behavior. For cavity configurations with $\tau_1 > 0.25$, the peak intensity always reduces with respect to the symmetric cavity. The reduction is stronger for greater asymmetries and reaches a minimum at $J_G \approx 0.54$ with a relative value of ≈ 0.6. For configurations with $\tau_1 < 0.25$, three different regimes can be identified: For small pump currents $J_G \lesssim 0.5$, the peak intensities show relative increases with values up to ≈ 1.06. For intermediate pump currents $0.5 \lesssim J_G \lesssim 1.0$, the pulse peak intensities reduce with the asymmetry and values down to ≈ 0.7 can be observed. For larger pump currents $J_G \gtrsim 1.0$, which are only observed for rather symmetric configurations, the behavior changes again and the peak intensities increase with the asymmetry. However, the relative improvements only reach values up to ≈ 1.01.

The relative pulse width changes show a qualitatively similar behavior for both asymmetry directions. For small pump currents $J_G \lesssim 0.6$ the relative pulse width increases. This effect is more pronounced for configurations with $\tau_1 > 0.25$, where

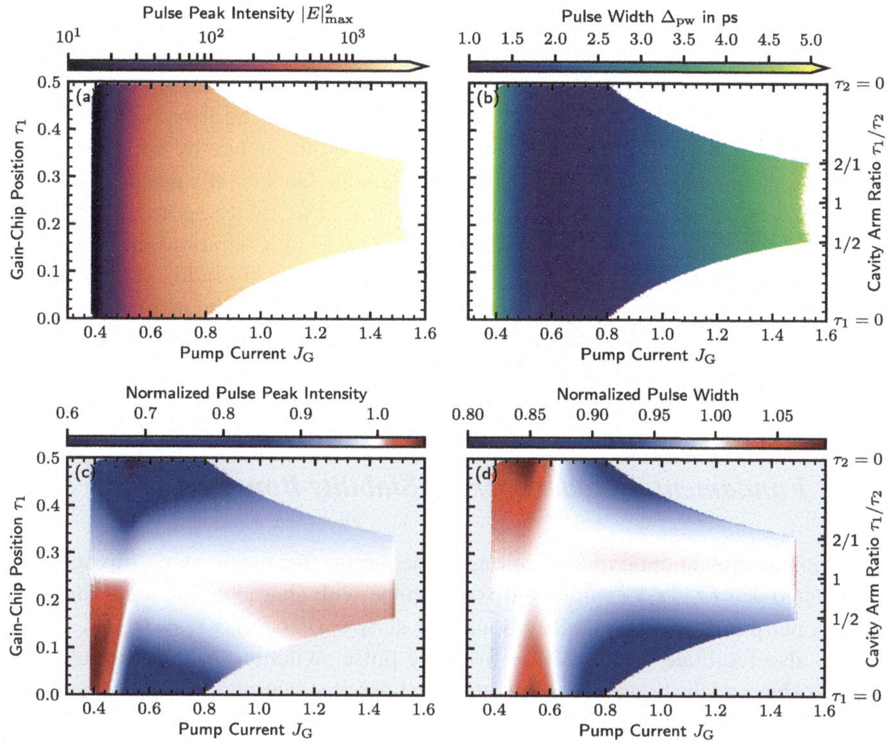

Fig. 5.22 Fundamental mode-locking performance in the pump current and gain-chip position parameter space (J_G, τ_1). Subfigures **a** and **b** show the pulse peak intensity and the pulse width, respectively. Subfigures **c** and **d** depict the data in (**a**) and (**b**) normalized to the symmetric laser cavity ($\tau_1 = 0.25$) for each pump current. Laser parameter: $Q_0 = -0.58$

the largest relative change with a value of 1.1 can be found additionally. A small exception is observed for configurations with $\tau_1 < 0.25$ just above the lasing threshold. At larger pump currents $J_G \gtrsim 0.6$, the behavior changes and the pulse width decreases with an increasing asymmetry. The reduction of the pulse width is more pronounced for configurations with $\tau_1 < 0.25$ and the minimum normalized pulse width is found close to the upper pump current stability boundary at $\tau_1 \approx 0.05$ with a value of ≈ 0.82. This reduction, however, decreases to a value of only ≈ 0.97 at the local pump current stability boundary maximum at $\tau_1 \approx 0.167$.

In conclusion, it is generally beneficial for the pulse performance to configure the laser in an asymmetric configuration with $\tau_1 < 0.25$. Only if the shortest absolute pulse width is to be achieved with no requirements for the peak intensity (obtained at $\tau_1 = 0.25$ and $J_G \approx 0.58$) a symmetric configuration can be recommended. If, furthermore, a great pump current tuning flexibility and strong peak powers are desired, a cavity configuration with $\tau_1 \approx 0.167$ is recommended, which corresponds to the lower local maximum of the upper pump current stability boundary.

Lastly, it is to be mentioned that some of the results must be interpreted carefully. As demonstrated in Sect. 5.3.1, the spectral filtering element, which models the combined gain spectrum of all optical components in a lumped element approach, has the greatest impact on the pulse shaping. To recall, in the investigated example (s. Fig. 5.8), the pulse peak intensity and pulse width change by approximately $\approx 15\%$ due to the interaction with the filtering element. Hence, only relative changes in Fig. 5.22, which exceed the 15%, can be attributed to cavity geometry induced changes of the gain and absorption dynamics. These considerations in particular weaken the results at large pump currents where the asymmetry induced peak intensity and pulse width changes are relatively small. Nonetheless, the results presented in this section exhibit a trend that is highly worthwhile to be investigated by experiments and by quantitatively more accurate models.

5.4.4 Fundamental Mode-Locking Stability Boundary

The region of fundamental mode-locking in the pump current and gain-chip position parameter space (J_G, τ_l) exhibits a distinct shape with characteristic local maxima of the upper pump current stability boundary at $\tau_l = 0.167$ and $\tau_l = 0.333$. These features also facilitate the transition from the pulse-switching to the leading-edge instability beyond the boundary. Motivated by those observations, this last subsection investigates the underlying mechanisms in detail and derives an analytic approximation of the upper stability boundary as a function of the cavity configuration.

To start out, Fig. 5.23 illustrates the three potential destabilizing mechanisms, which may occur for different cavity configurations τ_l. The pulse-switching instability ($J_G = 1.05$, $\tau_l = 0.1$) is shown in the left column (a), the leading-edge instability ($J_G = 1.5$, $\tau_l = 0.2$) in the middle column (b), and the second-order harmonic mode-locking leading-edge instability ($J_G = 1.4$, $\tau_l = 0.25$) in the right column (c). The top row ((a1) to (c1)) presents the intensity $|E|^2$ (blue), the gain G (green), and the absorber Q (red) dynamics and the bottom row ((a2) to (c2)) the corresponding net-gain $\mathcal{G}$ dynamics. The intensity $|E|^2$, the gain G, and the absorption Q have been normalized to 0.1. Moreover, the zero baselines of the gain and absorption have been shifted proportional to their cavity position with respect to the output coupler and are indicated by horizontal black dotted lines. The propagation through the laser cavity is illustrated by blue dashed lines. The intersection with the black dotted lines thus represents the interaction with the gain chip and the absorber chip. Additionally, black circles in (a2) to (c2) indicate semistable ($\mathcal{G} \approx 0$) net-gain windows for which perturbations would be neither amplified nor attenuated. The propagation of such perturbations is, furthermore, indicated by purple dashed lines in (a1) to (c1). Note that the perturbations are considered to be small and therefore leave no characteristic footprint in the gain dynamics, i.e., do not affect the gain recovery.

The pulse-switching instability ($\mathrm{FML_{PS}}$, (a)) is induced by the second local net-gain maximum, which occurs in between the pulse emissions as indicated by the black circle in (a2). The net-gain window is closed by the interaction of the gain and

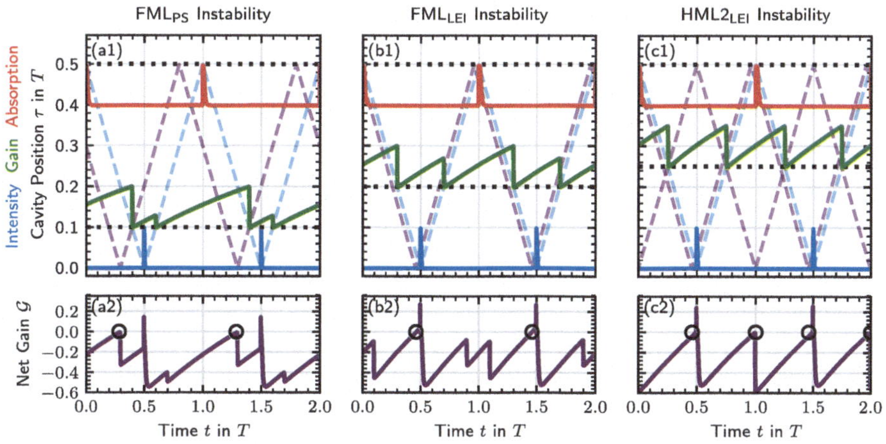

Fig. 5.23 Intensity and carrier dynamics of the fundamental mode-locking state close to the **a** pulse-switching instability ($J_G = 1.05$, $\tau_1 = 0.1$), **b** leading-edge instability ($J_G = 1.5$, $\tau_1 = 0.2$), and **c** second-order harmonic mode-locking leading-edge instability ($J_G = 1.4$, $\tau_1 = 0.25$). Subfigures **a1** to **c1**, respectively, show the temporal evolution of the intensity (blue lines), the gain (green lines), and the absorption (red lines), which have been normalized to 0.1. Their respective zero positions (black dotted lines) have been shifted proportional to their cavity positions. The propagation of the pulses is indicated by blue dashed lines and the propagation of the semistable perturbations by purple dashed lines. **a2**–**c2** plot the respective net gain $\mathcal{G}$. Black circles indicate semistable ($\mathcal{G} \approx 0$) net gain windows. Laser parameter: $Q_0 = -0.58$

the backward-traveling pulse. The resulting FML_{PS} dynamics beyond the stability boundary are presented in Fig. 5.19b. The leading-edge instability (FML_{LEI}, (b)), on the other hand, is generated, as the name indicates, by positive net gain at the leading edge of an existing pulse train. Hence, the destabilizing perturbation (purple dashed line in (b1)) travels right in front of the existing pulse train. The irregular FML_{LEI} dynamics resulting from this instability are presented in Fig. 5.19a. Lastly, the second-order leading-edge instability (HML2_{LEI}, (c)) is characterized by positive net gain at the leading edge of an existing pulse as well as the first, i.e., the earlier, local net-gain maximum in between pulse emissions. In the presented example, the first and second local net-gain maxima in between pulse emissions have merged due to the symmetric cavity configuration. The HML2_{LEI} dynamics, which result from the two simultaneously occurring net-gain instabilities, are presented in Fig. 5.6a.

Analytic Net-Gain Approximation

Based on those results, the three distinct irregular emission states, which can be observed beyond the stability boundary, can be attributed to the presented three distinct net-gain induced instabilities. Hence, the destabilizing mechanism can be entirely explained by the net-gain dynamics. This insight represents the foundation for the further analysis. The net gain, as given by Eq. (5.51), is reproduced here and reads

$$\mathcal{G}(t) = G(t - \tau_1) + G(t - \tau_1 - 2\tau_2) + 2Q(t - \tau_1 - \tau_2) + \ln(r_{OC}). \qquad (5.54)$$

The net gain thus dynamically depends on the gain G and the absorption Q at the respective delay times τ_1 and τ_2. Inspecting the time traces of the gain and absorption dynamics, as presented in Fig. 5.23, reveals the following features: Firstly, both the gain and the absorption are almost completely saturated upon the interaction with the mode-locked pulses. Secondly, the absorption recovers very quickly to its equilibrium value Q_0 in the absence of the pulses. Thirdly, the gain experiences an exponential relaxation towards its equilibrium value in the absence of the pulses, but does not reach that value.

Those features can be combined to yield an analytic approximation for the net gain for all times except at the trailing edge of the pulse, where the absorption is not yet recovered. In the first step, the dynamical absorption in Eq. (5.54) is simply replaced by the equilibrium value Q_0. In the second step, the gain recovery can be analytically solved by integrating the dynamical equation for the gain Eq. (5.39) in the absence of optical pulses, i.e., $|E|^2 \approx 0.0$. The result reads

$$G(t) = \frac{J_G}{\gamma_G} \left(1 - \exp(-\gamma_G t)\right), \qquad (5.55)$$

where the complete saturation $G \approx 0.0$ has been used as the initial condition. The net gain can then be written as

$$\mathcal{G} = G(\Delta_1) + G(\Delta_2) + 2Q_0 + \ln(r_{OC}), \qquad (5.56)$$

where Δ_1 and Δ_2 denote the times, which are available for the gain recovery. They describe the time intervals in between the gain-chip interaction with the considered perturbation and the last interaction with an existing high-power pulse. Δ_1 and Δ_2 depend on the emissions state and the cavity geometry, which is encoded by the gain-chip position τ_1. The previously introduced instabilities occur at the transition to positive net-gain values at the characteristic points. Hence, Eq. (5.56) is set to zero and solved for the critical pump current, which yields the expression

$$J_G^{\text{crit}} = \frac{\gamma_G \left(-\ln(r_{OC}) - 2Q_0\right)}{2 - \exp(-\gamma_G[\Delta_1 - \Delta_{\text{pw}}]) - \exp(-\gamma_G[\Delta_2 - \Delta_{\text{pw}}])}, \qquad (5.57)$$

where the gain-recovery intervals Δ_1 and Δ_2 are yet to be determined. In a last step, the effective recovery times have been adjusted for the finite pulse width Δ_{pw}, during which the electric field is not negligible and no gain recovery occurs.

Each of the three instabilities produces an independent branch of recovery intervals (Δ_1, Δ_2), which eventually leads to positive net-gain windows. In the example time series presented in Fig. 5.23, the recovery intervals (Δ_1, Δ_2) can be identified as the time differences between the intersection of the perturbation (purple dashed lines) with the gain chip (central black dotted line) and the preceding interaction of a high-power pulse with the gain. Crucially, the considered noise induced perturbations are small when they are generated, such that they can potentially be amplified twice by the same gain per round trip without causing any saturation.

The pulse-switching instability directly feeds off this mechanism. As Fig. 5.23a1 exemplifies, the backward-moving perturbation first passes the gain chip, is reflected at the output coupler, and then passes the gain chip in the forward direction right before the existing backward-moving pulse bleaches the gain G. Applying this rule, the gain-recovery intervals (Δ_1, Δ_2) that produce the pulse-switching instability (FML_{PS}) read

$$\left.\begin{aligned}\Delta_1 &= 1 - 4\tau_1\\ \Delta_2 &= 1 - 2\tau_1\end{aligned}\right\}\text{ for } \tau \leq 0.25 \quad \text{and} \quad \left.\begin{aligned}\Delta_1 &= 4\tau_1 - 1\\ \Delta_2 &= 2\tau_1\end{aligned}\right\}\text{ for } \tau > 0.25, \tag{5.58}$$

where the distinction of cases is due to the cavity configuration symmetry around $\tau_1 = 0.25$. Note that this construction allows the perturbation to be amplified twice by the long gain recovery, which amounts to $\Delta_1, \Delta_2 \approx 1$ for configurations with $\tau_1 \approx 0.0$ and $\tau_1 \approx 0.5$.

The leading-edge instability (FML_{LEI}), as highlighted in Fig. 5.23b1, is produced by a perturbation that sits right in front of the existing pulse. Hence, the characteristic recovery intervals follow the propagation times that determine the pulse amplification in Eq. (5.38) and read

$$\left.\begin{aligned}\Delta_1 &= 2\tau_1\\ \Delta_2 &= 1 - 2\tau_1\end{aligned}\right\}\text{ for } \tau \leq 0.25 \quad \text{and} \quad \left.\begin{aligned}\Delta_1 &= 1 - 2\tau_1\\ \Delta_2 &= 2\tau_1\end{aligned}\right\}\text{ for } \tau > 0.25. \tag{5.59}$$

Note that $\Delta_1 + \Delta_2 = 1$ always holds due to the construction. Lastly, the second-order leading-edge instability (HML2_{LEI}) is created by a perturbation that arrives at the gain chip in the backward direction right before the existing forward moving pulse bleaches the gain, as can be seen in Fig. 5.23c1. The resulting recovery intervals thus read

$$\left.\begin{aligned}\Delta_1 &= 2\tau_1\\ \Delta_2 &= 2\tau_1\end{aligned}\right\}\text{ for } \tau \leq 0.25 \quad \text{and} \quad \left.\begin{aligned}\Delta_1 &= 1 - 2\tau_1\\ \Delta_2 &= 1 - 2\tau_1\end{aligned}\right\}\text{ for } \tau > 0.25, \tag{5.60}$$

where Δ_1 and Δ_2 are always identical. Note that both recovery intervals drop to zero for the configurations $\tau_1 = 0.0$ and $\tau_1 = 0.5$, which causes Eq. (5.57) to diverge.

The resulting branches of recovery intervals (Δ_1, Δ_2) are plotted in Fig. 5.24a, where purple lines denote the pulse-switching instability (FML_{PS}), dark green lines the leading-edge instability (FML_{LEI}), and light green lines the second-order leading-edge instability (HML2_{LEI}). Solid lines denote Δ_1, which is by construction always the smaller among the two, and dashed lines denote Δ_2. The plot reveals three unique cavity configurations: The recovery intervals that belong to the leading-edge and second-order leading-edge instability are both equal for the symmetric cavity configuration $\tau_1 = 0.25$. The recovery intervals that belong to the leading-edge and

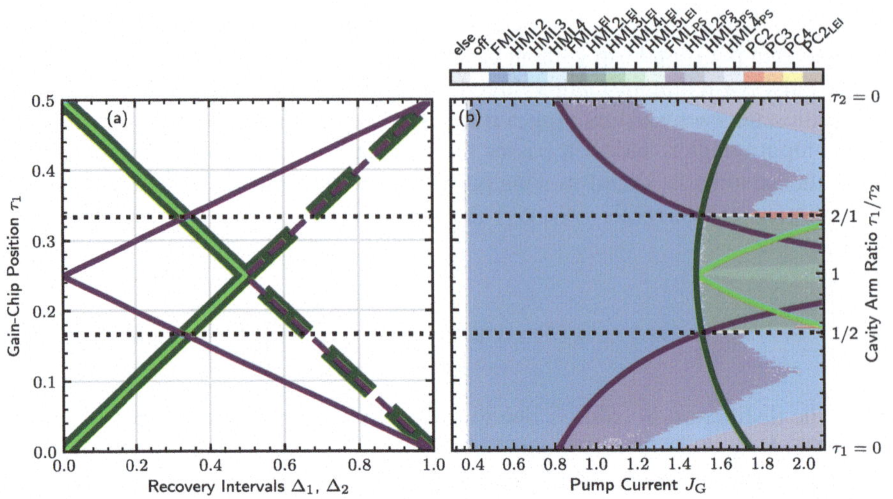

Fig. 5.24 Upper stability boundaries of the fundamental mode-locking state in the pump current and gain-chip position parameter space (J_G, τ_1). Subfigure **a** plots the branches of gain-recovery intervals Δ_1, Δ_2 (solid and dashed, respectively), which feed the net-gain approximation Eq. (5.57) and belong to the pulse-switching instability (purple), the leading-edge instability (dark green), and the second-order leading-edge instability (light green). Subfigure **b** plots the estimated stability boundaries (color code as in (**a**)) on top of the numerically obtained emission states for an effective pulse width $\Delta_{pw} = 0.05$. Horizontal black dotted lines indicate $\tau_1 = 0.167$ and $\tau_1 = 0.333$. Laser parameter: $Q_0 = -0.58$

the pulse-switching instability are both equal at the cavity configurations $\tau_1 \approx 0.167$ and $\tau_1 \approx 0.333$, which are highlighted by horizontal black dotted lines.

Using Eq. (5.57), the resulting stability boundary approximations are plotted in Fig. 5.24b on top of the numerically obtained emission states (s. Fig. 5.18) using the same color code as in (a). Important to note, the pulse width Δ_{pw} that enters Eq. (5.57) is not evaluated as the FWHM, but rather as the width for which the electric field dominates the dynamics of the gain (s. Eq. (5.39)). For the case at hand, this is fulfilled for $|E|^2 \gtrsim 0.1$, which yields an approximate pulse width of $\Delta_{pw} = 0.05$.[17]

As the comparison between the numerically obtained emission states and the analytic stability boundary approximations demonstrates, the latter almost perfectly approximates the former. The analytic approximation, moreover, explains the exact shape of the region of stable fundamental mode-locking as well as the respective emission states, which can be observed beyond the stability boundary. As can be seen in Fig. 5.24b, the fundamental mode-locking stability is determined by the lowest critical pump current J_G^{crit} among the three destabilizing mechanisms. For the rather asymmetric configurations, i.e., $\tau_1 \lesssim 0.167$ and $\tau_1 \gtrsim 0.333$, this happens to

[17] Similar to the intensity FWHM, which has been presented in Fig. 5.22b, the $|E|^2 \gtrsim 0.1$ width at the stability boundary also varies with the cavity configuration. Good results for the stability boundaries can be nonetheless obtained for a single-value approximation as long as the pulses are short compared to the round-trip time, i.e., $\Delta_{pw} \ll T$ holds.

be the pulse-switching instability (purple line). According to Fig. 5.24a, this can be attributed to the longest available recovery intervals Δ_1 and Δ_2, which require a smaller pump current to achieve positive net-gain windows.

For rather symmetric cavity configurations $0.167 \lesssim \tau_1 \lesssim 0.333$, the leading-edge instability provides the longest recovery intervals (Δ_1, Δ_2). Consequently, the fundamental mode-locking state is destabilized by the leading-edge instability (dark green line in Fig. 5.24b). The local pump current stability boundary maxima at $\tau_1 \approx 0.167$ and $\tau_1 \approx 0.333$ are generated by the intersection of the pulse-switching and leading-edge instability. The leading-edge instability, moreover, exhibits a global minimum for the symmetric cavity $\tau_1 = 0.25$, despite the constant sum $\Delta_1 + \Delta_2 = 1$ of the two recovery intervals. This property is caused by a monotonic and concave downward recovery of the gain G, which is more efficient, i.e., has a larger mean slope, for shorter recovery intervals and thus requires a smaller critical pump current to produce positive net-gain windows.

Lastly, the second-order leading-edge instability becomes only relevant for the symmetric cavity configuration $\tau_1 = 0.25$, where the associated recovery intervals (Δ_1, Δ_2) are identical to the leading-edge instability. For all other configurations, the critical pump current quickly grows and diverges towards $\tau_1 = 0.0$ and $\tau_1 = 0.5$. Hence, the leading-edge unstable second-order harmonic mode-locking state $\mathrm{HML2}_{\mathrm{LEI}}$ can only be observed right beyond the stability boundary for the symmetric configuration. Note that for further increasing pump currents J_G, a small cone of $\mathrm{HML2}_{\mathrm{LEI}}$ grows in the (J_G, τ_1) parameter space.

Parameter Dependence of the Stability Boundary

In conclusion, the stability boundaries of the fundamental mode-locking state can be understood in terms of the net-gain dynamics and can be quantitatively reproduced via a simple gain-recovery approximation. The resulting equation (5.57) for the critical pump current then only depends on the output coupler reflectivity r_{OC}, the unsaturated absorption Q_0, the gain recovery rate γ_G, the effective pulse width Δ_{pw}, and the gain-chip position τ_1 via the gain recovery intervals Δ_1 and Δ_2. Typical output coupler reflectivities, however, are close to unity [3, 4] and therefore only provide a small contribution.

The dependencies of the fundamental mode-locking stability boundary on the remaining laser parameters in the pump current and gain-chip position parameter space (J_G, τ_1) are presented in Fig. 5.25. The critical pump current J_G^{crit} is determined via Eq. (5.57) as the minimum among the three destabilizing mechanisms. Hence, the unsaturated absorption Q_0, the gain-recovery rate γ_G, and the pulse width Δ_{pw} parameterize the family of curves $J_G^{\mathrm{crit}}(J_G, \tau_1)$. Figure 5.25a–c each show seven such curves for representative values of the aforementioned parameters, where the fourth (red) curve always represents the previously used default parameter set (s. Fig. 5.24).

The unsaturated absorption Q_0, which is plotted in (a), is increased (towards negative values) in equidistant steps from $Q_0 = -0.18$ to $Q_0 = -1.08$. As Q_0 only appears linearly in the numerator of the governing equation (5.57) the respective curves $J_G^{\mathrm{crit}}(J_G, \tau_1)$ are merely shifted linearly towards larger pump currents.

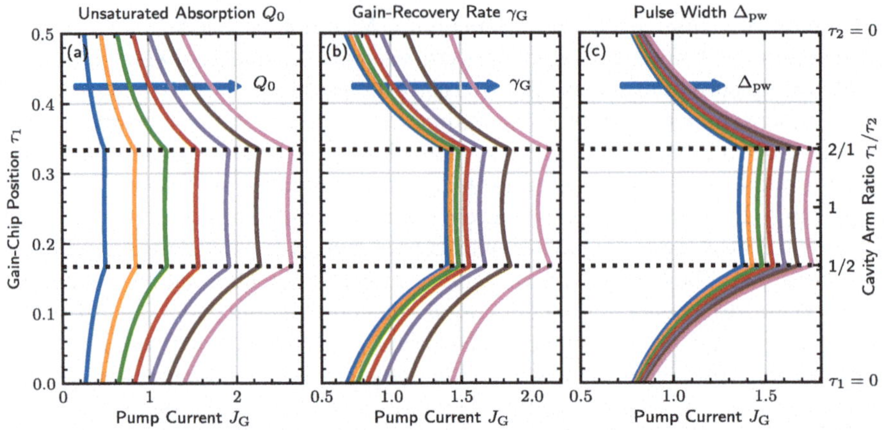

Fig. 5.25 Critical pump current J_G^{crit} (upper stability boundary) of the fundamental mode-locking state in the pump current and gain-chip position parameter space (J_G, τ_1). Subfigure **a** plots J_G^{crit} for a linearly increasing unsaturated absorption $Q_0 \in \{-0.18, \ldots, -1.08\}$. Subfigure **b** plots J_G^{crit} for a geometrically increasing gain-chip recovery rate $\gamma_G \in \{0.185, \ldots, 2.11\}$. Subfigure **c** plots J_G^{crit} for a linearly increasing pulse width $\Delta_{pw} \in \{0.0, \ldots, 0.1\}$. The red curve always represents the default parameter set

The gain-recovery rate, which is shown in Fig. 5.25b, is increased geometrically (constant multiplication factors) from $\gamma_G = 0.185$ to $\gamma_G = 2.11$. The gain-recovery rate enters the stability boundary approximation Eq. (5.57) linearly in the numerator and in the arguments of the exponential functions in the denominator, which implies a nonlinear response to changes of this parameter. As Fig. 5.25b reveals, an increasing recovery rate most dominantly shifts the critical pump current to larger values due to the factor of γ_G in the numerator, which describes the change of the gain equilibrium value with the gain-recovery rate. As a second-order effect, a fast, i.e., large, gain-recovery rate in the exponential functions leads to a faster and thus effectively 'more concave' gain-recovery process. This leads to more pronounced local maxima and a stronger curvature of the stability boundary at fast recovery rates (pink line in (b)). At small rates, the recovery is slow and effectively linear at the available recovery intervals Δ_1, Δ_2, which leads to a leading-edge instability boundary that is almost constant along gain-chip positions $0.167 \leq \tau_1 \leq 0.333$ (blue line in (b)).

Lastly, the effective pulse width Δ_{pw} is increased linearly with equidistant steps from $\Delta_{pw} = 0.0$ to $\Delta_{pw} = 0.1$ and the results are presented in Fig. 5.25 (c). The pulse width reduces the effective gain recovery intervals Δ_1 and Δ_2 and thus increases the pump current J_G that is needed to achieve positive net gain. Since the pulse width appears in the argument of the exponential functions in Eq. (5.57), the effect is nonlinear and, similarly to the gain-recovery rate, leads to a more pronounced curvature and stronger local maxima at larger pulse widths (pink curve in (c)). Moreover, the effect is weaker for smaller ratios $\Delta_{pw}/\Delta_{1,2}$, which occur for the rather asymmetric configurations.

To conclude, the unsaturated absorption Q_0, the gain-recovery rate γ_G, and the effective pulse width Δ_{pw} all three retain the characteristic shape of the region of fundamental mode-locking. The most dominant effect of all three parameters is their induced shift of the upper pump current stability boundary. Only secondary, the gain-recovery rate and the pulse width also nonlinearly affect the curvature of the boundary. Nevertheless, in all cases, the cavity configurations with $\tau_l \approx 0.167$ and $\tau_l \approx 0.333$ produce the local maxima in the stability boundary and thus provide the greatest pump current operating flexibility. This advantage increases for a strongly nonlinear gain recovery, e.g., as induced by faster recovery rates and broader pulses.

Summary and Applicability to More Complex Gain Systems

In summary, the stability of the fundamental mode-locking state can be qualitatively and quantitatively understood in terms of the net-gain dynamics. The FML state becomes unstable as soon as positive net-gain windows, which do not coincide with the emission of an existing pulse, appear. Depending on the temporal position of the additional positive net-gain window, the fundamental mode-locking state destabilizes via a pulse-switching, a leading-edge, or a second-order leading-edge instability.

The respective stability boundaries of the three destabilizing mechanisms can be analytically predicted by solving the gain recovery in the absence of optical pulses and using the results to construct the net gain as a function of the available gain-recovery intervals. The three instabilities each induce a branch of recovery times (Δ_1, Δ_2) that depends on the geometric cavity configuration and determines, which instability occurs at the lowest pump current and thereby limits the FML stability. The branches of recovery intervals (Δ_1, Δ_2) intersect at characteristic cavity configurations. Resulting from that, the stability boundaries intersect as well and thus create the characteristic shape of stable FML emission. The region of stable fundamental mode-locking emission in the pump current and cavity configuration parameter space retains its qualitative features, i.e. its shape, for all changes of the relevant laser parameters. These results are attributed to the concave downward gain-recovery process and the cavity configuration that determines the available gain-recovery intervals.

For that reason, the qualitative shape of the stable fundamental mode-locking region should generalize to any gain-chip active medium, which exhibits a concave downward recovery process. Hence, the largest pump current tuning-range should always be found independently from the gain-chip active medium at the characteristic gain-chip positions $\tau_l \approx 0.167$ and $\tau_2 \approx 0.167$.

For more complex gain-chip active media, the gain dynamics can likely not be solved analytically in the absence of optical pulses. This would prevent the simple approximation of the stability boundaries. Nevertheless, the stability boundary condition (net gain $\mathcal{G} \approx 0$) can be formulated as an ordinary-differential equation boundary problem. The numerical solution of this problem should be always much more inexpensive than simulating the complete set of equations that describe the full spatio-temporal laser dynamics. Hence, this approach could be easily implemented to determine the stability boundaries for gain-chips with quantum-dot active media [37, 38, 102–104], submonolayer quantum-dot active media [41, 88, 105, 106], and multi-level quantum-well active media [35, 79].

5.5 Chapter Conclusions

In this chapter, the emission dynamics of a passively mode-locked V-shaped external cavity semiconductor laser have been studied. To do so, a minimalistic delay-differential equation model has been derived from first principles to self-consistently describe the electric field, gain, and absorption dynamics of the laser. The model has been formulated such that the implementation of different gain media, e.g., semiconductor quantum wells or quantum dots, is straightforward. Unlike the Haus model [52, 97, 107, 108], instabilities and irregular pulse train dynamics are inherently within the scope of the model. Apart from the direct numerical integration of the equations, the model is suitable for path-continuation techniques [50, 52] and allows for analytical approximations [52].

Using the model, the experimentally observed sequence of emission states has been reproduced and understood in terms of the gain and net-gain dynamics as well as the cavity configuration. The observed emission states are characterized by equidistant interactions of the optical pulses with the gain chip. This optimizes the gain recovery and minimizes the non-radiative losses in the gain chip. The experimentally observed side pulses have been attributed to a leading-edge instability, which is predicted to only occur at slightly asymmetric cavity configurations.

The emission dynamics have been furthermore characterized as a function of the cavity configuration (gain-chip position τ_1). For rather symmetric cavity configurations, the equidistant gain interaction condition suppresses even-order harmonic mode-locking states and gives rise to the characteristic pulse-cluster mode-locking states. Rather asymmetric configurations restore the emission dynamics, which are known from ring-cavity models, where no pulse-cluster and only harmonic mode-locking states are observed. The symmetric configuration, in particular, represents a degenerate case with equal cavity arm lengths, which leads to a unique destabilizing mechanism of the stable emission states. The symmetric configuration has furthermore been found to not represent the optimal configuration in order to achieve high quality fundamental mode-locking: The best performance and operation flexibility has been found at $\tau_1 = 0.167$, i.e., a ratio of 1:2 of the cavity arms.

These results are further supported by an analytic approximation of the fundamental mode-locking upper pump current stability boundary. The procedure distinguishes three different positive net-gain destabilizing mechanisms, which lead to three different instabilities and thus to three qualitatively different irregular emission states beyond the stability boundary. The intersection of the different stability boundaries produces the characteristic shape of the stable fundamental mode-locking region. It has furthermore been discussed that the qualitative results of the analytic stability boundary approximation should generalize to every gain medium, which exhibits a monotonic and concave downward recovery process. As this applies to all conventional gain media, the optimal cavity configuration with respect to the pump current tuning range should always be found at $\tau_1 = 0.167$ or $\tau_1 = 0.333$, i.e., cavity arm ratios of 1:2 and 2:1. Results from the direct numerical integration have moreover suggested that the pulse cluster and harmonic mode-locking states are subject to the same positive net-gain induced destabilizing mechanisms. Hence, if those states are of particular interest, similar analytic approximations of the upper stability boundary could be constructed.

Open Questions and Outlook

The obvious open question addresses the stabilizing mechanism that produces the experimentally observed side pulses along stable periodic fundamental and pulse-cluster emission. Assuming that the delay-differential equation model has omitted the relevant physical mechanism due to the utilized approximations and assumptions, a number of extensions and modifications of the numerical model are possible: Easy to implement within the delay-differential equation framework, an effective multi-population charge-carrier model, as suggested in [35, 79], could provide the characteristic scattering time scales of the charge-carriers. If a more general electric field propagation scheme (s. Sect. 2.4) is used, analytic approximations of the quantum-well optical susceptibility [109–111] could be used to avoid the lumped-element gain filter approach and incorporate some of the intricacies of the asymmetric and complex gain spectrum. In the case that this does not suffice, the Maxwell-Bloch equations could be implemented either with the full polarization dynamics (s. Sect. 2.3) or with the effective polarization dynamics as introduced in Chap. 4. Note that only the former would be able to produce Rabi oscillations [112], which could become relevant due to the high pulse powers within the laser resonator. Ultimately, the highest order in the modeling hierarchy could be reached by coupling the Maxwell-Bloch equations to the full wave-equation as done in [64, 100, 101, 113]. Going beyond the traveling-wave approximations would furthermore allow to simulate the experimentally observed 100 fs pulse widths.

Independent of the experimentally observed side pulses, the above enumerated extensions and improvements of the numerical model could furthermore be utilized to study and understand the qualitative and quantitative limitations of the presented simple delay-differential equation model. With that knowledge, the pulse performance trends that have been observed for varied cavity configurations could be reevaluated in order to put the quantitative predictions regarding the optimal cavity configuration on firmer ground.

References

1. Kuznetsov M, Hakimi F, Sprague R, Mooradian A. High-power (> 0.5-W CW) diode-pumped vertical-external-cavity surface-emitting semiconductor lasers with circular TEM/sub 00/beams. IEEE Photon Technol Lett 9(8):1063–1065 (1997)
2. Keller U, Tropper AC (2006) Passively modelocked surface-emitting semiconductor lasers. Phys Rep 429:67–120
3. Tilma BW, Mangold M, Zaugg CA, Link SM, Waldburger D, Klenner A, Mayer AS, Gini E, Golling M, Keller U (2015) Recent advances in ultrafast semiconductor disk lasers. Light Sci Appl 4(e310). https://doi.org/10.1038/lsa.2015.83

4. Waldburger D, Link SM, Mangold M, Alfieri CGE, Gini E, Golling M, Tilma BW, Keller U (2016) High-power 100 fs semiconductor disk lasers. Optica 3(8):844–852. https://doi.org/10.1364/optica.3.000844
5. Moulton PF (1986) Spectroscopic and laser characteristics of Ti:Al_2O_3. J Opt Soc Am B 3(1):125–133. https://doi.org/10.1364/josab.3.000125
6. Sutter DH, Steinmeyer G, Gallmann L, Matuschek N, Morier-Genoud F, Keller U, Scheuer V, Angelow G, Tschudi T (1999) Semiconductor saturable-absorber mirror-assisted Kerr-lens mode-locked Ti:sapphire laser producing pulses in the two-cycle regime. Opt Lett 24(9):631–633. https://doi.org/10.1364/ol.24.000631
7. Geusic JE, Marcos HM, Van Uitert LG (1964) Laser oscillations in Nd-doped yttrium aluminum, yttrium gallium and gadolinium garnets. Appl Phys Lett 4(10):182–184. https://doi.org/10.1063/1.1753928
8. Spühler GJ, Südmeyer T, Paschotta R, Moser M, Weingarten KJ, Keller U (2000) Passively mode-locked high-power Nd:YAG lasers with multiple laser heads. Appl Phys B 71(1):19–25. https://doi.org/10.1007/pl00021154
9. Kafka JD, Baer T, Hall DW (1989) Mode-locked erbium-doped fiber laser with soliton pulse shaping. Opt Lett 14(22):1269–1271. https://doi.org/10.1364/ol.14.001269
10. Fermann ME, Hartl I (2009) Ultrafast fiber laser technology. IEEE J Sel Top Quantum Electron 15(1):191–206. https://doi.org/10.1109/jstqe.2008.2010246
11. Rafailov EU, Cataluna MA, Avrutin EA (2011) Ultrafast lasers based on quantum dot structures. WILEY-VCH, Weinheim. 978-3-527-40928-0
12. Norman JC, Mirin RP, Bowers JE (2021) Quantum dot lasers—history and future prospects. J Vac Sci Technol A 39(2). https://doi.org/10.1116/6.0000768
13. Wilcox KG, Tropper AC, Beere HE, Ritchie DA, Kunert B, Heinen B, Stolz W (1599) 4.35 kW peak power femtosecond pulse mode-locked VECSEL for supercontinuum generation. Opt Express 21(2):1599. https://doi.org/10.1364/oe.21.001599
14. Klopp P, Griebner U, Zorn M, Weyers M (2011) Pulse repetition rate up to 92 GHz or pulse duration shorter than 110 fs from a mode-locked semiconductor disk laser. Appl Phys Lett 98(7):071103. https://doi.org/10.1063/1.3554751
15. Butkus M, Viktorov EA, Erneux T, Maker G, Malcolm GPA, Rafailov EU (2013) 85.7 MHz repetition rate mode-locked semiconductor disk laser: fundamental and soliton bound states. Opt Express 21(21):25526–25531
16. Mangold M, Link SM, Klenner A, Zaugg CA, Golling M, Tilma BW, Keller U (2014) Amplitude noise and timing jitter characterization of a high-power mode-locked integrated external-cavity surface emitting laser. IEEE Photonics J 6(1):1500309. ISSN 1943-0655. https://doi.org/10.1109/jphot.2013.2295464
17. Gaafar MA, Rahimi-Iman A, Fedorova KA, Stolz W, Rafailov EU, Koch M (2016) Mode-locked semiconductor disk lasers. Adv Opt Photonics 8(3):370–400. https://doi.org/10.1364/aop.8.000370
18. Link SM, Klenner A, Keller U (2016) Dual-comb modelocked lasers: semiconductor saturable absorber mirror decouples noise stabilization. Opt Express 24(3):1889–1902. https://doi.org/10.1364/oe.24.001889
19. Zaugg CA, Klenner A, Sieber OD, Golling M, Tilma BW, Keller U (2013) Sub-100 MHz passively modelocked VECSEL. In: CLEO: 2013. IEEE, pp 1–2. Paper CW1G.6
20. Link SM, Waldburger D, Alfieri CGE, Golling M, Keller U (2017) Coherent beam combining and noise analysis of a colliding pulse modelocked VECSEL. Opt Express 25(16):19281–19290. https://doi.org/10.1364/oe.25.019281
21. Avrutin EA, Panajotov K (2019) Delay-differential-equation modeling of mode-locked vertical-external-cavity surface-emitting lasers in different cavity configurations. Materials 12(19):3224. https://doi.org/10.3390/ma12193224
22. Saraceno CJ, Emaury F, Schriber C, Hoffmann M, Golling M, Südmeyer T, Keller U (2014) Ultrafast thin-disk laser with 80 μJ pulse energy and 242 W of average power. Opt Lett 39(1):9–12

23. Udem T, Holzwarth R, Hänsch TW (2002) Optical frequency metrology. Nature 416:233–237. https://doi.org/10.1038/416233a
24. Klenner A, Schilt S, Südmeyer T, Keller U (2014) Gigahertz frequency comb from a diode-pumped solid-state laser. Opt Express 22(25):31008. https://doi.org/10.1364/oe.22.031008
25. Coddington I, Newbury N, Swann W (2016) Dual-comb spectroscopy. Optica 3(4):414–426. https://doi.org/10.1364/optica.3.000414
26. Link SM, Maas DJHC, Waldburger D, Keller U (2017) Dual-comb spectroscopy of water vapor with a free-running semiconductor disk laser. Science 356:1164. https://doi.org/10.1126/science.aam7424
27. Mayer AS, Klenner A, Johnson AR, Luke K, Lamont MRE, Okawachi Y, Lipson M, Gaeta AL, Keller U (2015) Frequency comb offset detection using supercontinuum generation in silicon nitride waveguides. Opt Express 23(12):15440–15451
28. Klenner A, Mayer AS, Johnson AR, Luke K, Lamont MRE, Okawachi Y, Lipson M, Gaeta AL, Keller U (2016) Gigahertz frequency comb offset stabilization based on supercontinuum generation in silicon nitride waveguides. Opt Express 24(10):11043–11053
29. Waldburger D, Mayer AS, Alfieri CGE, Nürnberg J, Johnson AR, Ji X, Klenner A, Okawachi Y, Lipson M, Gaeta AL, Keller U (2019) Tightly locked optical frequency comb from a semiconductor disk laser. Opt Express 27(3):1786. https://doi.org/10.1364/oe.27.001786
30. Aviles-Espinosa R, Filippidis G, Hamilton C, Malcolm G, Weingarten KJ, Südmeyer T, Barbarin Y, Keller U, Santos SICO, Artigas D, Loza-Alvarez P (2011) Compact ultrafast semiconductor disk laser: targeting GFP based nonlinear applications in living organisms. Biomed Opt Express 2(4):739. https://doi.org/10.1364/boe.2.000739
31. Voigt FF, Emaury F, Bethge P, Waldburger D, Link SM, Carta S, van der Bourg A, Helmchen F, Keller U (2017) Multiphoton in vivo imaging with a femtosecond semiconductor disk laser. Biomed Opt Express 8(7):3213. https://doi.org/10.1364/boe.8.003213
32. Hoogland S, Dhanjal S, Tropper AC, Roberts JS, Häring R, Paschotta R, Morier-Gemoud F, Keller U (2000) Passively mode-locked diode-pumped surface-emitting semiconductor laser. IEEE Photonics Technol Lett 12(9):1135–1137. ISSN 1941-0174. https://doi.org/10.1109/68.874213
33. Großmann M, Bek R, Jetter M, Michler P (2020) Stable fundamental and dual-pulse mode locking of red-emitting VECSELs. Laser Phys. Lett. 17(6). https://doi.org/10.1088/1612-202x/ab881f
34. Meyer JT, Lukowski ML, Hessenius C, Wright EM, Fallahi M (2020) High peak power, sub-ps green emission in a passively mode locked W-cavity VECSEL. Opt Express 28(4):5794. https://doi.org/10.1364/oe.384439
35. Waldburger D, Alfieri CGE, Link SM, Meinecke S, Jaurigue LC, Lüdge K, Keller U (2018) Multipulse instabilities of a femtosecond SESAM-modelocked VECSEL. Opt Express 26(17):21872–21886. https://doi.org/10.1364/oe.26.021872
36. Laurain A, Kilen I, Hader J, Ruiz Perez A, Ludewig P, Stolz W, Addamane S, Balakrishnan J, Koch SA, Moloney JV (2018) Modeling and experimental realization of modelocked VECSEL producing high power sub-100 fs pulses. Appl Phys Lett 113(12):121113. https://doi.org/10.1063/1.5042263
37. Hoffmann M, Barbarin Y, Maas DJHC, Golling M, Krestnikov IL, Mikhrin SS, Kovsh AR, Südmeyer T, Keller U (2008) Modelocked quantum dot vertical external cavity surface emitting laser. Appl Phys B 93:733–736. https://doi.org/10.1007/s00340-008-3267-0
38. Hoffmann M, Sieber OD, Wittwer VJ, Krestnikov IL, Livshits DA, Barbarin Y, Südmeyer T, Keller U (2011) Femtosecond high-power quantum dot vertical external cavity surface emitting laser. Opt Express 19(9):8108–8116. https://doi.org/10.1364/oe.19.008108
39. Nechay K, Mereuta A, Paranthoën C, Brévalle G, Levallois C, Alouini M, Chevalier N, Perrin M, Suruceanu G, Caliman A, Guina M, Kapon E (2019) InAs/InP quantum dot VECSEL emitting at 1.5 μm. Appl Phys Lett 115(17):171105. https://doi.org/10.1063/1.5125632
40. Finke T, Nürnberg J, Sichkovskyi V, Golling M, Keller U, Reithmaier JP (2020) Temperature resistant fast $In_xGa_{1-x}As/GaAs$ quantum dot saturable absorber for the epitaxial integration into semiconductor surface emitting lasers. Opt Express 28(14):20954. https://doi.org/10.1364/oe.396198

41. Alfieri CGE, Waldburger D, Nürnberg R, Golling M, Jaurigue LC, Lüdge K, Keller U (2018) Modelocking instabilities for high-gain semiconductor disk lasers based on active submonolayer quantum dots. Phys Rev Appl 10:044015. https://doi.org/10.1103/physrevapplied.10.044015
42. Jiang W, Shimizu M, Mirin RP, Reynolds TE, Bowers JE (1993) Electrically pumped mode-locked vertical-cavity semiconductor lasers. Opt Lett 18(22):1937–1939. https://doi.org/10.1364/ol.18.001937
43. Mulet J, Balle S (2005) Mode-locking dynamics in electrically driven vertical-external-cavity surface-emitting lasers. IEEE J Quantum Electron 41(9):1148–1156. ISSN 0018-9197. https://doi.org/10.1109/jqe.2005.853355
44. Chichkov NB, Yadav A, Kovalev AV, Smirnov SV, Herper M, Fedorova KA, Viktorov EA, Rafailov EU (2020) Pulse dynamics in SESAM-free electrically pumped VECSEL. Opt Express 28(9):13466. https://doi.org/10.1364/oe.385619
45. Keller U, Miller DAB, Boyd GD, Chiu TH, Ferguson JF, Asom MT (1992) Solid-state low-loss intracavity saturable absorber for Nd:YLF lasers: an antiresonant semiconductor Fabry-Perot saturable absorber. Opt Lett 17(7):505. https://doi.org/10.1364/ol.17.000505
46. Keller U, Weingarten KJ, Kärtner FX, Kopf D, Braun B, Jung ID, Fluck R, Hönninger C, Matuschek N, Aus-der Au J (1996) Semiconductor saturable absorber mirrors (SESAM's) for femtosecond to nanosecond pulse generation in solid-state lasers. IEEE J Sel Top Quantum Electron 2:435
47. Lagatsky AA, Rafailov EU, Sibbett W, Livshits DA, Zhukov AE, Ustinov VM (2005) Quantum-dot-based saturable absorber with p-n junction for mode-locking of solid-state lasers. IEEE Photonics Technol Lett 17(2):294–296. https://doi.org/10.1109/lpt.2004.839387
48. Liu X, Rafailov EU, Livshits D, Turchinovich D (2010) Quantum well saturable absorber mirror with electrical control of modulation depth. Appl Phys Lett 97(5):051103. https://doi.org/10.1063/1.3474799
49. Addamane SJ, Laurain A, Baker CW, Rotter TJ, Watt J, Reno JL, Balakrishnan G, Moloney JV (2021) Submonolayer quantum-dot based saturable absorber for femtosecond pulse generation. J Electron Mater 50(5). ISSN 1543-186X. https://doi.org/10.1007/s11664-021-08795-x
50. Hausen J, Meinecke S, Lingnau B, Lüdge K (2019) Pulse cluster dynamics in passively mode-locked semiconductor vertical-external-cavity surface-emitting lasers. Phys Rev Appl 11. https://doi.org/10.1103/physrevapplied.11.044055
51. Ippen EP (1994) Principles of passive mode locking. Appl Phys B 58:159–170. https://doi.org/10.1007/bf01081309
52. Hausen J, Meinecke S, Javaloyes J, Gurevich SV, Lüdge K (2020) Phase-incoherent photonic molecules in V-shaped mode-locked vertical-external-cavity surface-emitting semiconductor lasers. Phys Rev Appl 14(044059). https://doi.org/10.1103/physrevapplied.14.044059
53. van Tartwijk GHM, Agrawal GP (1998) Laser instabilities: a modern perspective. Prog Quantum Electron 22(2):43–122. https://doi.org/10.1016/s0079-6727(98)00008-1
54. Chow WW, Koch SW (1999) Semiconductor-laser fundamentals, 1st edn. Springer, Berlin Heidelberg, Berlin. 978-3-662-03880-2. https://doi.org/10.1007/978-3-662-03880-2
55. Lingnau B (2015) Nonlinear and nonequilibrium dynamics of quantum-dot optoelectronic devices. Springer theses. Springer, Switzerland. 978-3-319-25803-4. https://doi.org/10.1007/978-3-319-25805-8
56. Vladimirov AG, Pimenov AS, Rachinskii D (2009) Numerical study of dynamical regimes in a monolithic passively mode-locked semiconductor laser. IEEE J Quantum Electron 45(5):462–468. https://doi.org/10.1109/jqe.2009.2013363
57. New G (1974) Pulse evolution in mode-locked quasi-continuous lasers. IEEE J Quantum Electron 10(2):115. ISSN 0018-9197. https://doi.org/10.1109/jqe.1974.1145781
58. Avrutin EA, Marsh JH, Portnoi EL (2000) Monolithic and multi-gigahertz mode-locked semiconductor lasers: constructions, experiments, models and applications. IEE Proc Optoelectron 147(4):251. https://doi.org/10.1049/ip-opt:20000282
59. Vladimirov AG, Turaev DV (2005) Model for passive mode locking in semiconductor lasers. Phys Rev A 72(3). https://doi.org/10.1103/physreva.72.033808

60. Press WH, Flannery BP, Teukolsky SA, Vettering WT (2007) Numerical recipes, 3rd edn. Cambridge University Press, Cambridge
61. Haug H, Koch SW (2004) Quantum theory of the optical and electronic properties of semiconductors, 2nd edn. World Scientific, Singapore
62. Asada M, Kameyama A (1984) Gain and intervalence band absorption in quantum-well lasers. IEEE J Quantum Electron 20(1):745–753
63. Arakawa Y, Yariv A (1986) Quantum well lasers—gain, spectra, dynamics. IEEE J Quantum Electron 22:1887
64. Kilen I, Koch SW, Hader J, Moloney JV (2016) Fully microscopic modeling of mode locking in microcavity lasers. J Opt Soc Am B 33(1):75–80. https://doi.org/10.1364/josab.33.000075
65. Wilcox JZ, Peterson GL, Ou S, Yang JJ, Jansen M, Schechter D (1988) Gain- and threshold-current dependence for multiple-quantum-well lasers. J Appl Phys 64(11):6564–6567. https://doi.org/10.1063/1.342028
66. Makino T (1996) Analytical formulas for the optical gain of quantum wells. IEEE J Quantum Electron 32(3):493–501. https://doi.org/10.1109/3.485401
67. Avrutin EA, Russell BM (2009) Dynamics and spectra of monolithic mode-locked laser diodes under external optical feedback. IEEE J Quantum Electron 45(11):1456
68. Coldren LA, Corzine SW, Mashanovitch M (2012) Diode lasers and photonic integrated circuits. Wiley series in microwave and optical engineering, 2nd edn. Wiley
69. Schöll E (1988) Dynamic theory of picosecond optical pulse shaping by gain-switched semiconductor laser amplifiers. IEEE J Quantum Electron 24(2):435–442
70. Agrawal GP, Olsson NA (1989) Self-phase modulation and spectral broadening of optical pulses in semiconductor laser amplifiers. IEEE J Quantum Electron 25(11):2297–2306. https://doi.org/10.1109/3.42059
71. Tromborg B, Lassen HE, Olesen H (1994) Traveling wave analysis of semiconductor lasers: modulation responses, mode stability and quantum mechanical treatment of noise spectra. IEEE J Quantum Electron 30(4):939. https://doi.org/10.1109/3.291365
72. Bischoff S, Sørensen MP, Mørk J, Brorson SD, Franck T, Nielsen JM, Møller-Larsen A (1995) Pulse-shaping mechanism in colliding-pulse mode-locked laser diodes. Appl Phys Lett 67(26):3877–3879. https://doi.org/10.1063/1.115303
73. Bandelow U, Radziunas M, Sieber J, Wolfrum M (2001) Impact of gain dispersion on the spatio-temporal dynamics of multisection lasers. IEEE J Quantum Electron 37(2):183. ISSN 0018-9197. https://doi.org/10.1109/3.903067
74. Erneux T, Glorieux P (2010) Laser dynamics. Cambridge University Press, UK. https://doi.org/10.1017/cbo9780511776908
75. Otto C, Lüdge K, Vladimirov AG, Wolfrum M, Schöll E (2012) Delay induced dynamics and jitter reduction of passively mode-locked semiconductor laser subject to optical feedback. New J Phys 14. https://doi.org/10.1088/1367-2630/14/11/113033
76. Karin JR, Helkey RJ, Derickson DJ, Nagarajan R, Allin DS, Bowers JE, Thornton RL (1994) Ultrafast dynamics in field-enhanced saturable absorbers. Appl Phys Lett 64(6):676–678
77. Jones DJ, Zhang LM, Carroll JE, Marcenac D (1995) Dynamics of monolithic passively mode-locked semiconductor lasers. IEEE J Quantum Electron 31(6):1051–1058
78. Sieber OD, Hoffmann M, Wittwer VJ, Mangold M, Golling M, Tilma BW, Südmeyer T, Keller U (2013) Experimentally verified pulse formation model for high-power femtosecond VECSELs. Appl Phys B 113(1):133–145. ISSN 1432-0649. https://doi.org/10.1007/s00340-013-5449-7
79. Alfieri CGE, Waldburger D, Link SM, Gini E, Golling M, Eisenstein G, Keller U (2017) Optical efficiency and gain dynamics of modelocked semiconductor disk lasers. Opt Express 25(6):6402–6420. https://doi.org/10.1364/oe.25.006402
80. Rideout W, Yu B, LaCourse J, York PK, Beernink KJ, Coleman JJ (1990) Measurement of the carrier dependence of differential gain, refractive index, and linewidth enhancement factor in strained-layer quantum well lasers. Appl Phys Lett 56(8):706–708. https://doi.org/10.1063/1.102688
81. Ukhanov AA, Stintz A, Eliseev PG, Malloy KJ (2004) Comparison of the carrier induced refractive index, gain, and linewidth enhancement factor in quantum dot and quantum well lasers. Appl Phys Lett 84(7):1058–1060. https://doi.org/10.1063/1.1647688

82. Fordell T, Lindberg AM (2007) Experiments on the linewidth-enhancement factor of a vertical-cavity surface-emitting laser. IEEE J Quantum Electron 43(1):6–15
83. Pimenov AS, Viktorov EA, Hegarty SP, Habruseva T, Huyet G, Rachinskii D, Vladimirov AG (2014) Bistability and hysteresis in an optically injected two-section semiconductor laser. Phys Rev E 89:052903
84. Jaurigue LC, Nikiforov O, Schöll E, Breuer S, Lüdge K (2016) Dynamics of a passively mode-locked semiconductor laser subject to dual-cavity optical feedback. Phys Rev E 93(2). https://doi.org/10.1103/physreve.93.022205
85. Nikiforov O, Jaurigue LC, Drzewietzki L, Lüdge K, Breuer S (2016) Experimental demonstration of change of dynamical properties of a passively mode-locked semiconductor laser subject to dual optical feedback by dual full delay-range tuning. Opt Express 24(13):14301–14310. https://doi.org/10.1364/oe.24.014301
86. Agrawal GP, Dutta NK (1993) Semiconductor lasers. Van Nostrand Reinhold, New York
87. Wang J, Maitra A, Poulton CG, Freude W, Leuthold J (2007) Temporal dynamics of the alpha factor in semiconductor optical amplifiers. J Light Technol 25(3):891–900
88. Herzog B, Lingnau B, Kolarczik M, Kaptan Y, Bimberg D, Maaßdorf A, Pohl UW, Rosales R, Schulze JH, Strittmatter A, Weyers M, Woggon U, Lüdge K, Owschimikow N (2016) Strong amplitude-phase coupling in submonolayer quantum dots. Appl Phys Lett 109:201102. https://doi.org/10.1063/1.4967833
89. Engelborghs K, Luzyanina T, Roose D (2002) Numerical bifurcation analysis of delay differential equations using DDE-Biftool. ACM Trans Math Softw 28:1–21
90. Sieber J, Engelborghs K, Luzyanina T, Samaey G, Roose D (2014) DDE-BIFTOOL manual—bifurcation analysis of delay differential equations. arXiv:1406.7144
91. Vladimirov AG, Turaev DV (2004) A new model for a mode-locked semiconductor laser. Radiophys Quantum Electron 47:769–776. https://doi.org/10.1007/s11141-005-0015-8
92. Jaurigue LC (2017) Passively mode-locked semiconductor lasers: dynamics and stochastic properties in the presence of optical feedback, springer thesis edition. Springer, Cham. ISBN 978-3-319-58874-2. https://doi.org/10.1007/978-3-319-58874-2
93. Vladimirov AG, Rachinskii D, Wolfrum M (2011) Modeling of passively mode-locked semiconductor lasers. In Lüdge K (ed) Nonlinear laser dynamics—from quantum dots to cryptography. Reviews in nonlinear dynamics and complexity, chap 8. Wiley, Weinheim, pp 183–213. ISBN 978-3-527-41100-9
94. Haus HA (1975) Theory of mode locking with a fast saturable absorber. J Appl Phys 46(7):3049–3058. https://doi.org/10.1063/1.321997
95. Derickson DJ, Helkey RJ, Mar A, Karin JR, Wasserbauer JG, Bowers JE (1992) Short pulse generation using multisegment mode-locked semconductor lasers. IEEE J Quantum Electron 28(10):2186–2202
96. Thompson MG, Rae AR, Xia M, Penty RV, White IH (2009) InGaAs quantum-dot mode-locked laser diodes. IEEE J Quantum Electron 15(3):661–672
97. Haus HA (2000) Mode-locking of lasers. IEEE J Sel Top Quantum Electron 6(6):1173–1185. https://doi.org/10.1109/2944.902165
98. Otto C (2014) Dynamics of quantum dot lasers—effects of optical feedback and external optical injection. Springer Theses. Springer, Heidelberg. 978-3-319-03785-1. https://doi.org/10.1007/978-3-319-03786-8
99. Quarterman AH, Wilcox KG, Apostolopoulos V, Mihoubi Z, Elsmere SP, Farrer I, Ritchie DA, Tropper AC (2009) A passively mode-locked external-cavity semiconductor laser emitting 60-fs pulses. Nat Photonics 3(12):729–731. ISSN 1749-4885. https://doi.org/10.1038/nphoton.2009.216
100. Kilen I, Hader J, Moloney JV, Koch SW (2014) Ultrafast nonequilibrium carrier dynamics in semiconductor laser mode locking. Optica 1(4):192–197. https://doi.org/10.1364/optica.1.000192
101. McLaren S, Kilen I, Moloney JV (2020) Microscopic modeling of transverse mode instabilities in mode-locked vertical external-cavity surface-emitting lasers. Appl Phys Lett 116(3):031102. https://doi.org/10.1063/1.5134070

102. Lüdge K (2012) Nonlinear laser dynamics—from quantum dots to cryptography. Wiley, Weinheim. 978-3-527-41100-9
103. Lingnau B, Chow WW, Lüdge K (2014) Amplitude-phase coupling and chirp in quantum-dot lasers: influence of charge carrier scattering dynamics. Opt Express 22(5):4867–4879. https://doi.org/10.1364/oe.22.004867
104. Meinecke S, Drzewietzki L, Weber C, Lingnau B, Breuer S, Lüdge K (2019) Ultra-short pulse generation in a three section tapered passively mode-locked quantum-dot semiconductor laser. Sci Rep 9:1783. https://doi.org/10.1038/s41598-018-38183-1
105. Lingnau B, Lüdge K, Herzog B, Kolarczik M, Kaptan Y, Woggon U, Owschimikow N (2016) Ultrafast gain recovery and large nonlinear optical response in submonolayer quantum dots. Phys Rev B 94(1):014305. https://doi.org/10.1103/physrevb.94.014305
106. Hausen J, Herzog B, Nelde A, Meinecke S, Owschimikow N, Lüdge K (2021) Feedback induced locking in semiconductor lasers with strong amplitude-phase coupling. Phys Rev A 103(4):043511. https://doi.org/10.1103/physreva.103.043511. arXiv:2101.01623
107. Hausen J, Lüdge K, Gurevich SV, Javaloyes J (2020) How carrier memory enters the Haus master equation of mode-locking. Opt Lett 45(22):6210–6213. https://doi.org/10.1364/ol.406136
108. Perego AM, Garbin B, Gustave F, Barland S, Prati F, de Valcárcel GJ (2020) Coherent master equation for laser modelocking. Nat Commun 11(1):311. ISSN 2041-1723. https://doi.org/10.1038/s41467-019-14013-4
109. Mulet J, Mørk J (2006) Analysis of timing jitter in external-cavity mode-locked semiconductor lasers. IEEE J Quantum Electron 42(3):249. https://doi.org/10.1109/jqe.2006.869808
110. Javaloyes J, Balle S (2010) Mode-locking in semiconductor Fabry-Pérot lasers. IEEE J Quantum Electron 46(7):1023–1030
111. Javaloyes J, Balle S (2011) Anticolliding design for monolithic passively mode-locked semiconductor lasers. Opt Lett 36(22):4407–4409
112. Scully MO, Suhail Zubairy M (1997) Quantum optics, 1st edn. Cambridge University Press. ISBN 9780511813993 (online). https://doi.org/10.1017/cbo9780511813993
113. Kilen I, Koch SW, Hader J, Moloney JV (2017) Non-equilibrium ultrashort pulse generation strategies in VECSELs. Optica 4(4):412–417. https://doi.org/10.1364/optica.4.000412

Chapter 6
Summary and Outlook

In this thesis, the dynamics of passively mode-locked semiconductor lasers have been investigated. A special emphasis has been given to the spatio-temporal intracavity electric field and active medium gain dynamics. The results have been utilized to identify and suggest optimal laser configurations and operating conditions. The investigation has been split into two parts, which are both centered around a particular device: a monolithically integrated three-section tapered quantum-dot laser and a V-shaped external cavity semiconductor laser. Both analyses have been motivated and complemented by experimental characterizations, which have given rise to specific questions.

To construct effective laser models from first principles, the derivations of the traveling-wave equation framework for the electric field dynamics and the inhomogeneous Maxwell-Bloch equations for the coherent coupling to the gain medium have been presented. Moreover, two numerically efficient integration schemes have been developed. The first integration scheme can be applied for an arbitrary electric field source term and has thus been used for the three-section laser model, where the electric field is driven by an effective macroscopic polarization. The second integration scheme requires the electric field source to be linear in the electric field itself and has been used for the V-shaped laser model, where the polarization equations have been adiabatically eliminated.

In order to perform a quantitative analyses of the temporal pulse-train stability, the long-term timing jitter has been introduced. On that account, multiple established long-term timing jitter estimation methods have been reviewed and discussed in terms of their applicability to simulated passively mode-locked lasers. Additionally, a new pulse-period fluctuation autocorrelation based timing jitter estimation method has been developed. With the goal of reducing the computational costs of obtaining high-fidelity timing jitter estimates, the various methods have been benchmarked with respect to the calculated pulse train sample-set sizes. Both the pulse-period fluctuation autocorrelation method and the power spectral density based Kéfélian

S. Meinecke, *Spatio-Temporal Modeling and Device Optimization of Passively Mode-Locked Semiconductor Lasers*, Springer Theses,
https://doi.org/10.1007/978-3-030-96248-7_6

method have been demonstrated to perform better than the direct application of the time-domain definition by a factor of ≈ 100. Hence, the usage of the latter is strongly discouraged. Depending on the pulse train properties, recommendations regarding the most suitable timing jitter estimation method have furthermore been presented.

The first device under investigation has been a monolithically integrated tapered three-section semiconductor quantum-dot laser. Motivated by experimental characterizations, a numerical model has been developed, which couples the microscopically motivated quantum-dot charge-carrier dynamics via the Maxwell-Bloch equations to the intra-cavity electric field. The obtained model self-consistently describes the spatio-temporal laser dynamics and has been demonstrated to both qualitatively and quantitatively reproduce the measured emission dynamics. The excellent agreement validates the model and facilitates the further study and exploration of the laser dynamics.

To understand the pulse-shaping mechanism of the laser, a co-moving spatio-temporal representation of the electric field intensity and the gain dynamics has been developed. This representation has been proven highly effective, as it visualizes the evolution of pulse profile along the pulse propagation and thereby reveals the impact of the local interaction with the active medium. Utilizing this technique has demonstrated that the pulse-shaping in the three-section laser is contrary to the mechanisms reported in the literature: Pulses broaden in the absorber section and shorten in the gain sections due to the interplay between the saturable and the non-saturable gain and absorption.

The numerical model has been further employed to precisely identify and study the occurring instabilities and relate them to the experiment via their respective out-coupled emission characteristics. On that account, a pulse-switching instability has been found to limit the fundamental mode-locking state at large pump currents. This instability is induced by a positive net-gain window, which is created between the pulse emissions via the interactions of the forward and backward-traveling pulses. The instability has been shown to be characteristic for the devices geometry. The requirement for the spatio-temporal modeling has been highlighted by those results.

In the next step, the numerical model has been used to create detailed maps of the observable emission dynamics as well as maps of the pulse performance and the timing jitter of the fundamental mode-locking state. On this account, it has been shown that the pulse peak power, the pulse width, and the timing jitter can not be optimized simultaneously. However, a compromise among those figures of merit has been found at intermediate absorber reverse biases, which corresponds to the experimentally utilized biasing condition. The implications of the absorber reverse bias have further been unraveled by separately studying the induced effects. The investigation has demonstrated that the observed emission states critically depend on the interplay of the absorber redshift and absorber recovery rate, which are both determined by the applied reverse bias. Fundamental mode-locking only appears if both are properly tuned, which occurs at the aforementioned intermediate values.

Lastly, the numerical model has been utilized to explore the impact of the taper angle of the long gain section and the position of the saturable absorber section, and predict optimal configurations. The simulations have revealed that all pulse performance characteristics improve with an increasing taper angle. The absorber section, on the other hand, exhibits an intermediate optimal position within the laser cavity that maximizes both the pump current range for which fundamental mode-locking can be obtained as well as the achievable pulse peak power. It has been further demonstrated that these results can be interpreted and understood in terms of the pulse shaping and the gain and absorption recovery processes.

The second device under investigation has been a V-shaped external cavity semiconductor laser. Driven by experimental investigations, a minimalistic spatio-temporal delay-differential equation model has been derived, in order to study the effects, which are induced by the cavity geometry. Apart from the direct numerical integration of the equations, the model allows for analytical approximations and is suitable for path-continuation techniques. The latter, however, are rendered numerically unfeasible if the experimentally observed ultra-short optical pulses are to be studied. Deploying the numerical model, the experimentally observed sequence of emission states has been reproduced and understood in terms of the gain and the net-gain dynamics. Characteristic for the V-shaped laser, pulse-cluster mode-locking states, which contain multiple approximately identical pulses, appear at larger pump currents. Similar to the three-section laser, the emission dynamics critically depend on the cavity configuration and the resulting active medium interactions of the forward and backward-traveling pulses. The symmetric cavity configuration in particular gives rise to the pulse-cluster states and suppresses even-order harmonic mode-locking states. The experimentally observed side pulses have been attributed to a leading-edge instability, which is predicted to only occur at slightly asymmetric cavity configurations. The observed stable emission states have been characterized by equidistant interactions of the forward and backward-traveling pulses with the gain chip. This condition optimizes the gain recovery and minimizes the non-radiative losses.

Furthermore, the emission dynamics have been characterized as a function of the cavity configuration in terms of the gain-chip position. The simulations have revealed that rather asymmetric cavity configurations reestablish the even-order harmonic mode-locking states and suppress the pulse-cluster states. The governing mechanisms have been attributed to positive net-gain instabilities, which are characteristic for a given cavity configuration. On that account, the upper pump current stability boundary of the fundamental mode-locking state can be entirely understood in terms of three qualitatively different positive net-gain instabilities. Quantitatively, the critical pump current can be accurately constructed by approximating the net gain, which a perturbation experiences along its propagation. The three instabilities respond differently to changes of the cavity configuration and thereby generate a characteristic region of stable fundamental mode-locking. Based on those results, optimal fundamental mode-locking performance has been predicted for a cavity arm ratio of 1:2. These findings thus demonstrate the value of a net-gain analysis, which incorporates the spatio-temporal specifics of the laser dynamics.

Outlook

The fundamental equations of motion, i.e., the traveling-wave equation and the Maxwell-Bloch equations, which can be used to describe multi-mode laser dynamics, have been known for a number of decades [1, 2]. Detailed numerical investigations of passively mode-locked lasers, however, have been limited by the accessible computing powers. This has given rise to a number of minimalistic models [3–7], which employ strong approximations, but nonetheless capture many of the qualitative mechanisms and features of passive mode-locking. These models, however, often fail to reproduce and explain the details of experimental characterizations, and thus only carry limited quantitative predictive power.

Only recently, advances of semiconductor technologies have sufficiently reduced the computational costs, such that the spatio-temporal modeling of passively mode-locked lasers has become feasible [8–14]. Nevertheless, the simulation of such devices is still in the maturing process as this thesis demonstrates. Even with the availability of a small-scale cluster ($\approx$ 100 nodes), the numerical models must be carefully custom build for the specific device at hand, in order to achieve reasonable computation times, which allow to perform parameter studies and pulse-train stability analyses.

Although the computational costs can be expected to further decrease in the future, it is nonetheless highly desirable to further advance the simulation tools for multi-mode semiconductor lasers in the meantime. As this thesis demonstrates, the demands and expectations towards simulations may greatly vary. On the one hand, quantitatively exact simulations of mode-locking characteristics, pulse-shaping mechanisms, and parameter dependencies may be desired. On the other hand, qualitative descriptions and explanations of the dynamics may be sufficient.

To comprehensively include all cases, it would therefore be of great value to develop multi-mode laser simulation tools, which provide the flexibility to adapt the modeling depth to what is most appropriate for a given task. This would imply a modeling hierarchy, with a detailed microscopic description of the electric field and active medium on the one end, and an effective minimalist description on the other end. The ability to shift between the different levels within that hierarchy would directly reveal the validity of the approximations and assumptions, which are required to move to simpler levels, as well as their quantitative consequences on the pulse performance figures. This flexibility could then be used, e.g., to perform large parameter studies with a simple model and spot check the results using a more advanced model. Specific laser configurations and operating points could furthermore be investigated in detail with more advanced models, in order to improve the predictive power of the obtained results.

Irregardless of the modeling depth, the pulse-shaping mechanism can be studied by the means of the co-moving spatio-temporal representation, which has been introduced in this thesis. Similarly, a net-gain stability analysis can be performed at all modeling hierarchy levels. Equipped with those tools, the pulse generation can be visualized and understood in terms of the laser configuration and the active medium dynamics. Utilizing computer simulations thus may guide the development of new generations of efficient high-performance passively mode-locked lasers.

References

1. Fleck JA (1968) Emission of pulse trains by q-switched lasers. Phys Rev Lett 21:131–133. https://doi.org/10.1103/physrevlett.21.131
2. Schöll E (1988) Dynamic theory of picosecond optical pulse shaping by gain-switched semiconductor laser amplifiers. IEEE J Quantum Electron 24(2):435–442
3. New G (1974) Pulse evolution in mode-locked quasi-continuous lasers. IEEE J Quantum Electron 10(2):115. ISSN 0018-9197. https://doi.org/10.1109/jqe.1974.1145781
4. Haus HA (1975) Theory of mode locking with a slow saturable absorber. IEEE J Quantum Electron 11(9):736. https://doi.org/10.1109/jqe.1975.1068922
5. Haus HA (1975) Theory of mode locking with a fast saturable absorber. J Appl Phys 46(7):3049–3058. https://doi.org/10.1063/1.321997
6. Haus HA (2000) Mode-locking of lasers. IEEE J Sel Top Quantum Electron 6(6):1173–1185. https://doi.org/10.1109/2944.902165
7. Vladimirov AG, Turaev DV (2005) Model for passive mode locking in semiconductor lasers. Phys Rev A 72(3):033808. https://doi.org/10.1103/physreva.72.033808
8. Bandelow U, Radziunas M, Vladimirov AG, Hüttl B, Kaiser R (2006) 40GHz mode locked semiconductor lasers: theory, simulation and experiments. Opt Quantum Electron 38:495. https://doi.org/10.1007/s11082-006-0045-2
9. Mulet J, Mørk J (2006) Analysis of timing jitter in external-cavity mode-locked semiconductor lasers. IEEE J Quantum Electron 42(3):249. https://doi.org/10.1109/jqe.2006.869808
10. Avrutin EA, Russell BM (2009) Dynamics and spectra of monolithic mode-locked laser diodes under external optical feedback. IEEE J Quantum Electron 45(11):1456
11. Javaloyes J, Balle S (2010) Mode-locking in semiconductor Fabry-Pérot lasers. IEEE J Quantum Electron 46(7):1023–1030
12. Radziunas M, Vladimirov AG, Viktorov EA, Fiol G, Schmeckebier H, Bimberg D (2011) Pulse broadening in quantum-dot mode-locked semiconductor lasers: simulation, analysis, and experiments. IEEE J Quantum Electron 47(7):935–943
13. Rossetti M, Bardella P, Montrosset I (2011) Time-domain travelling-wave model for quantum dot passively mode-locked lasers. IEEE J Quantum Electron 47(2):139
14. Kilen I, Koch SW, Hader J, Moloney JV (2016) Fully microscopic modeling of mode locking in microcavity lasers. J Opt Soc Am B 33(1):75–80. https://doi.org/10.1364/josab.33.000075

The manufacturer's authorised representative in the EU is Springer Nature Customer Service Centre GmbH, Europaplatz 3, 69115 Heidelberg, Germany. If you have any concerns regarding our products, please contact ProductSafety@springernature.com

Printed and bound by CPI Group (UK) Ltd, Croydon, CR0 4YY
30/09/2026
02225498-0010